Diffusion and Reactions in Fractals and Disordered Systems

Fractal structures are found everywhere in Nature, and as a consequence anomalous diffusion has far-reaching implications for a host of phenomena. This book describes diffusion and transport in disordered media such as fractals, porous rocks, and random resistor networks. Part I contains material of general interest to statistical physics: fractals, percolation theory, regular random walks and diffusion, continuous-time random walks, and Lévy walks and flights. Part II covers anomalous diffusion in fractals and disordered media, while Part III serves as an introduction to the kinetics of diffusion-limited reactions. Part IV discusses the problem of diffusion-limited coalescence in one dimension. This book written in a pedagogical style is intended for upper-level undergraduates and graduate students studying physics, chemistry, and engineering. It will also be of particular interest to young researchers requiring a clear introduction to the field.

DANIEL BEN-AVRAHAM was born in 1957, in Sante Fe, Argentina, and obtained his Ph.D. in Physics from Bar-Ilan University in 1985. After a 2-year post-doctoral position in the Center of Polymer Studies at the University of Boston, he gained a permanent position at Clarkson University where he is now Associate Professor of Physics. Professor ben-Avraham has spent time as a Visiting Professor at various institutions including Heidelberg University, Bar-Ilan University, and the European Centre for Molecular Biology. He has published over 80 papers and contributed invited papers to several anthologies.

SHLOMO HAVLIN was born in 1942, in Jerusalem, Israel, and obtained his Ph.D. in 1972 from Bar-Ilan University. He stayed at Bar-Ilan University, progressing through the ranks of Research Associate, Lecturer, Senior Lecturer, and Associate Professor until in 1984 he became Professor and Chairman of the Department of Physics. He is now currently Dean of the Faculty of Exact Sciences. Since 1978 Professor Havlin has spent time as a Visiting Professor at numerous institutions including the University of Edinburgh, the National Institute of Health (USA), and Boston University. He is currently on the editorial boards of three journals. He is the author of over 400 papers and editor of ten books. He has given over 40 plenary and invited talks.

Diffusion and Reactions in Fractals and Disordered Systems

Daniel ben-Avraham
Clarkson University
and
Shlomo Havlin
Bar-Ilan University

PUBLISHED BY THE PRESS SYNDICATE OF THE UNIVERSITY OF CAMBRIDGE
The Pitt Building, Trumpington Street, Cambridge, United Kingdom

CAMBRIDGE UNIVERSITY PRESS
The Edinburgh Building, Cambridge CB2 2RU, UK
40 West 20th Street, New York NY 10011–4211, USA
477 Williamstown Road, Port Melbourne, VIC 3207, Australia
Ruiz de Alarcón 13, 28014 Madrid, Spain
Dock House, The Waterfront, Cape Town 8001, South Africa

http://www.cambridge.org

First published 2000
First paperback edition 2004

Typeface Times 11/14pt.   *System* LaTeX 2_ε [DBD]

A catalogue record for this book is available from the British Library

Library of Congress Cataloguing in Publication data
Ben-Avraham, Daniel, 1957–
Diffusion and reactions in fractals and disordered systems /
Daniel ben-Avraham and Shlomo Havlin.
p.   cm.   ISBN 0 521 62278 6 (hardback)
1. Diffusion.   2. Fractals.   3. Stochastic processes.   I. Havlin, Shlomo.   II. Title.
QC185.B46 2000
530.4′75–dc21   00-023591   CIP

ISBN 0 521 62278 6 hardback
ISBN 0 521 61720 0 paperback

to Akiva and Aliza

and
to Hava

Contents

Preface   *page* xiii

Part one: Basic concepts   1

1   Fractals   **3**
1.1   Deterministic fractals   3
1.2   Properties of fractals   6
1.3   Random fractals   7
1.4   Self-affine fractals   9
1.5   Exercises   11
1.6   Open challenges   12
1.7   Further reading   12

2   Percolation   **13**
2.1   The percolation transition   13
2.2   The fractal dimension of percolation   18
2.3   Structural properties   21
2.4   Percolation on the Cayley tree and scaling   25
2.5   Exercises   28
2.6   Open challenges   30
2.7   Further reading   31

3   Random walks and diffusion   **33**
3.1   The simple random walk   33
3.2   Probability densities and the method of characteristic functions   35
3.3   The continuum limit: diffusion   37
3.4   Einstein's relation for diffusion and conductivity   39
3.5   Continuous-time random walks   41
3.6   Exercises   43

3.7   Open challenges   44
3.8   Further reading   45

4   Beyond random walks   **46**
4.1   Random walks as fractal objects   46
4.2   Anomalous continuous-time random walks   47
4.3   Lévy flights and Lévy walks   48
4.4   Long-range correlated walks   50
4.5   One-dimensional walks and landscapes   53
4.6   Exercises   55
4.7   Open challenges   55
4.8   Further reading   56

       Part two: Anomalous diffusion   57

5   Diffusion in the Sierpinski gasket   **59**
5.1   Anomalous diffusion   59
5.2   The first-passage time   61
5.3   Conductivity and the Einstein relation   63
5.4   The density of states: fractons and the spectral dimension 65
5.5   Probability densities   67
5.6   Exercises   70
5.7   Open challenges   71
5.8   Further reading   72

6   Diffusion in percolation clusters   **74**
6.1   The analogy with diffusion in fractals   74
6.2   Two ensembles   75
6.3   Scaling analysis   77
6.4   The Alexander–Orbach conjecture   79
6.5   Fractons   82
6.6   The chemical distance metric   83
6.7   Diffusion probability densities   87
6.8   Conductivity and multifractals   89
6.9   Numerical values of dynamical critical exponents   92
6.10   Dynamical exponents in continuum percolation   92
6.11   Exercises   94
6.12   Open challenges   95
6.13   Further reading   96

7   Diffusion in loopless structures   **98**
7.1   Loopless fractals   98

7.2	The relation between transport and structural exponents	101
7.3	Diffusion in lattice animals	103
7.4	Diffusion in DLAs	104
7.5	Diffusion in combs with infinitely long teeth	106
7.6	Diffusion in combs with varying teeth lengths	108
7.7	Exercises	110
7.8	Open challenges	112
7.9	Further reading	113
8	**Disordered transition rates**	**114**
8.1	Types of disorder	114
8.2	The power-law distribution of transition rates	117
8.3	The power-law distribution of potential barriers and wells	118
8.4	Barriers and wells in strips ($n \times \infty$) and in $d \geq 2$	119
8.5	Barriers and wells in fractals	121
8.6	Random transition rates in one dimension	122
8.7	Exercises	124
8.8	Open challenges	125
8.9	Further reading	126
9	**Biased anomalous diffusion**	**127**
9.1	Delay in a tooth under bias	128
9.2	Combs with exponential distributions of teeth lengths	129
9.3	Combs with power-law distributions of teeth lengths	131
9.4	Topological bias in percolation clusters	132
9.5	Cartesian bias in percolation clusters	133
9.6	Bias along the backbone	135
9.7	Time-dependent bias	136
9.8	Exercises	138
9.9	Open challenges	139
9.10	Further reading	140
10	**Excluded-volume interactions**	**141**
10.1	Tracer diffusion	141
10.2	Tracer diffusion in fractals	143
10.3	Self-avoiding walks	144
10.4	Flory's theory	146
10.5	SAWs in fractals	148
10.6	Exercises	151
10.7	Open challenges	152
10.8	Further reading	153

Part three: Diffusion-limited reactions   155

11   Classical models of reactions   **157**
11.1   The limiting behavior of reaction processes   157
11.2   Classical rate equations   159
11.3   Kinetic phase transitions   161
11.4   Reaction–diffusion equations   163
11.5   Exercises   164
11.6   Open challenges   166
11.7   Further reading   166

12   Trapping   **167**
12.1   Smoluchowski's model and the trapping problem   167
12.2   Long-time survival probabilities   168
12.3   The distance to the nearest surviving particle   171
12.4   Mobile traps   174
12.5   Imperfect traps   174
12.6   Exercises   175
12.7   Open challenges   176
12.8   Further reading   177

13   Simple reaction models   **179**
13.1   One-species reactions: scaling and effective rate equations   179
13.2   Two-species annihilation: segregation   182
13.3   Discrete fluctuations   185
13.4   Other models   187
13.5   Exercises   189
13.6   Open challenges   189
13.7   Further reading   190

14   Reaction–diffusion fronts   **192**
14.1   The mean-field description   192
14.2   The shape of the reaction front in the mean-field approach   194
14.3   Studies of the front in one dimension   195
14.4   Reaction rates in percolation   196
14.5 $A + B_{static} \rightarrow C$ with a localized source of A particles   200
14.6   Exercises   201
14.7   Open challenges   202
14.8   Further reading   203

Part four: Diffusion-limited coalescence: an exactly solvable model 205

15	**Coalescence and the IPDF method**	**207**
15.1	The one-species coalescence model	207
15.2	The IPDF method	208
15.3	The continuum limit	211
15.4	Exact evolution equations	212
15.5	The general solution	213
15.6	Exercises	215
15.7	Open challenges	216
15.8	Further reading	216
16	**Irreversible coalescence**	**217**
16.1	Simple coalescence, $A + A \rightarrow A$	217
16.2	Coalescence with input	222
16.3	Rate equations	223
16.4	Exercises	227
16.5	Open challenges	228
16.6	Further reading	228
17	**Reversible coalescence**	**229**
17.1	The equilibrium steady state	229
17.2	The approach to equilibrium: a dynamical phase transition	231
17.3	Rate equations	233
17.4	Finite-size effects	234
17.5	Exercises	236
17.6	Open challenges	237
17.7	Further reading	237
18	**Complete representations of coalescence**	**238**
18.1	Inhomogeneous initial conditions	238
18.2	Fisher waves	240
18.3	Multiple-point correlation functions	243
18.4	Shielding	245
18.5	Exercises	247
18.6	Open challenges	247
18.7	Further reading	248
19	**Finite reaction rates**	**249**
19.1	A model for finite coalescence rates	249
19.2	The approximation method	250
19.3	Kinetics crossover	251
19.4	Finite-rate coalescence with input	254

19.5   Exercises   256
19.6   Open challenges   257
19.7   Further reading   257

Appendix A   *The fractal dimension*   258
Appendix B   *The number of distinct sites visited by random walks*   260
Appendix C   *Exact enumeration*   263
Appendix D   *Long-range correlations*   266

References   272
Index   313

Preface

Diffusion in disordered, fractal structures is anomalous, different than that in regular space. Fractal structures are found everywhere in Nature, and as a consequence anomalous diffusion has far-reaching implications for a host of phenomena. We see its effects in flow within fractured and porous rocks, in the anomalous density of states in dilute magnetic systems, in silica aerogels and in glassy ionic conductors, anomalous relaxation in spin glasses and in macromolecules, conductivity of superionic conductors such as hollandite and of percolation clusters of Pb on thin films of Ge and Au, electron–hole recombination in amorphous semiconductors, and fusion and trapping of excitations in porous membrane films, polymeric glasses, and isotropic mixed crystals, to mention a few examples.

It was Pierre Gilles de Gennes who first realized the broad importance of anomalous diffusion, and who coined the term "the ant in the labyrinth", describing the meandering of random walkers in percolation clusters. Since the pioneering work of de Gennes, the field has expanded very rapidly. The subject has been reviewed by several authors, including ourselves, and from various perspectives. This book builds upon our review on anomalous diffusion from 1987 and it covers the vast material that has accumulated since. Many questions that were unanswered then have been settled, yet, as usual, this has only brought forth a myriad of other questions. Whole new directions of research have emerged, most noticeably in the area of diffusion-limited reactions. The scope of developments is immense and cannot possibly be addressed in one volume. Neither do we have the necessary expertise. Hence, we have chosen once again to base the presentation mostly on heuristic scaling arguments.

The book is written for graduate students, and as an introduction to researchers wishing to enter the field. Much emphasis has been put on its pedagogical value. The end of each chapter includes exercises, open challenges, and references for further reading. The list of open challenges is not exhaustive. It is intended to inspire beginners (many of the challenges require computer programming, for which our youngsters show remarkable aptitude), and to educate the readers to identify new directions of research. Likewise, the references given are simply those that we had at hand. Many excellent works have been left out. Nothing is implied about their relative priority or importance. We merely wished to convey a general impression of the field's scope, and to provide with some starting points. In spite of our efforts, there are bound to be misprints, inaccuracies, and outright mistakes. Please alert us to their presence by sending messages to *benavraham@clarkson.edu* (D.b-A.), or to *havlin@ophir.ph.biu.ac.il* (S.H.).

The book is divided into four parts. Although they are closely related, they can be studied independently from one another. We ourselves have used different combinations in several graduate and upper-level undergraduate courses. Part I contains material of general interest to statistical physics: fractals, percolation theory, regular random walks and diffusion, continuous time random walks, and Lévy walks and flights. Part II expands on our previous review, covering anomalous diffusion in fractals and disordered media. Part III serves as an introduction to the kinetics of diffusion-limited reactions. (The classical case of reaction-limited kinetics is briefly reviewed in Chapter 11.) By and large, the approach used in Parts II and III is that of scaling. Diffusion-limited reactions are still poorly understood, so we believe that examples of exactly solvable models are particularly important. One such example is discussed in Part IV, where we attack the problem of diffusion-limited coalescence in one dimension with the method of inter-particle distribution functions.

We wish to thank our colleagues L. A. N. Amaral, J. S. Andrade, M. Barthelemy, S. Buldyrev, A. Bunde, M. A. Burschka, C. R. Doering, N. V. Dokholyan, A. L. Goldberger, P. Ch. Ivanov, P. R. King, J. Klafter, R. Kopelman, E. Koscielny-Bunde, P. A. Krapivsky, H. Larralde, Y. Lee, F. Leyvraz, R. Nossal, C.-K. Peng, V. Privman, S. Redner, H. E. Roman, S. Russ, M. Schwartz, S. Schwarzer, H. Sompolinsky, H. E. Stanley, H. Taitelbaum, G. M. Viswanathan, I. Webman, and G. Weiss for years of fruitful collaborations, and we gratefully acknowledge the input of our students. Special thanks are due to Jan Kantelhardt for the beautiful cover color prints, to Roi Elran for his help with the preparation of the figures, and to S. Capelin, J. Clegg, S. Holt, and T. Fishlock, of Cambridge University Press, for their patience and for help with the technical aspects of publishing. This book could not have been written without the help of our families and friends. We thank them for their continuous encouragement and support.

Daniel ben-Avraham

Shlomo Havlin

Part one

Basic concepts

The first part of the book includes introductory concepts and background necessary for the understanding of anomalous diffusion in disordered media.

Fractals might be familiar to most readers, but their importance in modeling disordered and random media, as well as certain characteristics of the trails made by diffusing particles, makes it worthwhile to spend some time reviewing the subject. In Chapter 1 we provide working definitions of fractals, fractal dimensions, self-affine fractals, and related ideas. More importantly, we describe several algorithms for determining whether a particular object is a fractal, and for finding its fractal dimension. In the early days of fractal theory much effort was spent on merely exploring the fractal properties of various natural objects and physical models, using precisely such algorithms, and they continue to be essential tools for the study of disordered phenomena.

Percolation, which is reviewed in Chapter 2, is perhaps the most important model of disordered media and of naturally occurring fractals. Percolation owes its enormous appeal to its simplicity (it can be defined and analyzed using only geometrical concepts), its remarkably wide range of applications, and its being one of the most basic models of critical phase transitions. Relevant to our purpose is the fact that studies of anomalous diffusion have traditionally focused on percolation systems, and the problem still attracts considerable interest. The percolation transition, on the other hand, gives us an excellent opportunity to introduce several useful concepts, such as critical exponents, scaling, and the upper critical dimension.

In Chapter 3 we present a brief introduction to random-walk theory. Discrete random walks (in regular lattices) are discussed first, then diffusion and the diffusion equation are obtained as limiting cases. In the course of the book, we shift freely between these discrete (random walk) and continuous (diffusion) representations. The method of generating functions is discussed in some detail, owing to its wide applicability in other realms of statistical physics. It also eases the introduction and discussion of continuous-time random walks (CTRWs).

Finally, in Chapter 4 we review other popular models of transport: Lévy walks, Lévy flights, and long-range correlated walks. These (and some instances of the CTRW) were originally introduced as models that exhibit anomalous transport kinetics even in *regular* lattices, and are therefore relatively easy to analyze, but eventually they came to be studied also in fractals and disordered lattices.

1

Fractals

Fractals model disorder in Nature far more successfully than do objects of classical geometry. In the now famous words of B. Mandelbrot: "Clouds are not spheres, mountains are not cones, coastlines are not circles, bark is not smooth, nor does lightning travel in a straight line" (Mandelbrot, 1982). We begin with a discussion of fractals and their most basic properties: self-similarity and symmetry under dilation, or scaling, and the fractal dimension and the ways to determine it. Our goal is to develop an intuitive understanding, and to provide some basic working tools.

1.1 Deterministic fractals

Deterministic fractals are idealized geometrical structures with the property that parts of the structure are similar to the whole. While this self-similarity is a general property of fractals, it is rather a vague definition, and the best way to understand what fractals really are is through examples. Some well-known deterministic fractals are shown in Figs. 1.1–1.3.

The Koch curve (Fig. 1.1) is constructed from a unit segment. The middle third section is replaced by two other segments of length $\frac{1}{3}$, making a tent shape, as seen in Fig. 1.1a. The same procedure is repeated for each of the four resulting segments (of length $\frac{1}{3}$). This process is iterated *ad infinitum*. The limiting curve is of infinite length, yet it is confined to a finite region of the plane. Thus, the Koch curve is somewhat "denser" than a regular curve of dimension $d = 1$, but certainly "sparser" than a two-dimensional object (its area is zero!). Intuitively, then, its dimension should be between one and two. If a regular object – such as a line segment, a square, or a cube, etc. – of dimension d is magnified by a factor b, the original object would fit b^d times in the magnified one. This consideration may serve as a working definition of the *fractal dimension*, d_f (see Appendix A for more rigorous definitions). In the Koch curve, magnified by a factor of three, there

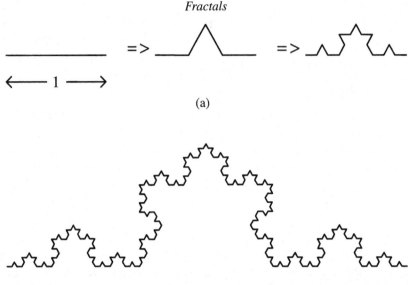

(a)

(b)

Fig. 1.1. The Koch curve. (a) Construction of the curve. The initiator is a unit segment. The generator replaces the middle third section by two similar sections, forming the shape of a tent. Two iterations of this process are shown. (b) The Koch curve after four iterations.

fit exactly four of the original curves. Therefore, its fractal dimension is given by $3^{d_f} = 4$, or $d_f = \ln 4/\ln 3 \simeq 1.262$.

Perhaps the most popular fractal is the Sierpinski gasket (Fig. 1.2). Here one begins with an equilateral triangle that is divided into four equal subunits, and the central subunit is discarded. Again, the process is repeated recursively. The resulting fractal dimension is given by $2^{d_f} = 3$, or $d_f = \ln 3/\ln 2 \simeq 1.585$.

The Sierpinski sponge (also known as the Menger sponge) (Fig. 1.3) is generated from a cube that is subdivided into $3 \times 3 \times 3 = 27$ smaller cubes. The small cube at the center and its six nearest neighbors are then discarded. The same is done with each of the remaining 20 cubes, and the process is iterated indefinitely. The limiting object has *zero* volume, but *infinite* surface area. This property is consistent with the fractal dimension of the sponge; $3^{d_f} = 20$, or $d_f = \ln 20/\ln 3 \simeq 2.727$, between two and three.

All deterministic fractal lattices are obtained in a similar way to the examples above. Construction begins from a genus, called the *initiator* (e.g., the unit segment in the case of the Koch curve, an equilateral triangle for the Sierpinski gasket, etc.) and proceeds with a set of operations that are repeated indefinitely in a recursive fashion. This set of operations is called the *generator*.

The generator may be one of two kinds. In one case, the initiator is replaced by *smaller* replicas of itself and the fractal builds inwardly, towards ever smaller

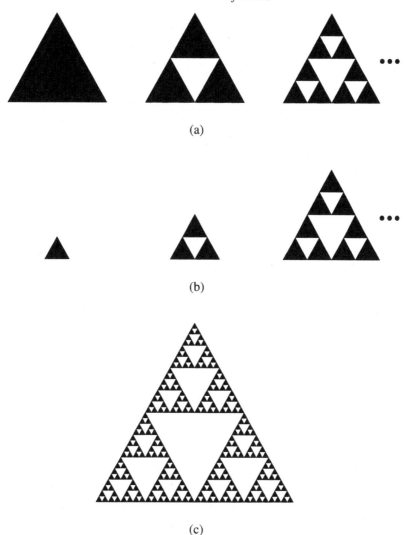

Fig. 1.2. The Sierpinski gasket. Two ways for generating this fractal are shown; (a) from the outside inwards, and (b) from the inside outwards. The initiator defines an upper cutoff length of the fractal, in case (a), and a lower cutoff length, in case (b). (c) The Sierpinski gasket, shown to five generations.

length scales. The resulting fractal has then an *upper cutoff* length – the length of the initiator – but no characteristic microscopic length scale. In the alternative approach, replicas of the initiator are assembled into a larger object and the fractal grows outwards. The lattice then has a *lower cutoff* length, but no characteristic large length scale. The two methods are illustrated in Fig. 1.2. Ideal fractal lattices possess no cutoff lengths (this may be achieved by combining the two kinds of

Fig. 1.3. The Sierpinski sponge. The initiator is a unit cube. The generator divides the unit cube into $3 \times 3 \times 3$ smaller cubes, and seven of these subunits are removed (the one at the center of the cube, and its six nearest neighbors). The Sierpinski sponge after three iterations is shown.

growth), but real-life objects, or fractals constructed in a computer, have both upper and lower cutoffs that represent the size of the fractal structure and the size of its elementary units, respectively.

1.2 Properties of fractals

A most important property of fractals is their *self-similarity*, or their symmetry under dilation. For example, if we examine the Koch curve (Fig. 1.1), or the *Koch snowflake*, as it is frequently called, we notice that there is a central object in the figure which is reminiscent of a snowman. To its right and left there are two other snowmen, each being an exact reproduction of the central snowman, only smaller by a factor of $\frac{1}{3}$. Both of these snowmen display even smaller replicas of themselves to their right and left, etc. In fact, if we look at the Koch curve at any given magnification (suppose we zoom in on the arm of the central snowman, say) we will see the same motif recurring again and again.

The same self-similarity property can be seen also in the Sierpinski gasket (Fig. 1.2). Each triangle subunit, when it is magnified properly, is similar to the whole gasket. Note the main difference between regular Euclidean space and fractal geometries: whereas regular space is symmetric under translation, in fractals this symmetry is violated. Instead, fractals possess a new symmetry, called scale invariance, i.e., invariance under dilation. This is the property of self-similarity. Self-similarity makes fractals useful in the study of phase transitions: the state of a system at the critical transition point is also characterized by a dilation symmetry, and may be thus modeled by fractals.

The fractal dimension d_f is clearly not sufficient as a full description of fractal objects. Another important characteristic of fractals is their ramification. A fractal is *finitely ramified* if any bounded subset of the fractal can be isolated by cutting a finite number of bonds, or intersections. Thus, the Sierpinski gasket and the Koch curve are finitely ramified, whereas the Sierpinski sponge is *infinitely ramified*. A large variety of physical processes that take place on finitely ramified fractals can be analyzed through an *exact* renormalization-group approach. This accounts for the enormous popularity of the Sierpinski gasket – it is one of the simplest fractals with this beautiful property, yet it is complex enough to yield interesting insights.

Many other measures and exponents may be defined to capture specific properties of fractals. For example, the concept of *lacunarity* is related to the degree of homogeneity of a fractal and to the extent that it is translation-invariant; the *fracton*, or *spectral* dimension, d_s is an exponent describing the scaling of the density of states associated with the Laplacian operator in a fractal; and the shortest path exponent, d_{min}, characterizes the length of the shortest, or *chemical*, path connecting two points in a fractal. Some of these properties will be required in our discussion and will be expanded upon as necessary.

1.3 Random fractals

Fractals need not be generated by stiff deterministic rules. Stochastic elements may be included in the generator. Consider for example the Sierpinski carpet – the surface of the Sierpinski sponge (Fig. 1.4a). It is obtained from a unit-square initiator. The generator divides this square into nine smaller cells and discards the central cell. In Fig. 1.4b we show the result obtained when the discarded cell is one of the nine cells, *chosen at random*. Clearly, the two figures are related, but the object in Fig. 1.4b is no longer exactly self-similar. Instead, we can argue that it is self-similar in a *statistical* sense: the distribution of holes looks similar at all length scales. Also, on average, the "mass" of this object (the black areas in the figure) increases by a factor of eight when space is dilated by a factor of three. Thus, the random carpet has the same fractal dimension $d_f = \ln 8/\ln 3$ as the deterministic carpet. Generally, the mass M of random fractals scales upon dilation by a factor b as

$$M(bL) = b^{d_f} M(L), \qquad (1.1a)$$

exactly as for deterministic fractals. Note that the solution of this functional equation is

$$M(L) = A L^{d_f}, \qquad (1.1b)$$

where A is a constant.

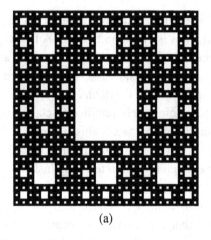

(a)   (b)

Fig. 1.4. Deterministic versus random fractals. (a) The *deterministic* Sierpinski carpet is generated from a unit square, subdivided into 3 × 3, by removing the *central* subunit. (b) The *random* Sierpinski carpet is obtained in a similar fashion, but in each generation the discarded subunit is chosen *randomly*.

Random fractals are useful as models for natural phenomena. Indeed, the fractal of Fig. 1.4b resembles a surface of a real sponge more closely than does the original Sierpinski carpet. A similar adaptation of the Koch curve, say, may provide an appropriate description of the coastline of Norway. Thanks to the pioneering work of Mandelbrot and others, we now know that natural objects are more likely to be fractals rather than not. How does Nature generate all these wonderful fractals? This is a very deep question and the subject of much recent research. In the next chapter we will discuss *percolation* – one of the most important and best understood phenomena giving rise to natural fractals. The path of a particle undergoing Brownian motion is yet another ubiquitous natural fractal that will concern us through most of this book.

The fractal dimensions of random fractals are usually found numerically, either from the scaling of mass with linear size (Eq. (1.1)) – the *sand-box algorithm* – or from the *box-counting algorithm*. In the sand-box method one calculates the mass within a radius L around a site belonging to the fractal. The fractal dimension is determined by averaging over many sites, and using Eq. (1.1b). In the box-counting algorithm the space embedding the fractal is subdivided into a hypercubic grid of cells of linear size ϵ. One then counts the number of cells (boxes) that contain parts of the fractal, $N(\epsilon)$. The procedure is repeated for several box sizes, and d_f is determined from the relation

$$N(\epsilon) \sim \epsilon^{-d_f}. \tag{1.2}$$

The scaling of mass may also be analyzed through the density–density correlation function:

$$c(r) = \frac{1}{V} \sum_{r'} \rho(r')\rho(r' + r). \tag{1.3}$$

Here $\rho(r') = 1$ if the site r' is part of the fractal and 0 otherwise, and $V = \sum_{r'} \rho(r')$ is a normalization factor. Simply, $c(r)$ is the average mass density around r from an arbitrary point belonging to the fractal. For isotropic self-similar fractals we expect that

$$c(r) = c(r) \sim r^{-\alpha}. \tag{1.4}$$

If the fractal is embedded in d-dimensional space, its mass within a linear size L is

$$M(L) \sim \int_0^L c(r) \, d^d r \sim L^{d-\alpha}, \tag{1.5}$$

and, since $M(L) \sim L^{d_f}$, it follows that

$$\alpha = d - d_f. \tag{1.6}$$

The measurement of d_f, by any of the techniques discussed so far, requires a digitalization of the available experimental data. It is often more convenient to obtain d_f directly from scattering experiments. The scattering intensity is proportional to the structure factor $S(q)$, which is simply the Fourier transform of the density $c(r)$. For isotropic random fractals

$$S(q) = S(q) \sim q^{-d_f}. \tag{1.7}$$

For natural fractals, the power-law dependence of mass upon distance is valid only within the cutoff length scales λ_- and λ_+, and Eq. (1.7) applies only for $1/\lambda_+ < q < 1/\lambda_-$.

1.4 Self-affine fractals

Until now we have considered only *isotropic* fractals, which display the same self-similar property in all directions. More generally, an object may possess an anisotropic dilation symmetry. In such a case we speak of *self-affinity*.

An example is the fractal shown in Fig. 1.5. The initiator is a unit square. The generator consists of subdividing the initiator into a grid of $b_1 \times b_2$ rectangular cells of which only n are retained ($b_1 b_2 - n$ cells are discarded). In our example $b_1 = 3$, $b_2 = 2$, and $n = 3$. This object scales differently under dilation along the horizontal (x) and vertical (y) directions: to recover the whole fractal one must magnify a subunit by a factor of three along the x-direction, and by a factor of two along the y-direction.

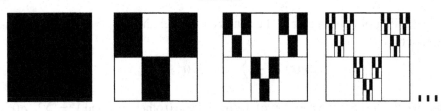

Fig. 1.5. A self-affine fractal. The rectangular initiator is divided into $b_1 \times b_2$ subunits, and n of the subunits are discarded. Three iterations of this process are shown. In this particular example $b_1 = 3$, $b_2 = 2$, and $n = 3$. Notice that the resulting fractal is disconnected. Designing connected self-affine fractals of a similar nature is an amusing challenge.

What is the fractal dimension of a self-affine object? If we try to apply the box-counting algorithm to the fractal of Fig. 1.5, together with Eq. (1.2), we find that the fractal dimension d_f seems to vary as a function of the length scale. On the one hand, as the size of the boxes decreases indefinitely,

$$d_{f,\text{local}} = \lim_{\epsilon \to 0} \frac{\ln N(\epsilon)}{\ln(1/\epsilon)} = \frac{\ln(nb_1/b_2)}{\ln b_1}. \tag{1.8}$$

This is a measure of the apparent *local* fractal dimension that we see as we zoom in to ever smaller length scales. On the other hand, for ever growing boxes,

$$d_{f,\text{global}} = \lim_{\epsilon \to \infty} \frac{\ln N(\epsilon)}{\ln(1/\epsilon)} = \frac{\ln(nb_2/b_1)}{\ln b_2}. \tag{1.9}$$

This *global* fractal dimension characterizes the object at the largest length scales. For an isotropic, self-similar fractal the local and global dimensions are the same and the distinction is unnecessary.

Another way of characterizing the fractal dimensions of self-affine objects focuses on the anisotropic scaling of the various spatial directions, rather than on the distinction between length scales. Thus, generalizing Eq. (1.1), we may write

$$M(b^{1/d_f^x} L_x, b^{1/d_f^y} L_y) = b M(L_x, L_y), \tag{1.10}$$

thus defining different dimensions for the x- and y-axes. In our example $M(b_1 L_x, b_2 L_y) = n M(L_x, L_y)$, and therefore $d_f^x = \ln n/\ln b_1$ and $d_f^y = \ln n/\ln b_2$.

Random affine fractals are also common in Nature. A mountainous landscape is the usual example, but more generally they tend to appear in phenomena involving surfaces and interfaces. The new science of *surface growth* profits immensely from the concept. The practical methods for measuring d_f of random isotropic fractals are easily generalized to affine fractals.

Fig. 1.6. The Cantor set in one dimension. The initiator is the unit segment, and the generator discards the middle third subsection. Successive iterations are shown beneath each other.

1.5 Exercises

1. The Cantor set is obtained from the unit segment $(0, 1)$ by removing the middle third $(\frac{1}{3}, \frac{2}{3})$ and then repeating the process recursively (Fig. 1.6). Find the fractal dimension of this set. Generalize to $d > 1$ dimensions. In $d = 2$ and 3 dimensions the Cantor set is identical to the Sierpinski carpet and the Sierpinski sponge, respectively. (Answer: $d_f = \ln(3^d - 1)/\ln 3$.)

2. Find the box dimension of the set $S = \{1, \frac{1}{2}, \frac{1}{3}, \ldots\}$ (see Appendix A). Is this set a fractal? (Answer: $\frac{1}{2}$; no.)

3. Generalize the Sierpinski gasket to $d > 2$ dimensions and compute the fractal dimensionality. Show in this way that a fractal may have integer d_f. (Answer: $d_f = \ln(d + 1)/\ln 2$; integer for $d = 2^n - 1$.)

4. Find examples of fractal lattices that have the same d_f but are different otherwise; for instance, finitely ramified versus infinitely ramified. Thus show that the fractal dimensionality does not characterize fractals uniquely.

5. Prove Eq. (1.7). (Hint: $S(q) = \int d^d r \, e^{iq \cdot r} c(r)$.)

6. Construct random Cantor sets on the computer, by randomly selecting the section omitted at each generation. Try the two different methods of construction (inward and outward). Measure the fractal dimensionality using the sand-box algorithm and the scaling of mass. Notice the influence of the lower and upper cutoffs.

7. Generalize the box-counting algorithm for measuring d_f^x and d_f^y of affine fractals. Generalize the method of density–density correlations for measuring $d_{f,local}$, $d_{f,global}$, d_f^x, and d_f^y.

8. Find a general relation between the two sets of exponents d_f^x, d_f^y and $d_{f,local}$, $d_{f,global}$. (Answer: $d_{f,local} = 1 + d_f^x - d_f^x/d_f^y$; $d_{f,global} = 1 + d_f^y - d_f^y/d_f^x$.)

9. Starting at (x, y), move to $(x+1, y-1)$ or to $(x+1, y+1)$ with equal probabilities. Repetition of this rule generates a random self-affine path. Simulate it on the computer and measure the various fractal dimensions. (Answer: $d_{f,local} = \frac{3}{2}$, $d_{f,global} = 1$, $d_f^x = 1$, and $d_f^y = 2$.)

1.6 Open challenges

The field of fractals has been very active since about 1980, in particular concerning diffusion-limited aggregation (Meakin, 1998), fractal surfaces (Barabási and Stanley, 1995), localization (Schreiber and Grussbach, 1991), and turbulence (Frisch, 1995; Meneveau and Sreenivasan, 1987; 1991), and finds applications in an impressive diversity of sciences: astrophysics (Labini *et al.*, 1998a; 1998b), chemistry, geochemistry, and biophysics (Birdi, 1993), physiology (West, 1990; West and Deering, 1994; Bassinthwaighte *et al.*, 1994), geology and geophysics (Turcotte, 1992), ecology (Sugihara and May, 1990), etc. Here are some of the many open problems.

1. What is the origin of fractals in Nature? See, for example, Bak (1996).
2. A question regarding the mere existence of fractals has been raised by Malcai *et al.* (1997).
3. Are there complex fractal dimensions? This question has been discussed by Sornette *et al.* (1996a).

1.7 Further reading

- About fractals in general and selected applications, see Mandelbrot (1977; 1982), Peitgen and Richter (1986), Peitgen *et al.* (1992), Feder (1988), Barnsley (1988), Takayasu (1990), Stanley and Ostrowsky (1990), Vicsek (1991), Gouyet (1992), Avnir (1992), Turcotte (1992), Bak (1996), Bunde and Havlin (1994; 1996), and Meakin (1998).
- Properties of fractals: ramification (Gefen *et al.*, 1981), lacunarity (Gefen *et al.*, 1980), the fracton dimension (Alexander and Orbach, 1982; Rammal and Toulouse, 1983), and the chemical dimension (Havlin and Nossal, 1984).
- On the effect of the cutoff lengths on the scaling of fractals, with applications to galaxies, see Amici and Montuori (1998).

2

Percolation

Random fractals in Nature arise for a variety of reasons (dynamic chaotic processes, self-organized criticality, etc.) that are the focus of much current research. Percolation is one such chief mechanism. The importance of percolation lies in the fact that it models critical phase transitions of rich physical content, yet it may be formulated and understood in terms of very simple geometrical concepts. It is also an extremely versatile model, with applications to such diverse problems as supercooled water, galactic structures, fragmentation, porous materials, and earthquakes.

2.1 The percolation transition

Consider a square lattice on which each bond is present with probability p, or absent with probability $1 - p$. When p is small there is a dilute population of bonds, and clusters of small numbers of connected bonds predominate. As p increases, the size of the clusters also increases. Eventually, for p large enough there emerges a cluster that spans the lattice from edge to edge (Fig. 2.1). If the lattice is infinite, the inception of the spanning cluster occurs sharply upon crossing a *critical threshold* of the bond concentration, $p = p_c$.

The probability that a given bond belongs to the incipient infinite cluster, P_∞, undergoes a phase transition: it is zero for $p < p_c$, and increases continuously as p is made larger than the critical threshold p_c (Fig. 2.2). Above and close to the transition point, P_∞ follows a power law:

$$P_\infty \sim (p - p_c)^\beta. \tag{2.1}$$

This phenomenon is known as the *percolation* transition. The name comes from the possible interpretation of bonds as channels open to the flow of a fluid in a porous medium (absent bonds represent blocked channels). At the transition point the fluid can percolate through the medium for the first time. The flow rate undergoes

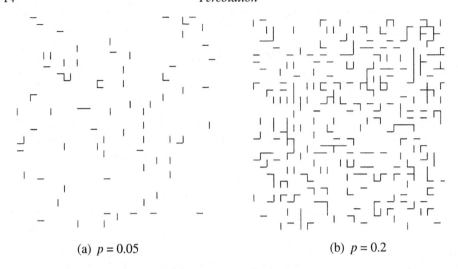

(a) $p = 0.05$ (b) $p = 0.2$

(c) $p = 0.5$

Fig. 2.1. Bond percolation on the square lattice. Shown are 40×40 square lattices, where bonds are present with probabilities $p = 0.05$ (a), 0.20 (b), and 0.50 (c). Notice how the clusters of connected bonds (i.e., the percolation clusters) grow in size as p increases. In (c) the concentration is equal to the critical concentration for bond percolation on the square lattice, $p_c = 0.5$. A cluster spanning the lattice (from top to bottom) appears for the first time. The bonds of this incipient infinite cluster are highlighted in bold.

a phase transition similar to that of P_∞. In fact, the transition is similar to all other continuous (second-order) phase transitions in physical systems. P_∞ plays the role of an *order parameter*, analogous to magnetization in a ferromagnet, and β is the *critical exponent* of the order parameter.

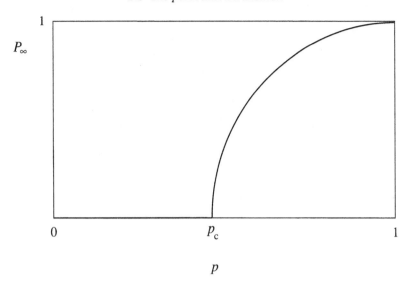

Fig. 2.2. A schematic representation of the percolation transition. The probability P_∞ that a bond belongs to the spanning cluster undergoes a sharp transition (in the thermodynamic limit of infinitely large systems): below a critical probability threshold p_c there is no spanning cluster, so $P_\infty = 0$, but P_∞ becomes finite when $p > p_c$.

There exists a large variety of percolation models. For example, the model above can be defined on a triangular lattice, or any other lattice besides the square lattice. In *site percolation* the percolating elements are lattice sites, rather than bonds. In that case we think of nearest-neighbor sites as belonging to the same cluster (Fig. 2.3). Other connectivity rules may be employed: in *bootstrap percolation* a subset of the cluster is connected if it is attached by at least two sites, or bonds. *Continuum percolation* is defined without resorting to a lattice – consider for example a set of circles randomly placed on a plane, where contact is made through their partial overlap (Fig. 2.4). Finally, one may consider percolation in different space dimensions. The percolation threshold p_c is affected by these various choices (Table 2.1), but critical exponents, such as β, depend only upon the space dimension. This insensitivity to all other details is termed *universality*. Clearly, critical exponents capture something very essential of the nature of the model at hand. They are used to classify critical phase transitions into *universality classes*.

Let us define some more of these important critical exponents. The typical length of finite clusters is characterized by the *correlation length* ξ. It diverges as p approaches p_c as

$$\xi \sim |p - p_c|^{-\nu}, \tag{2.2}$$

(a)

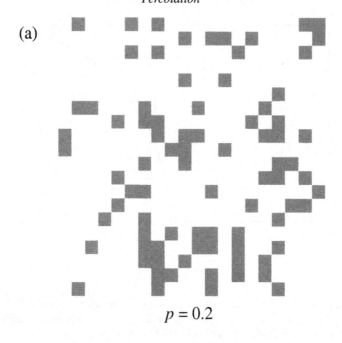

$p = 0.2$

(b)

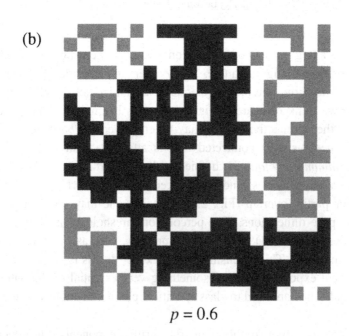

$p = 0.6$

Fig. 2.3. Site percolation on the square lattice. Shown are 20×20 square lattices with sites occupied (gray squares) with probabilities $p = 0.2$ (a) and 0.6 (b). Nearest-neighbor sites (squares that share an edge) belong to the same cluster. The concentration in (b) is slightly above p_c of the infinite system, hence a spanning cluster results. The sites of the "infinite" cluster are in black.

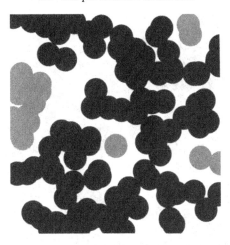

Fig. 2.4. Continuum percolation of circles on the plane. In this example the percolating elements are circles of a given diameter, which are placed *randomly* on the plane. Overlapping circles belong to the same cluster. As the concentration of circles increases the clusters grow in size, until a spanning percolating cluster appears (black circles). This type of percolation model requires no underlying lattice.

Table 2.1. *Percolation thresholds for several two- and three-dimensional lattices and the Cayley tree.*

Lattice	Percolation	
	Sites	Bonds
Triangular	$\frac{1}{2}$[a]	$2\sin(\pi/18)$[a]
Square	$0.592\,746\,0$[b,c]	$\frac{1}{2}$[a]
Honeycomb	$0.697\,043$[d]	$1 - 2\sin(\pi/18)$[a]
Face-centered cubic	0.198[e]	$0.120\,163\,5$[c]
Body-centered cubic	0.254[e]	$0.180\,287\,5$[c]
Simple cubic (first nearest neighbor)	$0.311\,605$[f,g]	$0.248\,812\,6$[c,h]
Simple cubic (second nearest neighbor)	0.137[i]	–
Simple cubic (third nearest neighbor)	0.097[i]	–
Cayley tree	$1/(z-1)$	$1/(z-1)$
Continuum percolation $d = 2$ (overlapping circles)	0.312 ± 0.005[j]	–
Continuum percolation $d = 3$ (overlapping spheres)	0.2895 ± 0.0005[k]	–

[a]Exact: Essam *et al.* (1978), Kesten (1982), Ziff (1992); [b]Ziff and Sapoval (1987); [c]Lorenz and Ziff (1998); [d]Suding and Ziff (1999); [e]Stauffer (1985a); [f]Strenski *et al.* (1991); [g]Acharyya and Stauffer (1998); [h]Grassberger (1992a); [i]Domb (1966); [j]Vicsek and Kertesz (1981), Kertesz (1981); and [k]Rintoul and Torquato (1997).

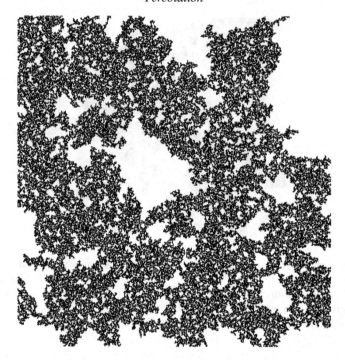

Fig. 2.5. An incipient infinite cluster. Shown is the spanning cluster in site percolation on the square lattice, as obtained from a computer simulation in a 400×400 square, with $p = 0.6$ (just above the percolation threshold). For clarity, occupied sites that do not belong to the spanning cluster have been removed, thus highlighting the presence of holes on all length scales – a characteristic feature of random fractals.

with the same critical exponent ν below and above the transition. The average mass (the number of sites in site percolation, or the number of bonds in bond percolation) of finite clusters, S, is analogous to the magnetic susceptibility in ferromagnetic phase transitions. It diverges about p_c as

$$S \sim |p - p_c|^{-\gamma}, \tag{2.3}$$

again with the same exponent γ on both sides of the transition. In the following sections we shall meet some more exponents and we shall see how they are related to each other.

2.2 The fractal dimension of percolation

The structure of percolation clusters can be well described by fractal concepts. Consider first the incipient infinite cluster at the critical threshold. An example is shown in Fig. 2.5. As is evident, the cluster contains holes on all length scales,

Fig. 2.6. A schematic representation of the infinite percolation cluster above p_c. The fractal features of the infinite cluster above the percolation threshold are represented schematically by repeating Sierpinski gaskets of length ξ, the so-called correlation length. There is self-similarity only at distances shorter than ξ, whereas on larger length scales the cluster is homogeneous (like a regular triangular lattice, in this drawing).

similar to the random Sierpinski carpet of Fig. 1.4b. In fact, with help of the box-counting algorithm, or other techniques from Chapter 1, one can show that the cluster is self-similar on all length scales (larger than the lattice spacing and smaller than its overall size) and can be regarded as a fractal. Its fractal dimension d_f describes how the mass S within a sphere of radius r scales with r:

$$S(r) \sim r^{d_f}. \tag{2.4}$$

$S(r)$ is obtained by averaging over many cluster realizations (in different percolation simulations), or, equivalently, averaging over different positions of the center of the sphere in a single infinite cluster.

Let us now examine percolation clusters off criticality. Below the percolation threshold the typical size of clusters is finite, of the order of the correlation length ξ. Therefore, clusters below criticality can be self-similar only up to the length scale of ξ. The system possesses a natural upper cutoff. Above criticality, ξ is a measure of the size of the *finite* clusters in the system. The incipient infinite cluster remains infinite in extent, but its largest holes are also typically of size ξ. It follows that the infinite cluster can be self-similar only up to length scale ξ. At distances larger than ξ self-similarity is lost and the infinite cluster becomes homogeneous. In other words, for length scales shorter than ξ the system is *scale invariant* (or self-similar) whereas for length scales larger than ξ the system is *translationally*

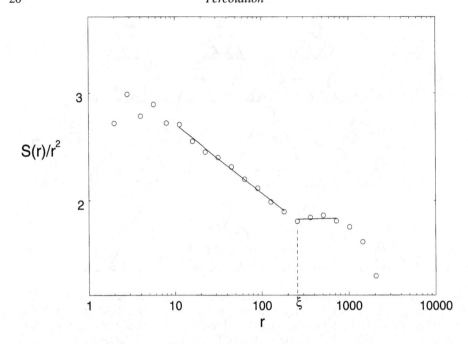

Fig. 2.7. The structure of the infinite percolation cluster above p_c. The dependence of the fractal dimension upon the length scale (Eq. (2.5)) is clearly seen in this plot of $S(r)/r^d$ ($d = 2$) versus r, for the infinite cluster in a 2500×2500 percolation system. The slope of the curve is $d_f - d$ for $r < \xi \approx 200$, and zero for $r > \xi$.

invariant (or homogeneous). The situation is cartooned in Fig. 2.6, in which the infinite cluster above criticality is likened to a regular lattice of Sierpinski gaskets of size ξ each. The peculiar structure of the infinite cluster implies that its mass scales differently at distances shorter and larger than ξ:

$$S(r) \sim \begin{cases} r^{d_f} & r < \xi, \\ r^d & r > \xi. \end{cases} \qquad (2.5)$$

Fig. 2.7 illustrates this crossover measured in a two-dimensional percolation system above p_c.

We can now identify d_f by relating it to other critical exponents. An arbitrary site, within a given region of volume V, belongs to the infinite cluster with probability S/V (S is the mass of the infinite cluster enclosed within V). If the linear size of the region is smaller than ξ the cluster is self-similar, and so

$$P_\infty \sim \frac{r^{d_f}}{r^d} \sim \frac{\xi^{d_f}}{\xi^d}, \qquad r < \xi. \qquad (2.6)$$

Using Eqs. (2.1) and (2.2) we can express both sides of Eq. (2.6) as powers of

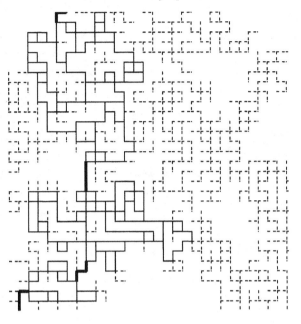

Fig. 2.8. Subsets of the incipient infinite percolation cluster. The spanning cluster (from top to bottom of the lattice) in a computer simulation of bond percolation on the square lattice at criticality is shown. Subsets of the cluster are highlighted: dangling ends (broken lines), blobs (solid lines), and red bonds (bold solid lines).

$p - p_c$:

$$(p - p_c)^\beta \sim (p - p_c)^{-\nu(d_f - d)}, \tag{2.7}$$

hence

$$d_f = d - \frac{\beta}{\nu}. \tag{2.8}$$

Thus, the fractal dimension of percolation is not a new, independent exponent, but depends on the critical exponents β and ν. Since β and ν are universal, d_f is also universal!

2.3 Structural properties

As with other fractals, the fractal dimension is not sufficient to fully characterize the geometrical properties of percolation clusters. Different geometrical properties are important according to the physical application of the percolation model.

Suppose that one applies a voltage on two sites of a metallic percolation cluster. The *backbone* of the cluster consists of those bonds (or sites) which carry the electric current. The remaining parts of the cluster which carry no current are

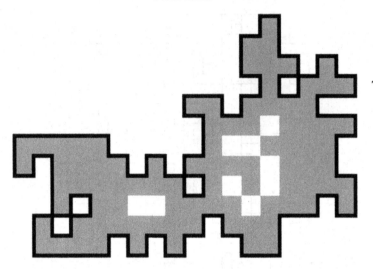

Fig. 2.9. The hull of percolation clusters. The external perimeter (the hull) is highlighted in bold lines in this computer simulation of a cluster of site percolation in the square lattice. The *total perimeter* includes also the edges of the internal "lakes" (not shown).

the *dangling ends* (Fig. 2.8). They are connected to the backbone by a single bond. The *red bonds* are those bonds that carry the total current; severing a red bond stops the current flow. The *blobs* are what remains from the backbone when all the red bonds are removed (Fig. 2.8). Percolation clusters (in the self-similar regime) are finitely ramified: arbitrarily large subsets of a cluster may always be isolated by cutting a finite number of red bonds.

The external perimeter of a cluster, which is also called the *hull*, consists of those cluster sites which are connected to infinity through an uninterrupted chain of empty sites (Fig. 2.9). In contrast, the *total perimeter* includes also the edges of internal holes. The hull is an important model for random fractal interfaces.

The fractal dimension of the backbone, d_f^{BB}, is smaller than the fractal dimension of the cluster (see Table 2.2). That is to say, most of the mass of the percolation cluster is concentrated in the dangling ends, and the fractal dimension of the dangling ends is equal to that of the infinite cluster. The fractal dimension of the backbone is known only from numerical simulations.

The fractal dimensions of the red bonds and of the hull are known from exact arguments. The mean number of red bonds has been shown to vary with p as $\langle N \rangle \sim (p - p_c)^{-1} \sim \xi^{1/\nu}$, hence the fractal dimension of red bonds is $d_{red} = 1/\nu$. The fractal dimension of the hull in $d = 2$ is $d_h = \frac{7}{4}$ – smaller than the fractal dimension of the cluster, $d_f = 91/48$. In $d \geq 3$, however, the mass of the hull is believed to be proportional to the mass of the cluster, and both have the same fractal dimension.

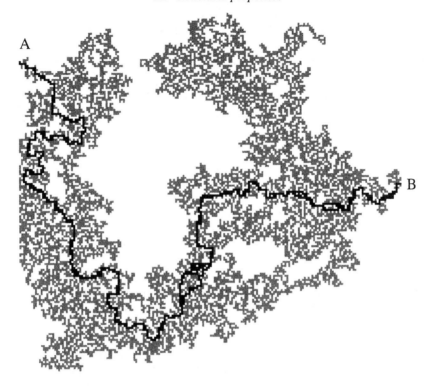

A

B

Fig. 2.10. Chemical distance. The chemical path between two sites A and B in a two-dimensional percolation cluster is shown in black. Notice that more than one chemical path may exist. The union of all the chemical paths shown is called the *elastic backbone*.

As an additional characterization of percolation clusters we mention the *chemical distance*. The chemical distance, ℓ, is the length of the shortest path (along cluster sites) between two sites of the cluster (Fig. 2.10). The *chemical dimension* d_ℓ, also known as the *graph dimension* or the *topological dimension*, describes how the mass of the cluster within a chemical length ℓ scales with ℓ:

$$S(\ell) \sim \ell^{d_\ell}. \tag{2.9}$$

By comparing Eqs. (2.4) and (2.9), one can infer the relation between regular Euclidean distance and chemical distance:

$$r \sim \ell^{d_\ell/d_f} \equiv \ell^{\nu_\ell}. \tag{2.10}$$

This relation is often written as $\ell \sim r^{d_{\min}}$, where $d_{\min} \equiv 1/\nu_\ell$ can be regarded as the fractal dimension of the minimal path. The exponent $d_{\min}$ is known mainly from numerical simulations. Obviously, $d_{\min} \geq 1$ (see Table 2.2). In many known deterministic fractals the chemical length exponent is either $d_\ell = d_f$ (e.g., for the Sierpinski gasket) or $d_\ell = 1$ (e.g., for the Koch curve). An example of an

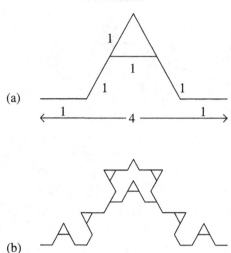

(a)

(b)

Fig. 2.11. The modified Koch curve. The initiator consists of a unit segment. Shown is the curve after one generation (a), and two generations (b). Notice that the shortest path (i.e., the chemical length) between the two endpoints in (a) is five units·long.

exception to this rule is exhibited by the modified Koch curve of Fig. 2.11. The fractal dimension of this object is $d_f = \ln 7/\ln 4$, while its chemical dimension is $d_\ell = \ln 7/\ln 5$ (or $d_{min} = \ln 5/\ln 4$).

The concept of chemical length finds several interesting applications, such as in the *Leath algorithm* for the construction of percolation clusters (Exercise 2), or in oil recovery, in which the first-passage time from the injection well to a production well a distance r away is related to ℓ. It is also useful in the description of propagation of epidemics and forest fires. Suppose that trees in a forest are distributed as in the percolation model. Assume further that in a forest fire at each unit time a burning tree ignites fires in the trees immediately adjacent to it (the nearest neighbors). The fire front will then advance one *chemical shell* (sites at equal chemical distance from a common origin) per unit time. The speed of propagation would be

$$v = \frac{dr}{dt} = \frac{dr}{d\ell} \sim \ell^{\nu_\ell - 1} \sim (p - p_c)^{\nu(d_{min} - 1)}. \qquad (2.11)$$

In $d = 2$ the exponent $\nu(d_{min} - 1) \approx 0.16$ is rather small and so the increase of v upon crossing p_c is steep: a fire that could not propagate at all below p_c may propagate very fast just above p_c, when the concentration of trees is only slightly bigger.

In Table 2.2 we list the values of some of the percolation exponents discussed above. As mentioned earlier, they are universal and depend only on the dimension-

Table 2.2. *Fractal dimensions of the substructures composing percolation clusters.*

d	2	3	4	5	6
d_f	$91/48^a$	2.53 ± 0.02^b	3.05 ± 0.05^c	3.69 ± 0.02^d	4
d_{min}	1.1307 ± 0.0004^e	1.374 ± 0.004^e	1.60 ± 0.05^f	1.799^g	2
d_{red}	$3/4^h$	1.143 ± 0.01^i	1.385 ± 0.055^j	1.75 ± 0.01^j	2
d_h	$7/4^k$	2.548 ± 0.014^i			4
d_f^{BB}	1.6432 ± 0.0008^l	1.87 ± 0.03^m	1.9 ± 0.2^n	1.93 ± 0.16^n	2
ν	$4/3^a$	0.88 ± 0.02^c	0.689 ± 0.010^p	0.571 ± 0.003^q	1/2
τ	$187/91^r$	2.186 ± 0.002^b	2.31 ± 0.02^r	2.355 ± 0.007^r	5/2

[a]den Nijs (1979), Nienhuis (1982); [b]Jan and Stauffer (1998). Other simulations (Lorenz and Ziff, 1998) yield $\tau = 2.189 \pm 0.002$; [c]Grassberger (1983; 1986); [d]Jan et al. (1985); [e]Grassberger (1992a). Earlier simulations (Herrmann and Stanley, 1988) yield $d_{min} = 1.130 \pm 0.004$ ($d = 2$); [f]calculated from $d_{min} = 1/\nu_\ell$; [g]Janssen (1985), from ϵ-expansions; [h]Coniglio (1981; 1982); [i]Strenski et al. (1991); [j]calculated from $d_{red} = 1/\nu$; [k]Sapoval et al. (1985), Saleur and Duplantier (1987); [l]Grassberger (1999a); [m]Porto et al. (1997b). Series expansions (Bhatti et al., 1997) yield $d_f^{BB} = 1.605 \pm 0.015$; [n]Hong and Stanley (1983a); [p]Ballesteros et al. (1997). They also find $\eta = 2 - \gamma/2 = 0.0944 \pm 0.0017$; [q]Adler et al. (1990); and [r]calculated from $\tau = 1 + d/d_f$. For the meaning of τ, see Section 2.4. Notice also that β and γ may be obtained from the other exponents, for example: $\beta = \nu(d - d_f)$, $\gamma = \beta(\tau - 2)/(3 - \tau)$.

ality of space, not on other details of the percolation model. Above $d = 6$ loops in the percolation clusters are too rare to play any significant role and they can be neglected. Consequently, the values of the critical exponents for $d > 6$ are exactly the same as for $d = 6$. The dimension $d = d_c = 6$ is called the *upper critical dimension*. The exponents for $d \geq d_c$ may be computed exactly, as we show in the next section.

2.4 Percolation on the Cayley tree and scaling

The Cayley tree is a loopless lattice, generated as follows. From a central site – the *root*, or *origin* – there emanate z branches. The end of each branch is a site, so there are z sites, which constitute the first shell of the Cayley tree. From each site of the first (chemical) shell there emanate $z - 1$ branches, generating $z(z - 1)$ sites, which constitute the second shell. In the same fashion, from each site of the ℓth shell there emanate $z - 1$ new branches whose endpoints are sites of the $(\ell + 1)$th shell (Fig. 2.12). The ℓth shell contains $z(z - 1)^{\ell-1}$ sites and therefore the Cayley tree may be regarded as a lattice of infinite dimension, since the number of sites grows exponentially – faster than any power law. The absence of loops in the

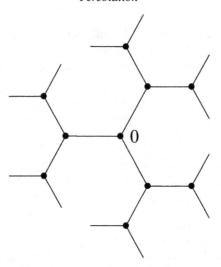

Fig. 2.12. The Cayley tree with $z = 3$. The chemical shells $\ell = 0$ (the "origin", 0), $\ell = 1$, and $\ell = 2$ are shown.

Cayley tree allows one to solve the percolation model (and other physics models) exactly. We now demonstrate how to obtain the percolation exponents for $d \geq 6$.

We must address the issue of distances beforehand. The Cayley tree cannot be embedded in any lattice of finite dimension, and so instead of Euclidean distance one must work with chemical distance. Because of the lack of loops there is only one path between any two sites, whose length is then by definition the chemical length ℓ. Above the critical dimension $d \geq d_c = 6$ we expect that correlations are negligible and that any path on a percolation cluster is essentially a random walk; $r^2 \sim \ell$, or

$$d_{\min} = 2, \tag{2.12}$$

(cf. Eq. (2.10)). This connects Euclidean distance to chemical distance.

Consider now a percolation cluster on the Cayley tree. Suppose that the origin is part of a cluster. In the first shell, there are on average $\langle s_1 \rangle = pz$ sites belonging to that same cluster. The average number of cluster sites in the $(\ell + 1)$th shell is $\langle s_{\ell+1} \rangle = \langle s_\ell \rangle p(z - 1)$. Thus,

$$\langle s_\ell \rangle = z(z - 1)^{\ell-1} p^\ell = zp[(z - 1)p]^{\ell-1}. \tag{2.13}$$

From this we can deduce p_c: when $\ell \to \infty$ the number of sites in the ℓth shell tends to zero if $p(z - 1) < 1$, whereas it diverges if $p(z - 1) > 1$; hence

$$p_c = \frac{1}{z - 1}. \tag{2.14}$$

For $p < p_c$, the density of cluster sites in the ℓth shell is $\langle s_\ell \rangle / \sum_{\ell=1}^{\infty} \langle s_\ell \rangle$.

Therefore the correlation length in chemical distance is (using Eqs. (2.13) and (2.14))

$$\xi_\ell = \frac{\sum_{\ell=1}^{\infty} \ell \langle s_\ell \rangle}{\sum_{\ell=1}^{\infty} \langle s_\ell \rangle} = \frac{p_c}{p_c - p}, \qquad p < p_c. \tag{2.15}$$

The correlation length in regular space is $\xi \sim \xi_\ell^{\nu_\ell}$, and therefore

$$\xi \sim (p_c - p)^{-1/2}, \tag{2.16}$$

or $\nu = \frac{1}{2}$. The mean mass of the finite clusters (below p_c) is

$$S = 1 + \sum_{\ell=1}^{\infty} \langle s_\ell \rangle = p_c \frac{1 + p}{p_c - p} = (p_c - p)^{-\gamma}, \tag{2.17}$$

which yields $\gamma = 1$ for percolation on the Cayley tree.

Consider next sn_s, the probability that a given site belongs to a cluster of s sites. The quantity n_s is the analogous probability *per cluster site*, or the probability distribution of cluster sizes in a percolation system. Suppose that a cluster of s sites possesses t perimeter sites (empty sites adjacent to the cluster). The probability of such a configuration is $p^s(1 - p)^t$. Hence,

$$n_s = \sum_t g_{s,t} p^s (1 - p)^t, \tag{2.18}$$

where $g_{s,t}$ is the number of possible configurations of s-clusters with t perimeter sites. In the Cayley tree all s-site clusters have exactly $2 + (z - 2)s$ perimeter sites, and Eq. (2.18) reduces to

$$n_s(p) = g_s p^s (1 - p)^{2+(z-2)s}, \tag{2.19}$$

where now g_s is simply the number of possible configurations of an s-cluster. We are interested in the behavior of n_s near the percolation transition. Expanding Eq. (2.19) around $p_c = 1/(z - 1)$ to lowest order in $p - p_c$ yields

$$n_s(p) \sim n_s(p_c) \exp[-(p - p_c)^2 s]. \tag{2.20}$$

To estimate $n_s(p_c)$ we need to compute g_s, which can be done through exact combinatorics arguments. The end result is that n_s behaves as a power law, $n_s(p_c) \sim s^{-\tau}$, with $\tau = \frac{5}{2}$. The above behavior of n_s is also typical of percolation in $d < 6$ dimensions. Generally,

$$n_s \sim s^{-\tau} f((p - p_c)s^\sigma), \tag{2.21}$$

where $f(x)$ is a *scaling function* that decays rapidly for large $|x|$. Thus n_s decays as $s^{-\tau}$ until some cutoff size $s_* \sim |p - p_c|^{1/\sigma}$, whereupon it quickly drops to zero. For percolation in the Cayley tree $f(x)$ is the exponential in Eq. (2.20), and so $\sigma = \frac{1}{2}$.

We will now use the scaling form of n_s to compute τ in yet another way. To this end we re-compute the mean mass of finite clusters, S, in terms of n_s. Since sn_s is the probability that an arbitrary site belongs to an s-cluster, $\sum sn_s = p$ ($p < p_c$). The mean mass of finite clusters is

$$S = \frac{\sum_s^\infty s\, sn_s}{\sum_s^\infty sn_s} \sim \frac{1}{p} \sum_s^{s_*} s^2 n_s \sim (p_c - p)^{-(3-\tau)/\sigma}, \qquad (2.22)$$

where we have used the scaling of n_s (and of the cutoff at s_*), and we assume that $\tau < 3$. By comparing this to Eq. (2.17) one obtains the scaling relation

$$\gamma = \frac{3 - \tau}{\sigma}. \qquad (2.23)$$

For percolation in the Cayley tree we see that $\tau = \frac{5}{2}$ (consistent with the assumption that $\tau < 3$).

Finally, let us compute the order-parameter exponent β. Any site in the percolation system is (a) empty, with probability $1 - p$, (b) occupied and on the infinite cluster, with probability pP_∞, or (c) occupied but not on the infinite cluster, with probability $p(1 - P_\infty)$. Therefore,

$$P_\infty = 1 - \frac{1}{p} \sum_s sn_s. \qquad (2.24)$$

For $p < p_c$ all clusters are finite, $\sum sn_s = p$, and $P_\infty = 0$. Above criticality $\sum sn_s$ is smaller than p, because there are occupied sites that belong to the infinite cluster. The correction comes from the upper cutoff of the sum at $s = s_*$; $\sum sn_s \sim \sum^{s_*} s^{1-\tau} \sim p - [\text{constant} \times (p - p_c)^{-(2-\tau)/\sigma}]$. We then find the scaling relation

$$\beta = \frac{\tau - 2}{\sigma}, \qquad (2.25)$$

and so $\beta = 1$ for percolation on the Cayley tree and for percolation in $d \geq 6$.

In closing this chapter, we would like to mention that there exist several variations of the percolation model that lie in different universality classes than regular percolation. These include directed percolation, invasion percolation, and long-range correlated percolation.

2.5 Exercises

1. Simulate percolation on the computer, following the simple-minded method of Section 2.1. Devise an algorithm to find out whether there is a spanning percolation cluster between any two sites, and to identify all the sites of the incipient percolation cluster.

2. *The Leath algorithm.* Percolation clusters can be built one chemical shell at a time, by using the Leath algorithm. Starting with an origin site (which represents the chemical shell $\ell = 0$) its nearest neighbors are assigned to the first chemical shell with probability p. The sites which were not chosen are simply marked as having been "inspected". Generally, given the first ℓ shells of a cluster, the $(\ell + 1)$th shell is constructed as follows: identify the set of nearest neighbors to the sites of shell ℓ. From this set discard any sites that belong to the cluster, or which are already marked as "inspected". The remaining sites belong to shell $(\ell + 1)$ with probability p. Remember to mark the newly inspected sites which were left out. Simulate percolation clusters at p slightly larger than p_c and confirm the crossover of Eq. (2.5).

3. Imagine an *anisotropic* percolation system in $d = 2$ with long range correlations, such that the correlation length depends on direction:

$$\xi_x \sim (p - p_c)^{-\nu_x}, \qquad \xi_y \sim (p - p_c)^{-\nu_y}.$$

Generalize the formula $d_f = d - \beta/\nu$ for this case. (Answer: $d_f^x = 1 + (\nu_y - \beta)/\nu_x$; $d_f^y = 1 + (\nu_x - \beta)/\nu_y$.)

4. From our presentation of the Cayley tree it would seem that the root of the tree is a special point. Show, to the contrary, that in an infinite Cayley tree all sites are equivalent!

5. Show that, in the Cayley tree, an s-cluster has exactly $2 + (z - 2)s$ perimeter sites. (Hint: prove it by induction.)

6. The exponent α is defined by the relation $\sum_s n_s \sim |p - p_c|^{2-\alpha}$. In thermodynamic phase transitions, α characterizes the divergence of the *specific heat*. Show that $2 - \alpha = (\tau - 1)/\sigma$.

7. The critical exponent δ characterizes the response to an external ordering field h. For percolation, it may be defined as $\sum_s s\, n_s e^{-hs} \sim h^{1/\delta}$. Show that $\delta = 1/(\tau - 2)$.

8. The exponents α, β, γ, and δ can all be written in terms of σ and τ. Therefore, any two exponents suffice to express the others. As an example, express α, δ, σ, and τ as functions of β and γ.

9. Percolation in one dimension may be analyzed exactly. Notice that only the subcritical phase exists, since $p_c = 1$. Analyze this problem directly and compare it with the limit of percolation in the Cayley tree when $z \to 2$.

10. Define the largest cluster in a percolation system as having rank $\rho = 1$, the second largest $\rho = 2$, and so on. Show that, at criticality, the mass of the clusters scales with rank as $s \sim \rho^{-d/d_f}$.

2.6 Open challenges

Percolation is the subject of much ongoing research. There remain many difficult theoretical open questions, such as finding exact percolation thresholds, and the exact values of various critical exponents. Until these problems are resolved, there is a point in improving the accepted numerical values of such parameters through simulations and other numerical techniques. Often this can be achieved using well-worn approaches, simply because computers get better with time! Here is a sample of interesting open problems.

1. The critical exponents β and ν are known exactly for $d = 2$, due to the relation of percolation to the one-state Potts model. However, no exact values exist for β and ν in $2 < d < 6$, nor for d_f^{BB} and d_{min} in all $1 < d < 6$. Also, is d_h truly equal to d_f in three-dimensional percolation, as is commonly assumed?

2. Series expansions (Bhatti *et al.*, 1997) suggest that d_{min} is nonuniversal: for the square lattice $d_{min} = 1.106 \pm 0.007$, whereas for the triangular lattice $d_{min} = 1.148 \pm 0.007$. Extensive numerical analysis is needed to clinch this issue. Numerical estimates for critical exponents in continuum percolation are currently markedly different than the universal values on lattices (Okazaki *et al.*, 1996; Balberg, 1998a; Rubin *et al.*, 1999). Simulations of larger systems, and closer to the transition point, are necessary to probe this issue.

3. Most values of the critical percolation thresholds are not known exactly. There is an ongoing search for expressions containing the various p_c which might be universal (Galam and Mauger, 1997; 1998; Babalievski, 1997; van der Marck, 1998).

4. Improve existing numerical algorithms for percolation, or invent new fresh ones. Researchers in the area keep finding better ways to perform the same old tricks! An example of this trend is the recent introduction of a new efficient algorithm for the identification of percolation backbones (Moukarzel, 1998). See also Babalievski (1998) and Stoll (1998).

5. Little is known about anisotropic percolation (see, for example, Dayan *et al.* (1991)). Transport properties of such models have not been studied.

6. It was commonly believed until recently that, for percolation in two-dimensional lattices at p_c, there exists exactly one incipient spanning cluster. New insight into this question was achieved by Aizenman (1997) when he proved that the number of incipient spanning clusters can be larger than one, and the probability of having at least k separate clusters in a system of size $L \times L$, $P_L(k)$, is bounded by $Ae^{-\alpha k^2} \leq P_L(k) \leq Ae^{-\alpha' k^2}$, where α and α' are constants. Numerical estimates for A, α, and α' were given by Shchur and Kosyakov (1998).

7. Percolation with long-range correlations has been studied by Prakash *et al.*

(1992), Makse *et al.* (1996), and Moukarzel *et al.* (1997). Makse *et al.* (1995) have applied the model to the study of the structure of cities. There remain many open questions.

8. The traditional percolation model assumes that one has only one kind of sites or bonds. Suppose for example that the bonds are of two different kinds: ε_1 and ε_2. One may then search for the path between two given points on which the sum of ε_i is minimal. This is the optimal-path problem (Cieplak *et al.*, 1994; 1996; Schwartz *et al.*, 1998). The relation of this problem to percolation is still open for research.

9. Are there other universal properties of percolation, in addition to the critical exponents and amplitude ratios? For example, results of recent studies by Cardy (1998), Aizenman (1997), and Langlands (1994) suggest that the crossing probability $\pi(\Omega)$ is a universal function of the shape of the boundary Ω of the percolation system.

10. Is there self-averaging in percolation, i.e., can ensemble averages be replaced by an average over one large (infinite) cluster? See De Martino and Giansanti (1998a; 1998b).

2.7 Further reading

- Reference books on percolation: Stauffer and Aharony (1994), and Bunde and Havlin (1996; 1999). For applications, see Sahimi (1994). A mathematical approach is presented by Essam (1980), Kesten (1982), and Grimmet (1989).

- Numerical methods for the generation of the backbone: Herrmann *et al.* (1984b), Porto *et al.* (1997b), Moukarzel (1998), and Grassberger (1999a). Experimental studies of the backbone in epoxy-resin–polypyrrol composites using image-analysis techniques can be found in Fournier *et al.* (1997).

- The fractal dimension of the red bonds: Coniglio (1981; 1982). Red bonds on the "elastic" backbone: Sen (1997). The fractal dimension of the hull: Sapoval *et al.* (1985) and Saleur and Duplantier (1987).

- Exact results for the number of clusters per site for percolation in two dimensions were presented by Kleban and Ziff (1998).

- Series-expansion analyses: Adler (1984). The renormalization-group approach: Reynolds *et al.* (1980). A renormalization-group analysis of several quantities such as the minimal path, longest path, and backbone mass has been presented by Hovi and Aharony (1997a). A recent renormalization-group analysis of the fractal dimension of the backbone, d_f^{BB}, to third-order in $\epsilon = 6 - d$ is given by Janssen *et al.* (1999).

- Forest fires in percolation: see, for example, Bak *et al.* (1990), Drossel and Schwabl (1992), and Clar *et al.* (1997).

- Continuum percolation: Balberg (1987). Experimental studies of continuum percolation in graphite–boron nitrides: Wu and McLachlan (1997). A recent study of percolation of finite-sized objects with applications to the transport properties of impurity-doped oxide perovskites: Amritkar and Roy (1998). Invasion percolation: Wilkinson and Willemsen (1983), Porto *et al.* (1997a), and Schwarzer *et al.* (1999). Directed percolation: Kinzel (1983), Frojdh and den Nijs (1997), and Cardy and Colaiori (1999).

- Percolation on fractal carpets: Havlin *et al.* (1983a) and Lin *et al.* (1997).

- A problem related to percolation that includes also long-range bonds, the "small-world network", has been studied by Watts and Strogatz (1998). They find that adding a very small fraction of randomly connected long-range bonds reduces the chemical distance dramatically. For interesting applications of the "small-world network" see Lubkin (1998).

- A new approach based on generating functions for percolation in the Cayley tree can be found in Buldyrev *et al.* (1995a).

- Applications of percolation theory and chemical distance to recovery of oil from porous media: Dokholyan *et al.* (1999), King *et al.* (1999), Lee *et al.* (1999), and Porto *et al.* (1999). Applications to ionic transport in glasses and composites: Roman *et al.* (1986), Bunde *et al.* (1994), and Meyer *et al.* (1996a). Applications to the metal–insulator transition: see, for example, Ball *et al.* (1994). Applications to fragmentation: see, for example, Herrmann and Roux (1990), Sokolov and Blumen (1999), and Cheon *et al.* (1999).

3

Random walks and diffusion

Random walks model a host of phenomena and find applications in virtually all sciences. With only minor adjustments they may represent the thermal motion of electrons in a metal, or the migration of holes in a semiconductor. The continuum limit of the random walk model is known as "diffusion". It may describe Brownian motion of a particle immersed in a fluid, as well as heat propagation, the spreading of a drop of dye in a glass of still water, bacterial motion and other types of biological migration, or the spreading of diseases in dense populations. Random-walk theory is useful in sciences as diverse as thermodynamics, crystallography, astronomy, biology, and even economics, in which it models fluctuations in the stock market.

3.1 The simple random walk

A *random walk* is a stochastic process defined on the points of a lattice. Usually, the time variable is considered discrete. At each time unit the "walker" steps from its present position to one of the other sites of the lattice according to a prescribed random rule. This rule is independent of the history of the walk, and so the process is Markovian.

In the simplest version of a random walk, the walk is performed in a hypercubic d-dimensional lattice of unit lattice spacing. At each time step the walker hops to one of its *nearest-neighbor* sites, with equal probabilities. Several configurations of simple random walks on a $d = 2$ lattice are shown in Fig. 3.1. After n steps the net displacement is

$$r(n) = \sum_{i=1}^{n} e_i, \qquad (3.1)$$

where e_i is a (unit) vector pointing to a nearest-neighbor site – it represents the ith step of the walk. Because $\langle e_i \rangle = 0$, the average displacement (averaged over many

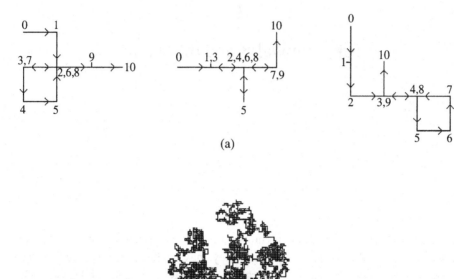

(a)

(b)

Fig. 3.1. Trails of simple random walks in two dimensions. (a) Configurations of ten-step walks. The step numbers are noted, to indicate instances when the walker retraces its own steps. (b) The trail resulting from a random walk of 20 000 steps.

realizations of the walk) is

$$\langle \mathbf{r}(n) \rangle = 0. \tag{3.2}$$

On the other hand, since $\langle \mathbf{e}_i \cdot \mathbf{e}_i \rangle = 1$, and $\langle \mathbf{e}_i \cdot \mathbf{e}_j \rangle = 0$ for $i \neq j$ (the steps are independent), the mean-squared displacement is

$$\langle r^2(n) \rangle = \left\langle \left(\sum_{i=1}^{n} \mathbf{e}_i \right)^2 \right\rangle = n + 2 \sum_{i>j}^{n} \langle \mathbf{e}_i \cdot \mathbf{e}_j \rangle = n. \tag{3.3}$$

More generally, let the lattice spacing be a and let the step time unit be τ; $t = n\tau$, then

$$\langle r^2(t) \rangle = (2d)Dt, \tag{3.4}$$

where $D = a^2/(2d)\tau$ is the *diffusion constant*. This is known as *regular* diffusion. Notice that the speed of the walker vanishes at long times: $v \sim \langle r^2 \rangle^{1/2}/t \sim 1/t^{1/2}$.

The mean-square displacement $\langle r^2(n) \rangle$ can be calculated from the probability density $P(r, t)$, the probability that the walker has displaced to r after time t: $\langle r^2(n) \rangle = \int r^2 P(r, t) \, d^d r$. In one dimension $P(x, t)$ can be calculated easily: assume that jumps to the right occur with probability p and jumps to the left with probability $1 - p$. If the walker moved m times to the right and $t - m$ times to the left its displacement is $x = m - (t - m) = 2m - t$. The probability for this is given by the binomial distribution

$$p(m, t) = \binom{t}{m} p^m (1 - p)^{t-m}.$$

On making the substitution $p = \frac{1}{2}$ and using the Sterling approximation $t! \cong (2\pi t)^{1/2}(t/e)^t$, and $x \ll t$, one obtains

$$P(x, t) \, dx \cong \frac{1}{(2\pi t)^{1/2}} e^{-x^2/(2t)} \, dx. \tag{3.5}$$

An interesting feature of regular diffusion is that its time dependence is universal, regardless of the dimension of the substrate, d. What happens if the walk takes place in a *fractal* substrate? Fractal lattices may be regarded, in some sense, as an extrapolation of regular Euclidean space of integer dimensions to spaces of noninteger dimension. Naively, then, one would expect regular diffusion to take place in fractals as well. However, this is not the case, and in fractal lattices one talks of *anomalous* diffusion. We shall return to this intriguing topic in Chapter 5.

3.2 Probability densities and the method of characteristic functions

We may be interested in a more general derivation of $P_n(r)$ – the probability density of being at r after n steps. We now tackle this question, following the method of characteristic functions. Consider a walk that takes place in R^d (continuous space) and the steps are drawn from the probability density $p(r')$, and assume that the walk starts at the origin. The simple random walk is a special case, in which $p(r') = [1/(2d)] \sum_i \delta(r' - ae_i)$.

The function

$$F_n(k) = \int P_n(r) e^{ik \cdot r} \, d^d r \tag{3.6}$$

is called the *characteristic function* of the probability density. It is in fact the Fourier transform of P, and hence it encompasses an equivalent amount of information. Similarly, the characteristic function of the step probability density is

$$\lambda(\boldsymbol{k}) = \int p(\boldsymbol{r}')e^{i\boldsymbol{k}\cdot\boldsymbol{r}'}\, d^d\boldsymbol{r}' \tag{3.7}$$

and is also known as the step *structure function*. Because the steps are independent, the process obeys the Markov property

$$P_{n+1}(\boldsymbol{r}) = \int P_n(\boldsymbol{r}')p(\boldsymbol{r} - \boldsymbol{r}')\, d^d\boldsymbol{r}'. \tag{3.8}$$

Taking the Fourier transform (applying the operator $\int d^d\boldsymbol{r}\, e^{i\boldsymbol{k}\cdot\boldsymbol{r}}$) results in

$$F_{n+1}(\boldsymbol{k}) = F_n(\boldsymbol{k})\lambda(\boldsymbol{k}), \tag{3.9a}$$

a recursive relation that yields

$$F_n(\boldsymbol{k}) = \lambda(\boldsymbol{k})^n. \tag{3.9b}$$

The probability density may now be obtained from the inverse transform

$$P_n(\boldsymbol{r}) = \frac{1}{(2\pi)^d} \int F_n(\boldsymbol{k})e^{-i\boldsymbol{k}\cdot\boldsymbol{r}}\, d^d\boldsymbol{k}. \tag{3.10}$$

Suppose that the step probability function has zero mean and finite variance:

$$\int \boldsymbol{r} p(\boldsymbol{r})\, d^d\boldsymbol{r} = 0, \qquad \int r^2 p(\boldsymbol{r})\, d^d\boldsymbol{r} = a^2 < \infty. \tag{3.11}$$

Then, at long times the distribution $P_n(\boldsymbol{r})$ tends to a Gaussian. This is a result of the *central-limit theorem*. (The central-limit theorem applies also for the case of nonzero mean; see Exercise 4.) Let us see how this happens in the simple special case of a walk on the line, $d = 1$. We may then work with scalar variables instead of the d-dimensional vectors $\boldsymbol{r}$ and $\boldsymbol{k}$. The structure function is, according to Eqs. (3.7) and (3.11),

$$\lambda(k) = 1 - \tfrac{1}{2}k^2 a^2 + \mathcal{O}(k^2), \tag{3.12}$$

and so the characteristic function is

$$F_n(k) = e^{-nk^2 a^2/2 + n\mathcal{O}(k^2)}. \tag{3.13}$$

For very long times ($n \gg 1$) the main contribution to $F_n(k)$ comes from $|k| < 1/n^{1/2}$, and $n\mathcal{O}(k^2) \sim 1/n^{1/2}$ may be neglected. We then have

$$P_n(r) = \frac{1}{2\pi} \int_{-\infty}^{\infty} e^{-nk^2 a^2/2 - ikr}\, dk = \frac{1}{\sqrt{2\pi a^2 n}}e^{-r^2/(2a^2 n)}. \tag{3.14}$$

This may be written in terms of the time variable $t = n\tau$ as

$$P(r, t) = \frac{1}{\sqrt{4\pi Dt}}e^{-r^2/(4Dt)}, \tag{3.15}$$

which is the same result as Eq. (3.5), since $D = \frac{1}{2}$ in our case. The probability density in d dimensions is essentially the same, the only difference being that the normalization factor is then $(4\pi Dt)^{-d/2}$.

Figure 3.2a shows $P(r, t)$ for walks on the line at several values of t. The different curves share a universal shape, which is evident in the plot of $t^{1/2}P(r, t)$ versus $r/t^{1/2}$ of Fig. 3.2b. The collapsing of all curves into one is made possible by the appropriate *scaling* of distance with time $r \sim t^{\alpha}$ ($\alpha = \frac{1}{2}$).

3.3 The continuum limit: diffusion

So far we have discussed random walks with a discrete time variable. (We have considered both discrete and continuous steps.) The limit of continuous time and continuous space is known as *diffusion*. We shall now discuss this limit.

Consider a discrete random walk (both in space and in time) in a one-dimensional lattice of equal spacing. At each time step the walker hops to the nearest site to its right or left, with equal probability $\frac{1}{2}$. The probability of being at site m at the nth step, $P_n(m)$, satisfies the equation (Fig. 3.3):

$$P_{n+1}(m) = \tfrac{1}{2}P_n(m - 1) + \tfrac{1}{2}P_n(m + 1),$$

or

$$P_{n+1}(m) - P_n(m) = \tfrac{1}{2}P_n(m - 1) - P_n(m) + \tfrac{1}{2}P_n(m + 1). \tag{3.16}$$

This difference equation offers an alternative approach to the method of characteristic functions. It is far easier, however, to deal with differential equations, and this motivates the passage to the continuum limit.

On making the change of variables $x = ma$ and $t = n\tau$, and taking the limit $a \to 0$ and $\tau \to 0$, the probability $P_n(m)$ is replaced by the probability density $P(x, t)$. Provided that we keep $a^2/\tau = 2D$ constant in the limiting process, Eq. (3.16) becomes

$$\frac{\partial}{\partial t}P(x, t) = D\frac{\partial^2}{\partial x^2}P(x, t). \tag{3.17a}$$

This is the diffusion equation in one dimension. Notice that (3.15) is a solution of the diffusion equation (3.17a), with initial condition $P(x, 0) = \delta(x)$. Indeed, in the long-time limit there is virtually no difference between discrete random walks and diffusion. We shall therefore use these terms loosely throughout the remainder of this book.

The generalization of diffusion to higher dimensions is straightforward. One obtains

$$\frac{\partial}{\partial t}P(r, t) = D\nabla^2 P(r, t), \tag{3.17b}$$

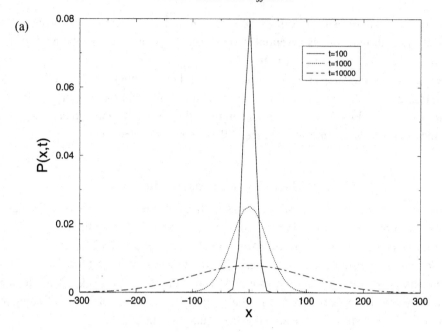

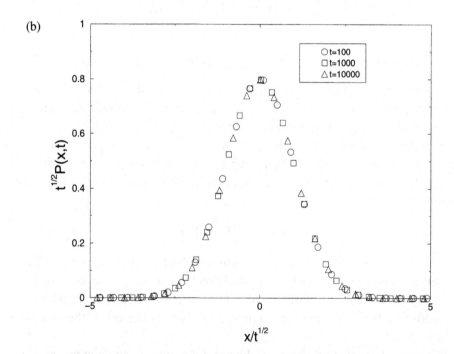

Fig. 3.2. The probability distribution function of a simple RW. (a) The distribution function $P(x, t)$ is shown for $t = 100$, 1000, and 10 000 steps. (b) The various curves collapse into a universal function (a Gaussian) upon rescaling of the distance variable, $x \rightarrow x/t^{1/2}$.

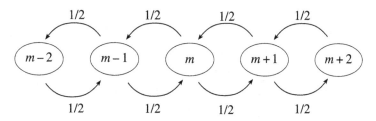

Fig. 3.3. Hopping rates of a simple RW in one dimension: Eq. (3.16) is written from inspection of this diagram.

where ∇^2 is the d-dimensional Laplacian operator.

3.4 Einstein's relation for diffusion and conductivity

Consider an asymmetric walk, in which the probabilities of stepping to the right and left are $\frac{1}{2}+\epsilon$ and $\frac{1}{2}-\epsilon$, respectively. This results in an overall drift of the walk to the right. Following the same procedure as above, one obtains the diffusion equation

$$\frac{\partial}{\partial t}P(x,t) = -v\,\frac{\partial}{\partial x}P(x,t) + D\,\frac{\partial^2}{\partial x^2}P(x,t), \qquad (3.18)$$

where $D = a^2/(2\tau)$, as before, and $v = \langle \Delta x \rangle / \Delta t = 2\epsilon a/\tau$ is the drift velocity of the walker. The diffusive motion is then simply superposed on the constant drift. Hence Eq. (3.18) can also be written as a *continuity equation*;

$$\frac{\partial}{\partial t}P(x,t) = -\frac{\partial}{\partial x}J(x,t), \qquad (3.19)$$

where $J \equiv vP - D\,\partial P/\partial x$ has the meaning of *probability current*. The continuity equation expresses the conservation of probability, resulting from the fact that walkers cannot be created or destroyed.

In Fig. 3.4 we show numerical data for $\langle r^2 \rangle$ of diffusion with a drift. Notice that the problem may be solved exactly (cf. Exercise 4), but it provides an opportunity for presenting a further example of a scaling analysis.

Let the walkers now be electric charges e, in a chunk of metal (Fig. 3.5). The charges undergo diffusion, characterized by some diffusion constant D. In the presence of an electric field E, the charges attain a constant terminal velocity v given by Ohm's law: $nev = -\sigma E$, where n is the density of charges per unit volume and σ is the dc conductivity of the medium. If the metal is restricted to the half-space $x > 0$, and the field E is in the positive x-direction, the charges will

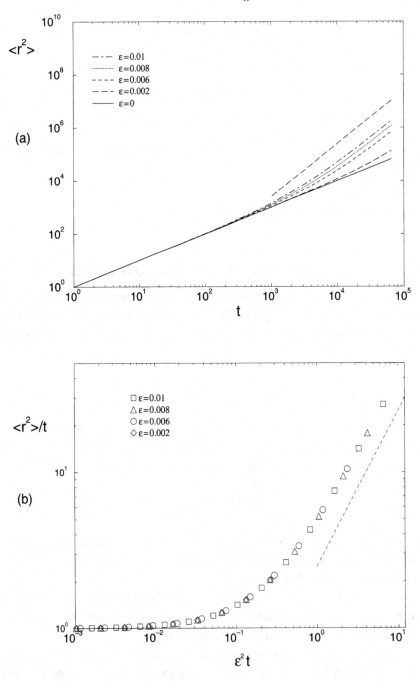

Fig. 3.4. A simple RW with drift. (a) The mean-square displacement for various values of the asymmetry parameter ϵ. The asymptotic slope is denoted by the broken line of slope 1. (b) Scaling analysis of the displacement. The analytical solution of the problem yields $\langle r^2 \rangle = t + \epsilon^2 t^2$.

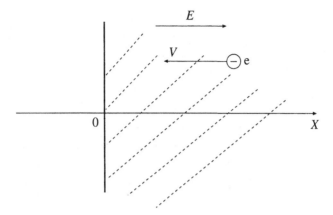

Fig. 3.5. Charges in an electric field. The electric field E imparts to the electrons a drift velocity in the negative x-direction. The current is zero at the metal's surface (at $x = 0$), since the charges are confined to the metal.

attain a stationary state. This may be obtained from Eq. (3.18), with $\partial P/\partial t = 0$:

$$\frac{\sigma E}{ne}\frac{\partial P}{\partial x} + D\frac{\partial^2 P}{\partial x^2} = 0; \qquad x > 0, \tag{3.20}$$

and the boundary condition $J(x = 0) = 0$ – since there is no flux of charges into the empty region $x < 0$. The solution is $P(x) = \text{constant} \times \exp[-\sigma Ex/(neD)]$. On the other hand, if the metal is at temperature T, the charges would arrive at thermal equilibrium, characterized by the Boltzmann distribution $P_{\text{eq}}(x) = \text{constant} \times \exp[-Eex/(k_B T)]$. On comparing the two results, one concludes that

$$\sigma = \frac{e^2 n}{k_B T} D. \tag{3.21}$$

This is known as the *Einstein relation* for conductivity and diffusion. There are similar remarkable relations between other *macroscopic* transport parameters and the *microscopic* coefficient of diffusion, D.

3.5 Continuous-time random walks

Montroll and Weiss (1965) introduced the concept of continuous-time random walks (CTRWs) as a way to render time continuous, without appealing to the diffusion limit. The model achieves much more than this modest goal. Some forms of CTRW are fundamentally different than the classical diffusion model, and the theory has numerous important applications.

Imagine a random walk on a lattice, starting at the origin, but such that the steps are taken at *random times*. Let $\psi(t)$ be the probability density for the waiting time between successive steps. We assume that the waiting times for different jumps are statistically independent and are all characterized by $\psi(t)$. The probability that the waiting time between steps is greater than t is

$$\Psi(t) = \int_t^\infty \psi(t')\,dt'. \tag{3.22}$$

Define also $\psi_n(t)$ as the probability density that the nth jump occurs at time t. Clearly, $\psi_1(t) = \psi(t)$, and, since the waiting times between steps are independent,

$$\psi_{n+1}(t) = \int_0^t \psi_n(t')\psi(t-t')\,dt'. \tag{3.23}$$

Using the Laplace transform

$$\hat{\psi}(s) = \int_0^\infty \psi(t)e^{-st}\,dt \tag{3.24}$$

(and likewise for other functions), we find

$$\hat{\psi}_n(s) = \hat{\psi}(s)^n, \qquad \hat{\Psi}(s) = \frac{1 - \hat{\psi}(s)}{s}. \tag{3.25}$$

Given now $P_n(r)$ – the probability of being at r at the nth step – one can express $P(r,t)$ as

$$P(r,t) = \sum_{n=0}^\infty P_n(r) \int_0^t \psi_n(t')\Psi(t-t')\,dt'. \tag{3.26}$$

The integral represents the probability that, once the walker had arrived at site r at time $t' < t$, it would *remain* there until time t. Taking the Laplace transform, we obtain

$$\hat{P}(r,s) = \sum_{n=0}^\infty P_n(r)\hat{\psi}(s)^n \frac{1 - \hat{\psi}(s)}{s}. \tag{3.27}$$

Finally, we take the Fourier transform, and, using the result of Eqs. (3.6) and (3.9), the infinite sum may be carried out explicitly:

$$\hat{P}(k,s) = \frac{1 - \hat{\psi}(s)}{s[1 - \hat{\psi}(s)\lambda(k)]}. \tag{3.28}$$

One can now invert the double transform to obtain the distribution itself. The moments are somewhat easier to obtain. For example, in one dimension:

$$\langle \hat{r}(s) \rangle = -i \left. \frac{\partial \hat{P}(k,s)}{\partial k} \right|_{k=0} = \frac{\mu_1 \hat{\psi}(s)}{s[1 - \hat{\psi}(s)]}, \tag{3.29}$$

and

$$\widehat{\langle r^2(s)\rangle} = (-i)^2 \left.\frac{\partial^2 \hat{P}(k, s)}{\partial k^2}\right|_{k=0} = \frac{\mu_2 \hat{\psi}(s)}{s[1 - \hat{\psi}(s)]} + \frac{2\mu_1^2 \hat{\psi}(s)^2}{s[1 - \hat{\psi}(s)]^2}, \tag{3.30}$$

where $\mu_n \equiv \int r^n p(r)\, dr$ are the moments of the step distribution function.

The special case of $\psi(t) = \delta(t - \tau)$ (τ constant) reduces to the regular random walks discussed earlier and confirms the CTRW formalism. In fact, any $\psi(t)$ that falls off fast enough yields similar regular behavior. However, interesting anomalous behavior may be obtained with slow-decaying ψ, as will be shown in the following chapter.

3.6 Exercises

1. Compute $\lambda(k)$ of Eq. (3.7) for a one-dimensional random walk (RW).
2. Consider a RW in a d-dimensional hypercubic lattice with lattice spacing a. In a time step τ the walker steps to one of the $2d$ nearest sites with probability $\epsilon/(2d)$, or stays put with probability $1 - \epsilon$. Write a master equation (similar to (3.16)) and obtain the continuum limit, Eq. (3.17). What is the diffusion coefficient $D(\epsilon)$? (Answer: $D = \epsilon a^2/(2d\tau)$.)
3. Verify that $P(r, t)$ of Eq. (3.15) solves the diffusion equation (3.17), by direct substitution. Obtain the solution explicitly, for when the initial condition is $P(x, 0) = \delta(x)$.
4. *Asymmetric RWs.* Consider an asymmetric random walk in one dimension, with $\int rp(r)\, dr = b > 0$ and $\int (r - b)^2 p(r)\, dr = a^2$. Apply the method of characteristic functions and show that $P(r, t)$ of Eq. (3.15) generalizes to $P(r, t) = (4\pi Dt)^{-1/2} \exp[(r - vt)^2/(4Dt)]$, where $D = a^2/(2\tau)$ and $v = b/\tau$. Verify that this distribution is a solution to Eq. (3.18).
5. *Persistent RWs.* A random walker in one dimension has probability α of continuing in the same direction as the previous step (and probability $\beta = 1 - \alpha$ of reversing direction). Let the probability that the walker is at m at step n be $P_n(m)$, when it is coming from $m - 1$; and $Q_n(m)$, when it is coming from $m + 1$. Generalize the characteristic-function approach of Section 3.2 by considering the state vector $(P_n(m), Q_n(m))$. The structure function may then be written as a matrix. Work out the long-time asymptotic limit and show that it is similar to a regular random walk, without persistence.
6. Derive recursion relations for $P_n(m)$ and $Q_n(m)$ of the persistent RW. Pass to the continuum limit in the usual way, writing $\alpha = 1 - \tau/(2T)$, where T is a constant. For $U = P + Q$, obtain the *telegrapher's equation*

$$\frac{\partial^2 U}{\partial t^2} + \frac{1}{T}\frac{\partial U}{\partial t} = c^2 \frac{\partial^2 U}{\partial x^2}.$$

7. Motion in a fluid medium is hindered by a drag force proportional to the speed of motion: $F_{drag} = -\gamma M v$. M is the mass of the moving object, and γ is the macroscopic *drag coefficient*. Derive the Einstein relation between γ and the microscopic diffusion coefficient D. (Hint: consider motion under an external constant force, such as gravity.) (Answer: $\gamma^{-1} = M D/(k_B T)$.)

8. The regular RW is a special case of the CTRW, with a waiting-time distribution function $\psi(t) = \delta(t - \tau)$. Verify this statement by computing $\langle r \rangle$, $\langle r^2 \rangle$, and $P(r, t)$ through the CTRW formalism.

9. Given a CTRW with $\psi(t) = (1/T) \exp(-t/T)$, where T is a constant, compute $\langle r \rangle$ and $\langle r^2 \rangle$ as functions of time. Show that the probability density is $P(r, t) = \exp(-t/T) I_r(t/T)^d$, where $I_r(z) = [1/(2\pi)] \int_{-\pi}^{\pi} \exp(-ir\theta - z\cos\theta)\, d\theta$ is a Bessel function of imaginary argument. Obtain the limit $t \to \infty$ of $P(0, t) \sim \sqrt{T/(2\pi t)}$. Compare this result with the simple random walk.

10. Generalize the approach of Section 3.5 to higher dimensions, and obtain formulae analogous to (3.29) and (3.30).

3.7 Open challenges

1. The number of distinct sites visited by a walker in a lattice up to time t, $S(t)$, is an important quantity for many applications (Appendix B). The full asymptotic distribution of $S(t)$ is known to be Gaussian in three dimensions, with a mean t and variance $t \log t$; and also in higher dimensions it is Gaussian. The scaling of the first moment is known in all dimensions: $\langle S(t) \rangle \sim t^{1/2}$ ($d = 1$), $t/\log t$ ($d = 2$), and t ($d \geq 3$). The second moment of $S(t)$ has also been derived (Larralde and Weiss, 1995). However, the full distribution of $S(t)$ remains unknown.

2. The problem of self-attracting (or self-repelling) random walks has not yet been resolved. In this case the RW has a higher (or lower) probability of visiting previously visited sites. Open problems include the derivation of the mean-square displacement and the number of distinct sites. The scaling of these quantities with time is controversial (Lee, 1998; Sapozhnikov, 1998). The limit in which the probability of revisiting previously visited sites is zero is the well-known problem of self-avoiding walks (SAWs).

3. The paths generated by random walks on lattices are intricate geometrical objects of which little is known. For example, the scaling of the chemical distance with the Euclidean distance has not been studied. A partial study of random-walk paths in two dimensions was performed by Movshovitz and Havlin (1988). The question of the distribution of loops in random-walk trails has recently been introduced by Wolfling and Kantor (1999).

4. The problem of N diffusing walkers when a fraction p of the walkers can die and a fraction q gives birth has been studied by Meyer *et al.* (1996b). The mean displacement of the walkers with respect to the center of mass has been found for $p = q$ to reach a plateau as a function of time, i.e., the walkers tend to cluster. The number of distinct sites visited by this group (the territory) has not been studied. Moreover, the known solution (Meyer *et al.*, 1996b) is valid only when p and q are fixed fractions, rather than probabilities.

3.8 Further reading

- The theory of random walks: Chandrasekhar (1943), Spitzer (1976), Barber and Ninham (1970), Weiss and Rubin (1983), Montroll and Shlesinger (1984), Weiss (1994), and Hughes (1995). Random walks in biology: Berg (1993). Brownian motion: Lavenda (1985).

- The number of distinct sites visited by N diffusing particles in d dimensions has been studied analytically by Larralde *et al.* (1992a; 1992b).

- Selected applications of random walks: optical imaging (Bonner *et al.*, 1987) and amorphous disordered systems (Bunde and Havlin, 1996).

- Diffusion in condensed matter: Kärger *et al.* (1998).

- Random walks contributed to the field of crystallography (Klug, 1958). The development of this theory has been recognized in the awarding of the Nobel Prize to Hauptman and Karle in 1985.

- An introduction to transport models based on CTRWs: Scher and Lax (1973) and Scher and Montroll (1975).

4

Beyond random walks

Random walks normally obey Gaussian statistics, and their average square displacement increases linearly with time; $\langle r^2 \rangle \sim t$. In many physical systems, however, it is found that diffusion follows an anomalous pattern: the mean-square displacement is $\langle r^2 \rangle \sim t^{2/d_w}$, where $d_w \neq 2$. Here we discuss several models of anomalous diffusion, including CTRWs (with algebraically long waiting times), Lévy flights and Lévy walks, and a variation of Mandelbrot's fractional-Brownian-motion (FBM) model. These models serve as useful, tractable approximations to the more difficult problem of anomalous diffusion in disordered media, which is discussed in subsequent chapters.

4.1 Random walks as fractal objects

The trail left by a random walker is a complicated random object. Remarkably, under close scrutiny it is found that the trail is self-similar and can be thought of as a fractal (Exercise 1). The ubiquity of diffusion in Nature makes it one of the most fundamental mechanisms giving rise to random fractals.

The fractal dimension of a random walk is called the *walk dimension* and is denoted by d_w. If we think of the sites visited by a walker as "mass", then the mass of the walk is proportional to time. We can then write

$$M \sim t \sim r^{d_w},\tag{4.1}$$

where r is the typical distance covered after time t. The mean-square displacement is then given by

$$\langle r^2(t) \rangle \sim t^{2/d_w}.\tag{4.2}$$

For regular diffusion $d_w = 2$, but in fractals $d_w \neq 2$ and one then talks of *anomalous diffusion*. Next, we shall use the CTRW model to illustrate a different cause for anomalous diffusion: slow-decaying waiting times between consecutive steps.

4.2 Anomalous continuous-time random walks

We have seen that, when the waiting-time density function of a CTRW has a characteristic time scale (i.e., a finite first moment), the mean-square displacement is proportional to t and the probability density is Gaussian. Here we consider the case in which the waiting-time distribution possesses no characteristic time scale and show that this leads to anomalous diffusion. Consider a CTRW with a power-law waiting-time distribution

$$\psi(t) \sim A t^{-(\gamma+1)}, \qquad 0 < \gamma \le 1. \tag{4.3}$$

(The exponent γ, used here, and β, in Section 4.3, should not be confused with the γ and β of percolation theory and critical phase transitions.) Such a situation occurs for example when at each lattice site the walker is subject to a potential V, distributed exponentially, $P(V) = V_0 \exp(-V/V_0)$. In a three-dimensional lattice a random walker hardly ever revisits a previously visited site and therefore the waiting times between steps are practically independent, just like in a CTRW. The probability that the walker exits a well of depth V is proportional to the Boltzmann factor $w \sim \exp(-\beta V)$, where $\beta = 1/(k_B T)$. The distribution of transition rates would then be $\phi(w) = P(V) \, dV/dw \sim w^{(1/V_0 - \beta)/\beta}$, and hence the waiting time – which is proportional to $1/w$ – is distributed as in Eq. (4.3), with $\gamma = k_B T/V_0$. (The restriction on γ can always be met by lowering the temperature.) Such a system may model the thermal relaxation of complex molecules such as large proteins, or it may mimic dynamic processes in glasses.

The mean-square displacement can be found in an analogous way to Eq. (3.30):

$$\langle r^2(t) \rangle \sim t^\gamma. \tag{4.4}$$

By comparing this to (4.2) we conclude that diffusion is anomalous, with $d_w = 2/\gamma$. It is also instructive to derive the probability density of the walk, $P(r, t)$, in the long-time asymptotic limit. One finds

$$\log P(r, t) \sim -\left(\frac{r}{t^{1/d_w}}\right)^\delta, \qquad \delta = \frac{d_w}{d_w - 1}, \qquad r/t^{1/d_w} \ll 1. \tag{4.5}$$

This generalizes the result for regular diffusion, where $\delta = 2$ (Eq. (3.15)).

As with other fractal objects, the fractal dimension d_w does not fully characterize random walks. For example, it is not completely clear whether δ depends solely on d_w, or whether there are cases in which δ is a new independent exponent. Many other properties of random walks are studied with different applications in mind.

Examples are the *span* of the walk, which is the distance between the leftmost and rightmost points visited by a walker; *first-passage times*, i.e., the time until a certain set of points is visited for the first time; and *survival probabilities* – the chances that a walker has avoided a given subset of sites up to a certain time. A useful quantity in many applications is the *number of distinct sites* visited by a random walker (Appendix B).

4.3 Lévy flights and Lévy walks

In many natural realizations it was found that diffusion is enhanced and $\langle r^2 \rangle$ scales faster than linearly with time. One of the important models for enhanced diffusion is Lévy flights and its generalization, called Lévy walks (Shlesinger *et al.*, 1987).

Consider a random walker that at each time step t jumps in some random direction (taken from a uniform distribution) to a distance r, taken from a power-law distribution;

$$p(r) \sim 1/r^{1+\beta}. \tag{4.6}$$

It is implied by Eq. (4.6) that the second moment of $p(r)$ diverges for $\beta < d$ (d is the system's dimension), or that the Lévy jumps have no characteristic length scale. Figure 4.1 shows some typical 1000-step Lévy flights. It can be shown that, for $\beta < d$, the probability density of the walker being at r at time t is not Gaussian but rather follows the *Lévy* distribution

$$P(r, t) \sim t/r^{1+\beta}, \tag{4.7}$$

and the mean-square displacement diverges. The form of the probability density, when it is transformed from real space r to Fourier k-space, is

$$\tilde{P}(k, t) = \exp(-t|k|^{\beta}). \tag{4.8}$$

The fractal dimension of the sites visited by a Lévy flight is $d_f = \beta$ for $\beta < 2$ and $d_f = 2$ for $\beta \geq 2$. The largest jump after t trials, $r_{\max}$, can be estimated by comparing the distance probability-density function with a uniformly distributed variable u: $p(r)\,dr = du/t$, or $r^{-\beta} \sim u$. The minimal value of u after t trials is $1/t$, hence $r_{\max} \sim t^{1/\beta}$. This relation also represents the first-passage time for a Lévy flight to reach a distance r; $t \sim r^{\beta}$. In either case, the scaling implies a fractal dimension of the Lévy flight of $d_f = \beta$ ($\beta < 2$). Similar analytical arguments are used in Chapter 8.

Lévy flights have been found useful for describing a wide range of physical phenomena, including chaotic diffusion in Josephson junctions (Geisel *et al.*, 1985) and turbulent diffusion (Shlesinger and Klafter, 1986). Lévy flights were observed recently in several biological systems: foraging ants and *Drosophila* flies perform

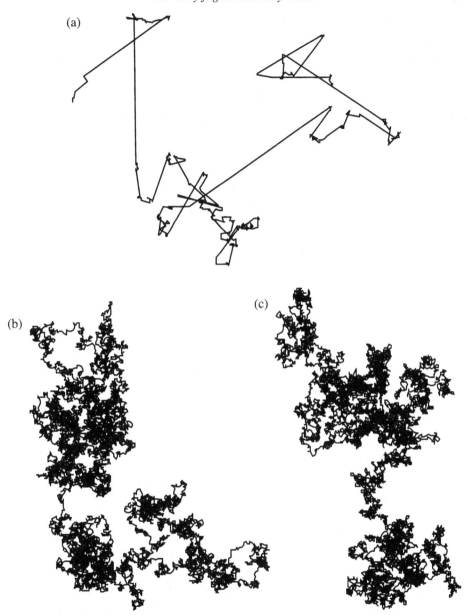

Fig. 4.1. Trajectories of typical Lévy flights in $d = 2$ for several values of β: (a) $\beta = 2$, (b) $\beta = 3$, and (c) $\beta = 6$.

Lévy flights (Cole, 1995); large birds such as the wandering albatross follow a power-law distribution of flight-time intervals (Viswanathan *et al.*, 1996); and a model for evolution of DNA based on Lévy flights has been suggested by Buldyrev *et al.* (1993). Lévy flights have been found useful even in economics (Mantegna and Stanley, 1995).

When the time to make a step r is assumed to be proportional to r (constant velocity) the resulting model is called a *Lévy walk*. One way to introduce this velocity is by using a coupled spatio-temporal probability density $\psi(r, t)$ for the random walker performing a displacement r at time t (Klafter *et al.*, 1996):

$$\psi(r, t) = \psi(r|t)p(r). \tag{4.9}$$

Here $\psi(r|t)$ is the conditional probability that a jump of size r takes a time t. For random walks moving with a velocity v, $\psi(r|t) = \delta(t - |r|/v)$. Note that in general the velocity need not be constant. Finite velocities lead to a finite mean-square displacement at any given time t. For constant velocity it can be shown that the mean-square displacement depends on β as

$$\langle r^2 \rangle = \begin{cases} t^2, & 0 < \beta < 1, \\ t^2/\ln t, & \beta = 1, \\ t^{3-\beta}, & 1 < \beta < 2, \\ t \ln t, & \beta = 2, \\ t, & \beta > 2. \end{cases} \tag{4.10}$$

An interesting example of nonconstant velocity is provided by turbulent diffusion. It was discovered by Richardson in 1926 that the mean-square separation $\langle r^2 \rangle$ between two particles in a turbulent flow increases as t^3. Since generally $\langle r^2 \rangle \sim Dt$, it follows that $D \sim t^2 \sim r^{4/3}$. The relation $D \sim r^{4/3}$ was actually suggested by Richardson. From $\langle r^2 \rangle \sim v^2 t^2$ also follows that for turbulence $v^2(r) \sim r^{2/3}$, and the Fourier transform yields $v^2(k) \sim k^{-5/3}$. Indeed, Kolmogorov had discovered this property of the turbulent spectrum in 1941.

4.4 Long-range correlated walks

An alternative way to generalize the random-walk model is by introducing long-range correlations similar to those of the FBM model (Mandelbrot, 1982; Havlin *et al.*, 1988; 1989; Voss, 1991; Peng *et al.*, 1991; Makse *et al.*, 1996). In the calculation of the mean-square displacement of a simple random walk, Eq. (3.3), it was assumed that the single steps e_i are uncorrelated, i.e., $\langle e_i \cdot e_j \rangle = \delta_{ij}$. If one assumes instead that e_i and e_j are correlated, anomalous diffusion may result. When the correlations are of short-range nature then Eq. (3.3) is still valid at large times, but it needs to be modified for long-range correlations. Assume for example that e_i and e_j are correlated in the power-law form;

$$\langle e_i \cdot e_j \rangle \approx \frac{A}{|i - j|^\gamma}. \tag{4.11}$$

(a)

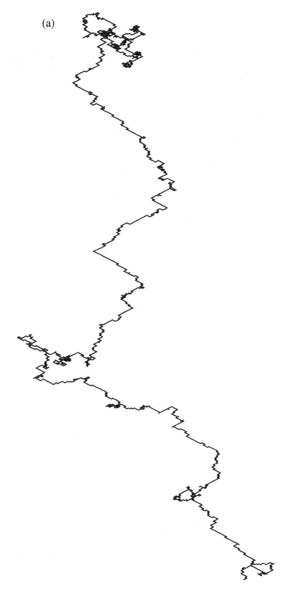

Fig. 4.2. Trajectories of typical long-range correlated walks in $d = 2$, obeying Eq. (4.11), for (a) $\gamma = 0.5$, (b) $\gamma = 0.8$, and (c) $\gamma = 0.3$. (Continued overleaf.)

Now Eq. (3.3) becomes

$$\langle r^2(n) \rangle = \left\langle \left(\sum_{i=1}^{n} e_i \right)^2 \right\rangle = n + 2 \sum_{i>j}^{n} \langle e_i \cdot e_j \rangle \approx n + Bn^{2-\gamma}.$$

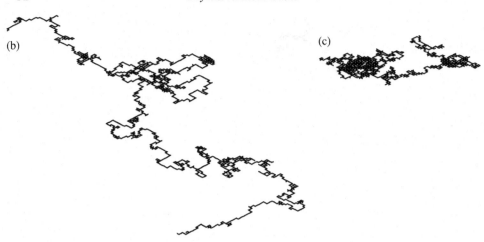

(b)

(c)

Fig. 4.2. Continued.

The exponent $2 - \gamma$ comes from the double sum over i and j of the second term. Thus, for $\gamma < 1$ and for large n the second term dominates and

$$\langle r^2(n) \rangle \sim n^{2-\gamma}, \tag{4.12}$$

and d_{w} is

$$d_{\mathrm{w}} = \frac{2}{2 - \gamma}, \qquad \gamma < 1. \tag{4.13}$$

For $\gamma > 1$ the first term dominates and Eq. (3.3) holds. The above arguments are valid also for the case in which the steps of the walk are taken from a distribution with finite second moment. For a method to generate walks with long-range correlations, Eq. (4.11), using the inverse Fourier transform, see Appendix D. Examples of such walks in $d = 2$ are shown in Fig. 4.2.

The cause of anomalous diffusion cannot be determined from the dependence of the displacement on time. Although Lévy walks can be shown to have long-range correlations, as in Eq. (4.11), the inverse statement is not always true. For example, the inverse Fourier method for the generation of long-range correlations of Appendix D does not give rise to a Lévy distribution of the persistent segments. Instead, their distribution seems to be stretched exponential (see Open challenge 2). An example of such a situation is flow in the presence of layered random fields, for which $\langle r^2 \rangle \sim t^{3/2}$ but the probability density $P(r, t)$ is Gaussian rather than Lévy (Redner, 1990; Zumofen *et al.*, 1990; Araujo *et al.*, 1991a).

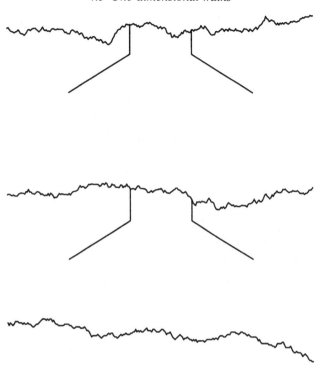

Fig. 4.3. A landscape created by a simple one-dimensional random walk. The x-axis represents the time (number of steps, n) and the y-axis the walk's displacement. Successive figures illustrate the self-affinity property. At each stage a section along the x-axis is magnified by a factor of 5 while the y-axis is magnified by a factor of $5^{1/2}$, and the landscape preserves its overall appearance.

4.5 One-dimensional walks and landscapes

Consider a simple random walk on the y-axis;

$$y(n) = \sum_{i=1}^{n} e_i, \qquad e_i = \pm 1. \tag{4.14}$$

If one plots $y(n)$ as a function of n (which we take to be the x-axis), the resulting trail is a self-affine landscape, or interface (Exercise 1.9, and Fig. 4.3). The curve is self-affine because dilation of the x-axis by a factor b requires that the y-axis be magnified by a different factor, $b^{1/2}$, if the shape is to be preserved. The *roughness*, $w(\ell)$, is defined as the root-mean-square fluctuations of y inside a window of size ℓ. In the case of correlated walks it scales as (Eq. (4.12))

$$w(\ell) \equiv \langle y^2(\ell) - \bar{y}^2(\ell) \rangle^{1/2} \sim \ell^{(2-\gamma)/2} \sim \ell^{\alpha}. \tag{4.15}$$

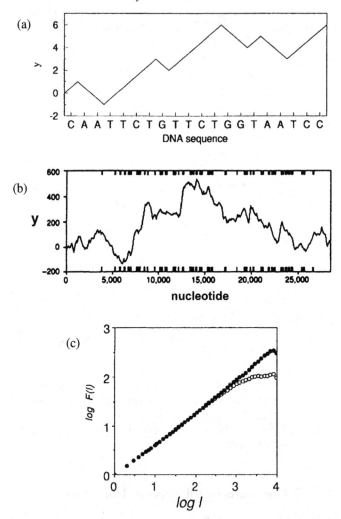

Fig. 4.4. Long-range correlations in DNA. (a) A DNA-walk landscape generated from the rule that the walker steps up for a pyrimidine (C or T) and down for a purine (A or G). After Buldyrev *et al.* (1995b). (b) The DNA-walk representation of the intron-rich human β-cardiac myosin heavy-chain gene sequence. Heavy bars correspond to the coding regions of the gene. After Peng *et al.* (1992a). (c) Double-logarithmic plots of the mean-square fluctuation function $F(\ell)$ as a function of the linear distance ℓ along the DNA chain for the embryonic skeletal myosin heavy-chain gene (○) and the "intron-spliced" sequence (●). After Peng *et al.* (1992b).

The roughness $w(\ell)$ is a useful parameter in the study of interfaces (Barabási and Stanley, 1995). The exponent α is called the roughness exponent. An interesting application of rough landscapes is for a geometrical representation of sequences of DNA (Fig. 4.4), and for the detection of long-range correlations between its nucleotides (Peng *et al.*, 1992a; 1992b). A value of $\alpha = \frac{1}{2}$ indicates short-range correlations, or no correlation. Long-range correlations in DNA are evidenced by the observed roughness exponent of $\alpha \approx 0.6$–0.7. Similar approaches have been applied for the detection of long-range correlations in a wide range of systems, including human writings (Shenkel *et al.*, 1993), heartbeat fluctuations (Peng *et al.*, 1993), and weather fluctuations (Koscielny-Bunde *et al.*, 1998). For methods that distinguish between polynomial trends and power-law correlations see Appendix D.

4.6 Exercises

1. *Self-similarity of RWs.* Consider a simple RW in one dimension, with a Gaussian step probability density; $p(r) = (\pi a^2)^{-1/2} \exp(-r^2/a^2)$. Now, rescale the time step, i.e., regard the original walk as a new RW in which the time between steps is $\tau' = N\tau$ $(N \gg 1)$. Show that the probability density of the rescaled step is also Gaussian. How should one rescale space, so that the RW is invariant (that is, $P'(r', t')$ and $p'(r')$ are similar to $P(r, t)$ and $p(r)$ of the original walk)? Argue that, in the long-time asymptotic limit, the same self-similarity is observed for other forms of $p(r)$.

2. Calculate the number of distinct sites visited by a Lévy flight in a one-dimensional system and in a d-dimensional system. See Gillis and Weiss (1970).

3. Use Eqs. (3.30) and (4.3) to derive the anomalous behavior cited in Eq. (4.4).

4. Derive the behavior of Lévy walks cited in Eq. (4.10).

4.7 Open challenges

1. The form of $P(r, t)$ in the CTRW model in the regime $r \ll t^{1/d_w}$ is not fully understood. See Weissman *et al.* (1989).

2. Equation (4.12) represents a self-affine landscape. Define a persistence length as the number of steps in a sequence of steps taken in the same direction. When the steps e_i are uncorrelated, the distribution of persistence lengths ℓ can be easily shown to decay exponentially as $P(\ell) = (1/a)e^{-a\ell}$. The question is, what is the functional form of $P(\ell)$ for the case of long-range correlated steps according to Eq. (4.9)?

3. The question of Lévy flights in which the size of the jumps $r(x)$ depends on the lattice site x (quenched disorder) and is taken from a Lévy distribution has been studied in $d = 1$ by Kutner and Maass (1998). Many questions remain unanswered. It is plausible that, for $d \geq 2$, the transport exponent will be the same as that for the case of annealed Lévy flights. This is because a random

walk in $d \geq 2$ rarely revisits the same sites. See also similar considerations in Section 8.4.

4.8 Further reading

- Lévy flights have been observed in two-dimensional fluid flow in a rotating vessel and in other interesting experimental setups by Swinney and coworkers (Solomon *et al.*, 1993; 1994; Weeks *et al.*, 1996).
- Lévy flights and walks: Shlesinger *et al.* (1987), Klafter *et al.* (1996), and Araujo *et al.* (1991b). The number of distinct sites visited by Lévy flights: Gillis and Weiss (1970). *N* Lévy flights: Berkolaiko *et al.* (1996), Berkolaiko and Havlin (1997), and Appendix B. Several mathematical aspects of Lévy stable distributions are discussed by Samorodnitsky and Taqqu (1994). Applications to the migration of contaminants in ground-water systems: Berkowitz and Scher (1998), Margolin *et al.* (1998), and Berkowitz *et al.* (2000).
- Long-range correlated walks: Peng *et al.* (1991) and Makse *et al.* (1996). *Optimal paths* (i.e., paths with the lowest possible energy) in random-energy landscapes were found to be long-range correlated, with $\gamma = \frac{2}{3}$, and are believed to be in the same universality class as directed percolation (Cieplak *et al.*, 1994; 1995; Schwartz *et al.*, 1999).
- Anomalous diffusion in layered velocity fields: Redner (1990), Zumofen *et al.* (1990), Araujo *et al.* (1991a), ben-Avraham *et al.* (1992), and Ben-Naim *et al.* (1992).
- Fractional Brownian motion: Mandelbrot (1982) and Voss (1991).

Part two

Anomalous diffusion

This part of the book explores the subject of anomalous diffusion in fractals and disordered media. Our goal here is twofold: to introduce various approaches to the problem – exact, as well as approximate; and to become intimately acquainted with the phenomenon of anomalous diffusion, its characteristics, and its causes.

Chapter 5 discusses diffusion in the Sierpinski gasket. Anomalous diffusion is demonstrated through simulations and exact enumerations, and through an exact renormalization of the first-passage time. The important relation between diffusion and conductivity (the Einstein relation), introduced in Section 3.4, is used to rederive the anomalous diffusion exponent in yet another way. The chapter closes with a discussion of probability-density-distribution functions and of fractons and spectral dimensions.

In Chapter 6 we present a summary of diffusion in percolation clusters. The question of diffusion in the incipient infinite cluster versus diffusion in *all* of the clusters is analyzed through scaling and simulations. The chapter describes also the attempt of the Alexander–Orbach conjecture to connect between static and dynamic exponents, diffusion in chemical space, and the multifractality of conductivity and diffusion in percolation clusters.

Chapter 7 discusses diffusion in loopless fractals: Eden trees, combs, etc. In this important case some exact results may be derived, including general relations between static and dynamic exponents, which shed some light on the more general situation in which loops are relevant.

Diffusion in *regular* lattices but with disordered transition rates is also anomalous, and is surveyed in Chapter 8. The subject is richer than that of diffusion in fractals (in fact, the latter may be regarded as a special case), and includes the famous Sinai problem.

Chapter 9 deals with the complicated question of biased diffusion (topological and Cartesian). The focus is mostly on the relatively tractable case of biased diffusion in combs, and the results are used for the interpretation of simulations in percolation and other random fractals.

Finally, in Chapter 10 we review incidents of diffusion with excluded-volume interactions: tracer diffusion and self-avoiding walks. Tracer diffusion describes the motion of a tagged random walker that interacts with other walkers through hard-core repulsion. Self-avoiding walks are an elementary model for linear polymers, and can be mapped onto the $n = 0$ vector model for phase transitions. Together they serve as examples of what might change when interactions are added to the picture.

5

Diffusion in the Sierpinski gasket

In previous chapters, we have seen examples of anomalous diffusion in the CTRW model, in Lévy flights, and in long-range correlated walks. Diffusion in fractal lattices is also anomalous. Here we consider nearest-neighbor random walks in the Sierpinski gasket, for which an exact solution is possible. We analyze the problem and solve it following several different approaches. The analysis not only illustrates anomalous diffusion in a simple way, but also stresses important aspects of diffusion theory, such as the relation to conductivity and elasticity.

5.1 Anomalous diffusion

Imagine a random walk in the Sierpinski gasket. At each step the walker moves randomly to one of the four nearest-neighbor sites *on the gasket*, with equal probabilities. We require the mean-square displacement after n steps, $\langle r^2(n) \rangle$.

Naively, one should think that, since diffusion is regular ($\langle r^2(n) \rangle \sim n$) in all integer dimensions, so would be the case for fractals, since fractals may be regarded as mere extrapolations of regular space to noninteger dimensions. Surprisingly, this turns out to be wrong! Perhaps the best way to convince oneself of this fact is to perform numerical simulations of random walks on the Sierpinski gasket. By averaging the mean-square displacement over several different runs (the points of origin may also be chosen randomly) one obtains a direct estimate of $\langle r^2(n) \rangle$. It is an instructive exercise! Alternatively, an exact enumeration of all possible walks may be performed numerically for arbitrarily large n, and starting from different origins. The technique is described in Appendix A.

In Fig. 5.1 we show results obtained from exact enumeration. One finds that there is little sensitivity to the starting point. At large times the plot of $\log n$ versus $\log \langle r^2(n) \rangle^{1/2}$ follows a straight line. The slope of the curve is the fractal dimension of the walks, which is found numerically to be $d_w = 2.33 \pm 0.01$. This is, as expected, in very good agreement with results from simulations. The accuracy

The Sierpinski gasket

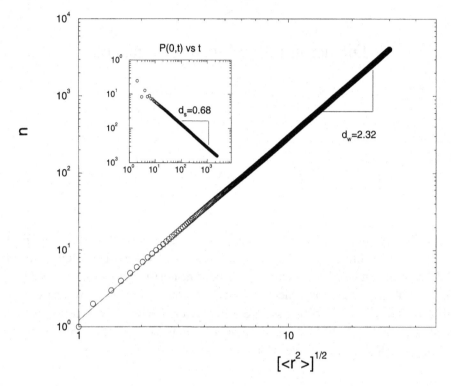

Fig. 5.1. The fractal dimension of RWs on the Sierpinski gasket. A plot of $\log n$ versus $\log \langle r^2(n) \rangle^{1/2}$, obtained from an exact enumeration of walks carried out for 3000 steps. The resulting curve converges quickly, after just four or five steps, to a straight line of slope $d_w = 2.33 \pm 0.04$, regardless of the starting point. The inset shows the probability of being at the origin after n steps.

may be increased considerably, but it is not necessary, for we shall soon compute d_w exactly. At any rate, the result is convincingly different than that for regular diffusion, for which $d_w = 2$.

What makes diffusion in the Sierpinski gasket anomalous? We could rephrase the question: we have proven that diffusion in all dimensions is regular (Section 3.1) – where does the argument fail when it is applied to the Sierpinski gasket? The answer is that now the correlations between different steps are no longer zero; $\langle e_i \cdot e_j \rangle \neq 0$. In Euclidean lattices all sites are equivalent and it is easy to prove that correlations are zero. In fractals, however, the different sites are not equivalent. Regard the Sierpinski gasket as embedded in a regular two-dimensional triangular lattice. At each node of the gasket two bonds of the embedding triangular lattice are missing, but which bonds are missing varies from one gasket site to the next. As a consequence $e_i \cdot e_j$ depends on *where* the walker is at times i and j. This makes diffusion on fractals non-Markovian.

One could imagine that the correlations may average out. For small $|i - j|$, however, it is easy to convince oneself (by enumerating all walks) that this is not the case, and that $\langle e_i \cdot e_j \rangle \neq 0$. The same is true for large $|i - j|$, though less evident. The nonzero correlations are ultimately responsible for the anomalous behavior. The exact approaches presented in Sections 5.2 and 5.3 will help build up further intuition for the rise of anomalous diffusion in fractals.

It is generally true that $d_w > 2$ for anomalous diffusion in fractals. The number of fractal lattice sites within a radius L is L^{d_f}. On the other hand, a walker must perform $t \sim L^{d_w}$ steps to attain a r.m.s. displacement L. It follows that, if $d_w > d_f$, each site within the radius L would be visited several times (the number of visits per site would increase with the radius as $L^{d_w - d_f}$). In this case the trail of the walk is "compact" – the number of *distinct* sites visited equals the volume L^{d_f}. In the opposite case of $d_w < d_f$, the walker cannot visit all lattice sites within L because their number exceeds the number of steps. The trail is tenuous: only a vanishing fraction $L^{d_w - d_f}$ of the sites is visited.

The compact walks are also called *recurrent*, because of their property that, in the long-time limit, a walker returns to the origin with probability unity. For *nonrecurrent* walks with $d_w < d_f$ the probability of return to the origin vanishes with increasing time. In regular Euclidean space $d_w = 2$ and $d_f = d$, hence diffusion in one dimension is recurrent and diffusion in $d \geq 3$ is nonrecurrent. The case of $d = 2$ is marginal, $d_w = d_f$, and is accompanied by logarithmic divergences.

5.2 The first-passage time

We now compute d_w for the Sierpinski gasket, following an exact real-space renormalization-group procedure. We focus on T, the mean first-passage time taken to traverse the lattice from the vertex at the apex to one of the remaining two vertices at the bottom (Fig. 5.2a). T can be related to T', the equivalent first-passage time in a lattice rescaled by a factor of two (Fig. 5.2b).

Let A and B be the first-passage times from the inner vertices of the rescaled lattice to the lower O vertices (Fig. 5.2b). Then, exploiting the finite ramification of the gasket, we see that

$$
\begin{aligned}
T' &= T + A, \\
A &= T + \tfrac{1}{4}A + \tfrac{1}{4}B + \tfrac{1}{4}T', \\
B &= T + \tfrac{1}{2}A.
\end{aligned}
\tag{5.1}
$$

For example, suppose that a walker is in one of the nodes denoted by A in Fig. 5.2b. If it gets first to the vertex at the top, it will then take an additional time T' to exit

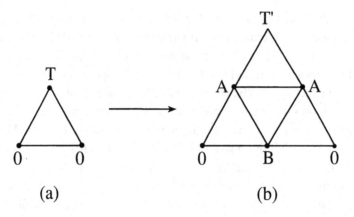

(a)   (b)

Fig. 5.2. Rescaling of first-passage times: (a) The mean first-passage time for traversing a Sierpinski gasket from the top apex to either of the vertices at the base of the gasket (marked by 0) is T. (b) When the lattice is rescaled by a factor of two, the first passage time is T', which is clearly larger than T. A and B are the first-passage times for exit through 0, for walks originating from the junctions between the rescaled subunits. A, B, and most importantly, T' may be related to T exactly (Eq. (5.1)).

through O. If instead it gets first to the A node, it will take an extra time A to exit through O. If it gets first to the B node, it will take an extra time B to exit. Finally, if it gets first to O, no additional time is necessary, for it would have already exited the unit. Since each of these possibilities happens with equal probability, and since it takes an average time T to get to any of the nearest nodes, we have

$$A = \tfrac{1}{4}(T + T') + \tfrac{1}{4}(T + A) + \tfrac{1}{4}(T + B) + \tfrac{1}{4}T,$$

which is the second equation in (5.1). The remaining equations are obtained in a similar fashion.

The solution of Eqs. (5.1) is $T' = 5T$ (and $A = 4T$ and $B = 3T$). We can then compute d_w from the way it relates time T to length L:

$$T \sim L^{d_w} \tag{5.2}$$

(cf. Eq. (4.2)). Upon rescaling of space by a factor of two, $L \to L' = 2L$, time rescales as $T \to T' = 5T$. Since Eq. (5.2) holds also for the rescaled problem; $T' \sim (L')^{d_w}$, we conclude that

$$d_w = \frac{\log 5}{\log 2} \cong 2.322 \tag{5.3}$$

for random walks on the Sierpinski gasket. This compares very nicely with the numerical results of Monte Carlo simulations and exact enumeration.

5.3 Conductivity and the Einstein relation

One of the main reasons for the interest in diffusion in fractals and disordered media is its relevance to other physical properties of the medium. Here we wish to discuss the connection between diffusion and electric conductivity.

The basic relation is the Einstein relation derived in Section 3.4, $\sigma = ne^2 D/(k_B T)$ (Eq. (3.21)). In anomalous diffusion the diffusion constant D is not really a constant, but actually changes as diffusion proceeds:

$$D \equiv \frac{\langle r^2(t) \rangle}{t} \sim \frac{L^2}{t}, \tag{5.4}$$

where L is the typical distance covered at time t. The carrier density n is proportional to the density of the substrate, and hence in a fractal $n \sim M/L^d \sim L^{d_f - d}$. The conductivity of a fractal is expected to scale as a power of its linear size,

$$\sigma \sim L^{-\tilde{\mu}}, \tag{5.5}$$

where $\tilde{\mu}$ is the *conductivity exponent*. On inserting these expressions into the Einstein relation we find

$$t \sim L^{2 - d + \tilde{\mu} + d_f},$$

or, since $t \sim L^{d_w}$,

$$d_w = 2 - d + d_f + \tilde{\mu} = d_f + \tilde{\zeta}. \tag{5.6}$$

In the last expression $\tilde{\zeta}$ is the *resistance exponent* which denotes the scaling of resistance with length:

$$R \sim L^{\tilde{\zeta}}. \tag{5.7}$$

The relation between $\tilde{\mu}$ and $\tilde{\zeta}$ is obtained from the relation between conductivity and resistance. Conductivity is defined through the equation $j = \sigma E$, where j is the current density and E is the electric field. Resistance is given by Ohm's law, $I = V/R$, where I is the total current and V is the potential gap, $V \sim EL$. In d dimensions the current density is $j \sim I/L^{d-1}$ (current per unit "area"). Comparison of the equations for σ and R yields $\tilde{\zeta} = 2 - d + \tilde{\mu}$, as in Eq. (5.6). Notice that, for regular space, conductivity is an *intensive* quantity and $\tilde{\mu} = 0$, but generally for fractals $\tilde{\mu} > 0$.

We now exploit the relation between resistance and random walks (Eq. (5.6)) to compute d_w indirectly. Suppose that a current I is injected into the top vertex of a Sierpinski gasket and is drawn out from the other two vertices (Fig. 5.3a). We wish to compute the resistance R between the top vertex and a vertex at the bottom. Let

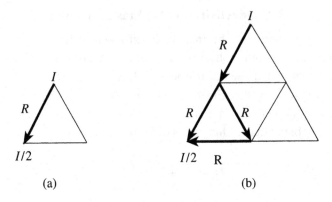

Fig. 5.3. Rescaling of the resistance of the Sierpinski gasket. A current I is injected at the apex of the gasket and collected from the lower vertices. (a) The effective resistance between the apex and a vertex at the bottom is R. (b) In a gasket rescaled by a factor of two the current will flow from vertex to vertex along the path indicated by the arrows. This suggests the rescaling of resistance of Eq. (5.8).

R' be the analogous resistance in a lattice rescaled by a factor of two. Owing to symmetry the current flows as indicated in Fig. 5.3b. From this we conclude that

$$R' = R + (R \parallel 2R) = \tfrac{5}{3}R, \qquad (5.8)$$

where the vertical bars denote combination of resistances in parallel. From the scaling of resistance with length (Eq. (5.7)) we find

$$\tilde{\zeta} = \frac{\log \tfrac{5}{3}}{\log 2}. \qquad (5.9)$$

Using Eq. (5.6) and the fact that the fractal dimension of the Sierpinski gasket is $d_{\mathrm{f}} = \log 3/\log 2$, we recover $d_{\mathrm{w}} = \log 5/\log 2$. This method can be applied to calculate $\tilde{\zeta}$ and d_{w} for finitely ramified fractals.

The remarkable relation between resistance and random walks provides us with a most useful technique for estimating $\tilde{\zeta}$ in fractals. One simply simulates diffusion or performs exact enumeration in the fractal in question and measures d_{w}. The fractal dimensionality which is also necessary for the computation of $\tilde{\zeta}$ (Eq. (5.7)) is related to the probability of the walker returning to the origin (see next section) and can too be measured from the same diffusion simulations. This method has been used to obtain reliable estimates of $\tilde{\zeta}$ and $\tilde{\mu}$ for percolation clusters, lattice animals, diffusion-limited aggregates, and other random substrates.

5.4 The density of states: fractons and the spectral dimension

The Laplacian operator ∇^2 of the diffusion equation

$$\frac{\partial P}{\partial t} = D \nabla^2 P$$

appears in many other physics problems. For example, in the Hamiltonian $\mathcal{H} = -[\hbar^2/(2m)]\nabla^2$ of the Schrödinger equation for a free particle:

$$i\hbar \frac{\partial}{\partial t}\psi = \mathcal{H}\psi$$

as well as in heat equations, in Poisson's equation for electric potential, etc.

As a specific example, consider the vibrational modes of an elastic fractal network consisting of particles connected by harmonic springs. In fractals the vibrational modes are called *fractons* rather than phonons. In the isotropic case in which the spring constants are assumed to be scalars one obtains the equations of motion

$$\frac{d^2 U_i(t)}{dt^2} = \sum_j k_{ij}[U_j(t) - U_i(t)], \tag{5.10}$$

where U_i is the displacement of the ith site, and the sum runs over all nearest neighbors j of site i. (For simplicity, we assume that all particles have unit mass.) Diffusion in a discrete fractal lattice would be described by a similar equation, except that $U_i(t)$ is then replaced by $P_i(t)$ – the probability of being at site i at time t – and there is a first-order time derivative instead of the second-order derivative on the LHS of Eq. (5.10). Also, in the case of diffusion k_{ij} represents the jump frequency for jumps from site i to site j. Notice that, when all the k_{ij} are equal, the RHS of (5.10) is in fact a discrete form of $\nabla^2 U$.

Equation (5.10) may be solved by a standard classical-mechanics approach. On making the substitution $U_i(t) = U_i(w)\exp(-i\omega t)$ one obtains N linear homogeneous equations for the N unknowns $U_i(w)$ (N is the total number of sites). When $k_{ij} = k_{ji}$, i.e., for *unbiased* diffusion, the equations yield positive real eigenvalues: $\omega_n^2 \geq 0$ ($n = 1, 2, \ldots, N$). The general solution of (5.10) is $U_i(t) = \Re\{\sum_{n=1}^{N} C_n \Psi_i^n \exp(-i\omega_n t)\}$, where $(\Psi_1^n, \ldots, \Psi_N^n)$ is an orthogonal set of eigenvectors and the C_n are complex constants that are determined from initial conditions. For example, if at $t = 0$ only the j_0th particle is displaced, $U_i(0) = U_{j_0}\delta_{ij_0}$, we get

$$U_{i+j_0}(t) = U_{j_0}(0)\Re\left(\sum_{n=1}^{N} (\Psi_{j_0}^n)^* \Psi_{i+j_0}^n \exp(-i\omega_n t)\right).$$

For the corresponding diffusion problem, when the walks originate at $i = j_0$, the solution is

$$P_{i+j_0}(t) = \Re \left(\sum_{n=1}^{N} (\Psi_{j_0}^n)^* \Psi_{i+j_0}^n \exp(-\epsilon_n t) \right),$$

where $\epsilon_n = \omega_n^2$ is the energy associated with the vibration mode n. The average probability of finding the walker at distance r from the origin at time t is then

$$\langle P(r,t) \rangle = \Re \left(\sum_{n=1}^{N} \Phi(r,n) \exp(-\epsilon_n t) \right),$$

where

$$\Phi(r,n) = \frac{1}{N} \sum_{j_0=1}^{N} \frac{1}{N(r)} \sum_{i=1}^{N(r)} (\Psi_{j_0}^n)^* \Psi_{i+j_0}^n.$$

Notice the double average: over all $N(r)$ points a distance r away from j_0, and over the starting point j_0. In particular, for $r = 0$,

$$\langle P(0,t) \rangle = \frac{1}{N} \sum_{n=1}^{N} \exp(-\epsilon_n t),$$

since the eigenvectors are normalized, $\Phi(0,n) = 1/N$. Passing to the continuum limit $N \to \infty$, one may rewrite the last equation as

$$\langle P(0,t) \rangle = \int d\epsilon \, \rho(\epsilon) \exp(-\epsilon t), \qquad (5.11)$$

where $\rho(\epsilon)$ is the energy density of states.

The probability of returning to the origin at time t may also be estimated from the following argument. At time t the r.m.s. displacement is $L \sim t^{1/d_w}$. The probability that a walker is in any of the L^{d_f} sites within that radius is approximately uniform, so $P(0,t) \sim 1/L^{d_f} \sim 1/t^{d_f/d_w}$. Using this relation in Eq. (5.11), we finally find

$$\rho(\epsilon) \sim \epsilon^{d_f/d_w-1} \equiv \epsilon^{d_s/2-1}. \qquad (5.12a)$$

The vibrational density of states $g(\omega)$ then follows from $\rho(\epsilon)\, d\epsilon = g(\omega)\, d\omega$ and $\epsilon = \omega^2$:

$$g(\omega) \sim \omega^{d_s-1}. \qquad (5.12b)$$

In regular Euclidean space the two densities are $\rho(\epsilon) \sim \epsilon^{d/2-1}$ and $g(\omega) \sim \omega^{d-1}$. In fractals, the vibrational modes are called *fractons* instead of phonons. We see that, in the case of fractals, the effective dimension which controls the density of states is

$$d_s = 2d_f/d_w \qquad (5.13)$$

rather than d. The exponent d_s is known as the *fracton* dimension, or also as the *spectral* dimension. (We use the subscript s to avoid confusion with the fractal dimension d_f.)

The relation between diffusion and the density of states can also be derived from the following scaling argument. The density of states is an *extensive* quantity (proportional to the volume of the system), thus, upon rescaling of length $L \to bL$, the density of states rescales as

$$\rho_{bL}(\epsilon) = b^{d_f} \rho_L(\epsilon). \tag{5.14}$$

From the Schrödinger equation (we could use any phenomenon involving the ∇^2 operator) we see that the energy – the eigenvalue of $\mathcal{H}$ – is inversely proportional to time, which scales as $t \sim L^{d_w}$, so

$$\epsilon_{bL} = b^{-d_w} \epsilon_L. \tag{5.15}$$

The density of states in the original system and that in the rescaled system are related through probability theory: $\rho_{bL}(\epsilon_{bL}) \, d\epsilon_{bL} = \rho_L(\epsilon_L) \, d\epsilon_L$, or, using Eq. (5.15),

$$\rho_{bL}(\epsilon_{bL}) = b^{d_w} \rho_L(\epsilon_L). \tag{5.16}$$

The solution of Eqs. (5.14)–(5.16) is $\rho(b\epsilon) = b^{d_f/d_w-1} \rho(\epsilon)$, in agreement with Eq. (5.12a). For fractons one can follow the same line of reasoning, except that instead of Eq. (5.15) we now have $\omega_{bL} \sim b^{-d_w/2} \omega_L$ (because of the second-order time derivative, as opposed to the first-order derivative in Schrödinger's equation). The result is identical with Eq. (5.12b).

Both the Schrödinger equation and the problem of fractons have been treated exactly in the Sierpinski gasket. The analyses confirm the relation to diffusion discussed above.

5.5 Probability densities

In random-walk theory probability densities play a central role. Indeed, let $P(r, t)$ be the probability density that a walker that started at the origin at time $t = 0$ is at point r at time t. In regular space random walks are Markovian and $P(r, t)$ determines the diffusion process completely. Diffusion in fractals is in general non-Markovian and, although $P(r, t)$ does not then provide a complete characterization it remains a quantity of primary importance. For example, knowledge of it allows the computation of the moments

$$\langle r^k(t) \rangle = \int r^k P(r, t) \, d^d r \tag{5.17}$$

the probability of return to the origin is simply $P(0, t)$ and from it one may obtain the fracton dimension d_s; and $P(r, t)$ plays a prominent role in the Flory theory of self-avoiding walks.

For regular diffusion $P(r, t)$ is a Gaussian distribution (Eq. (3.15)). In the case of anomalous diffusion, one expects that $P(r, t)$ would have a similar scaling and functional form. A reasonable assumption is

$$P(r, t) = \frac{r^{d_f - d}}{t^{d_s/2}} \Phi\left(\frac{r}{t^{1/d_w}}\right). \tag{5.18}$$

The fundamental scaling of $r \sim t^{d_w}$ is contained in the argument of Φ. The prefactor of $r^{d_f - d}$ accounts for the expected scaling of P with fractal density. (Consider for example the case in which P is homogeneous in space.) Finally, the prefactor of $t^{-d_s/2}$ follows from normalization; $\int P \, d^d r = 1$, if $d_s = 2d_f/d_w$.

For $r \gg t^{1/d_w}$ we expect some form of exponential decay, which may be incorporated in the function Φ (Fig. 5.4):

$$P(r, t) = \frac{r^{d_f - d}}{t^{d_s/2}} \left(\frac{r}{t^{1/d_w}}\right)^g \exp\left[-a\left(\frac{r}{t^{1/d_w}}\right)^\delta\right], \tag{5.19}$$

where δ is a *shape exponent*. The smaller δ the sharper the peak of the distribution. It has been suggested that $g = \delta/2 - d_f$ (Acedo and Yuste, 1998). Notice that Eqs. (5.18) and (5.19) agree with regular diffusion (where $d_f = d$, $d_w = 2$, and $\delta = 2$).

As a particularly simple example consider diffusion in linear polymers, modeled by self-avoiding walks (SAWs). SAWs have a random fractal structure but are topologically linear, similar to the Koch curve. Therefore, the process of diffusion along a SAW is normal, just like in one-dimensional space. Using the length ℓ measured along the SAW (the chemical length), the probability density $\phi(\ell, t)$ of diffusion in the SAW is

$$\phi(\ell, t) = \frac{1}{(4\pi Dt)^{1/2}} \exp\left(-\frac{\ell^2}{4Dt}\right). \tag{5.20}$$

The relation between ℓ and distance in regular space, r, is given through the fractal dimensionality of the SAW, $r^{d_f} \sim \ell$. However, a better description is obtained from the probability density $P(r|\ell)$ that an ℓ-step SAW section is of length r:

$$P(r|\ell) = C\ell^{-dv}\left(\frac{r}{\ell^v}\right)^{g_{SAW}} \exp\left[-a\left(\frac{r}{\ell^v}\right)^u\right], \qquad u = \frac{1}{1 - v}. \tag{5.21}$$

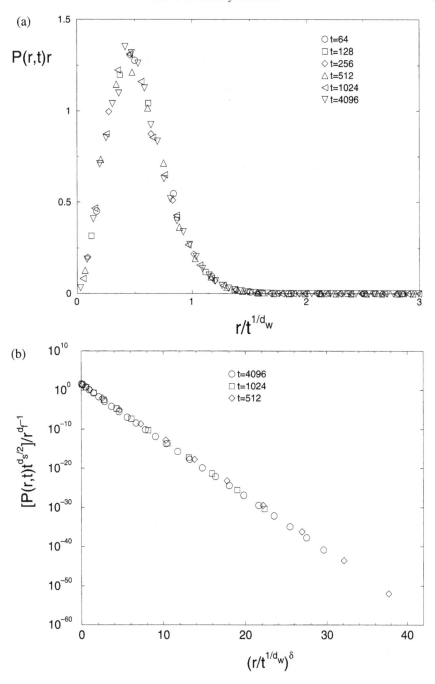

Fig. 5.4. $P(r, t)$ of random walks in the Sierpinski gasket. (a) The universal scaling of $P(r, t)$ for walks of differing time spans is shown. (b) A plot of $\log[P(r, t)t^{d_s/2}]$ versus $(r/t^{1/d_w})^\delta$ for walks of various lengths. The data collapse confirms the scaling form of Eq. (5.19). The slope of the linear fit is $\delta = 1.76 \pm 0.05$, in agreement with $\delta = d_w/(d_w - 1) = 1.756$. Note that the plots are for $P(r, t)$ rather than $\mathbf{P}(\mathbf{r}, t)$.

Here $\nu = 1/d_f$ is the end-to-end exponent of the SAW. We can now write

$$P_{SAW}(\mathbf{r}, t) = \int_0^\infty P(\mathbf{r}|\ell)\phi(\ell, t)\, d\ell,$$

and, using a steepest-descent approximation, one obtains

$$P_{SAW}(\mathbf{r}, t) \sim \exp\left[-b\left(\frac{r}{t^{1/d_w}}\right)^\delta\right], \qquad \delta = \frac{d_w}{d_w - 1}. \tag{5.22}$$

In the final form of Eq. (5.22) we have used the fact that diffusion in a SAW has walk dimension $d_w = 2d_f$. Equation (5.22) has the same form as Eq. (5.19). Interestingly, it also agrees with the probability density of anomalous diffusion of CTRWs (cf. Eq. (4.5)).

The probability density of diffusion in the Sierpinski gasket has been studied in great detail. Most authors agree that the data are well described by (5.19) and some theoretical arguments support this conclusion. Barlow and Perkins (1988) have rigorously proven that $\delta = d_w/(d_w - 1)$ for walks in the Sierpinski gasket, though it has been suggested that $\delta = d_w$ might better describe short distances, $r < t^{1/d_w}$.

A consequence of the scaling form of $P(\mathbf{r}, t)$, Eq. (5.18), is that the moments may be characterized by a *gap exponent*: $\langle r^k \rangle \sim t^{\alpha k}$, $\alpha = 1/d_w$. (Notice that the gap exponent is independent of δ.) There is at present no proof that the probability density should scale as postulated. Indeed, we now know of several phenomena in disordered media displaying *multifractality*, for which moments do not possess a gap exponent. It turns out that the probability density of anomalous diffusion in random fractals has multifractal features. This has been shown by examining the moments $\langle P^q(r, t) \rangle$, for which the average is taken over different random configurations (Havlin and Bunde, 1989; Bunde *et al.*, 1990). A similar multifractal property arises for $P(\mathbf{r}, t)$ of diffusion in percolation clusters (Chapter 6).

5.6 Exercises

1. Consider the *branching Koch curve* (BKC) of Fig. 5.5. Produce this fractal lattice in a computer to 4–5 generations. Simulate random walks (up to time $t = 100$) starting from different points on the lattice and measure d_w. Perform exact enumerations (with different origins) and measure d_w.
2. Write down renormalization-group equations for the first-passage time from edge to edge in the BKC. Derive d_w from these equations. (Answer: $d_w = \log(40/3)/\log(3)$.)
3. Compute the resistance exponent $\tilde{\zeta}$ of the BKC from a renormalization-group

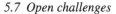

(a)

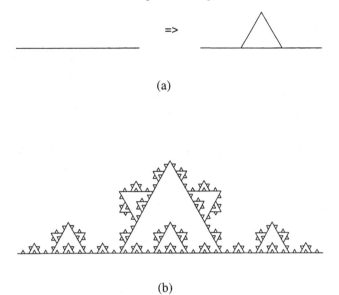

(b)

Fig. 5.5. A branching Koch curve. (a) The unit-segment initiator and the generator rule. (b) The branching Koch curve after four iterations.

approach. Combine your result with d_f of the BKC to rederive d_w, using the Einstein relation.

4. The renormalization-group approach can be successfully used for the computation of d_w, and also of $\tilde{\zeta}$, for regular one-dimensional space (try it!). Why can one not apply the same method to two dimensions?

5. Compute the spectral dimension of the BKC and also obtain it from the probability of return to the origin in your simulations of random walks. (Answer: $2\log 5/\log(40/3)$.)

6. Check out the scaling of $P(r, t)$, Eq. (5.18), in your simulations (or exact enumeration) of random walks in the BKC. It is important to start from many different origins! Estimate the shape exponent δ and compare it with $d_w/(d_w - 1)$.

7. Perform the steepest-descent approximation leading to Eq. (5.22).

5.7 Open challenges

1. Exact results on anomalous diffusion are known only for finitely ramified fractals. Examples of exact analyses for infinitely ramified fractals would be most useful.

2. In certain situations, the spectral dimension d_s which follows from the scaling

of $P(0, t) \sim t^{-d_s/2}$ does not agree with the one from $g(\omega) \sim \omega^{d_s-1}$ (Burioni and Cassi, 1996; Burioni *et al.*, 1999). The problem needs further elucidation.

3. The question of $P(r, t)$ in fractals needs more researching. Is there always scaling (Eq. (5.18)) in the long-time asymptotic limit? If so, what is Φ? It is also not known what happens in the short-time regime.

4. How does anomalous diffusion in affine fractals look? What is then the scaling (if any) of $P(r, t)$?

5. The distribution of the number of *distinct* sites visited by a walker (each site is counted only once, even after multiple visits) in fractals is a widely open problem, including in the Sierpinski gasket.

6. Most research into diffusion in fractals has focused on simple random walks. What happens, for example, with persistent random walks in fractals? How is their persistence length affected by the fractal substrate?

5.8 Further reading

- Anomalous diffusion (or conductivity) in exact fractals: Dhar (1978a; 1978b), Mandelbrot (1982), Given and Mandelbrot (1983), Gefen *et al.* (1981; 1983a; 1984), Havlin and ben-Avraham (1982a), and Gould and Kohin (1984). In random fractals: Kirkpatrick (1979), Havlin and ben-Avraham (1983), Pandey and Stauffer (1983), Hong *et al.* (1984), Majid *et al.* (1984), Wilke *et al.* (1984), Havlin *et al.* (1984a; 1984b; 1985d), Meakin and Stanley (1983), Meakin *et al.* (1984), Dhar and Ramaswamy (1985), Derrida and Vannimenus (1982), Derrida *et al.* (1984), Zabolitzky (1984), and Reis (1996a). The first-passage time and number of distinct sites visited by many walkers on fractals: Yuste (1997; 1998) and Havlin *et al.* (1992). Mathematical approaches: diffusion in the Sierpinski gasket (Barlow *et al.*, 1990; Barlow and Bass, 1990; 1992), and renormalization of finitely ramified fractals (Metz, 1995).

- Einstein's relation: Alexander and Orbach (1982), Havlin (1985), Cates (1985), and Gefen and Goldhirsch (1987).

- Fractons: Alexander *et al.* (1978; 1981), Alexander and Orbach (1982), Rammal and Toulouse (1982; 1983), Domany *et al.* (1983), and Teitel and Domany (1985). Localization of fractons, and electrons: Bunde and Havlin (1996) and Kantelhardt and Bunde (1997). Diffusion of electrons, and multifractality of the energy spectra: Huckestein and Schweitzer (1994) and Ketzmerick *et al.* (1997). Vibrations of fractal drums: Sapoval *et al.* (1991) and Even *et al.* (1999).

- $P(r, t)$: Fisher (1966), Fisher and Burford (1967), Domb (1969), McKenzie (1976), McKenzie and Moore (1971), Mandelbrot (1982), Havlin and ben-Avraham (1982b), Havlin and Nossal (1984), Havlin *et al.* (1985b; 1985c), O'Shaughnessy and Procaccia (1985a), Banavar and Willemson (1984), and

Guyer (1984). Exact results for $P(r, t)$ on random walks for $r \ll t^{1/d_w}$: Rabinovich *et al.* (1996). Dependence of $P(r, t)$ on the number of configurations: Drager and Bunde (1996).

6

Diffusion in percolation clusters

In an insightful pioneering work de Gennes (1976a; 1976b) pondered the problem of a random walker in percolation clusters, which he described as "the ant in the labyrinth". Similar ideas were presented at the time by Brandt (1975), Kopelman (1976), and Mitescu and Roussenq (1976). de Gennes' "ants" triggered intensive research on diffusion in disordered media. Here we describe the more important aspects of the subject. A brief account of percolation theory has been presented in Chapter 2.

6.1 The analogy with diffusion in fractals

As discussed in Chapter 2, percolation clusters may be regarded as random fractals. Below the critical threshold, $p < p_c$, the clusters are finite. The largest clusters have a typical size of the order of the (finite) correlation length $\xi(p)$, and they possess a fractal dimension $d_f = d - \beta/\nu$ (Eq. (2.8)).

At criticality, $p = p_c$, there emerges an infinite percolation cluster that may be described as a random fractal with the same dimension $d_f = d - \beta/\nu$. The inception of the infinite cluster coincides with the divergence of the correlation length, $\xi \sim |p - p_c|^{-\nu}$. Along with the incipient infinite cluster there exist clusters of finite extent. The finite clusters may be regarded as fractals possessing the usual percolation dimension d_f. They are not different, practically, than the clusters that form below the percolation threshold, other than that there is no limit to their typical size – since the global correlation length diverges.

Above criticality, $p > p_c$, the percolation system still consists of the incipient infinite cluster and of finite clusters. However, the typical size of the largest finite clusters is again limited to the finite correlation length $\xi(p)$, exactly like below criticality, and the clusters are fractals with the usual dimension d_f. The infinite cluster is different than the one at criticality. It resembles a fractal only within

74

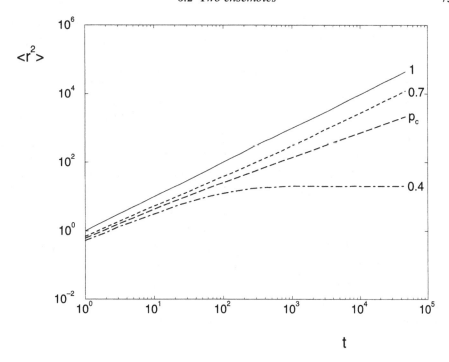

Fig. 6.1. Crossover behavior for diffusion in the incipient infinite percolation cluster. A plot of $\langle r^2(t) \rangle$ versus t, for 46 500-step walks in the percolation cluster with various values of p. Above criticality, $p > p_{\text{c}}$, diffusion is normal. The solid line with slope 1 ($d_{\text{w}} = 2$) is shown for comparison. Below criticality the walk attains a finite span, dictated by the finite size of the clusters.

distances shorter than $\xi(p)$; at distances larger than ξ it behaves like regular space (cf. Eq. (2.5) and Figs. 2.6 and 2.7).

Diffusion in percolation clusters can then be explained as diffusion in fractals. In view of the situation pictured above, there are three different diffusion regimes. For $p > p_{\text{c}}$ the infinite cluster is homogeneous at large length scales, and hence in the long-time asymptotic limit diffusion is regular, with $d_{\text{w}} = 2$. At criticality, $p = p_{\text{c}}$, the incipient infinite cluster is self-similar at all length scales and diffusion is anomalous, with $d_{\text{w}} > 2$. Finally, for $p < p_{\text{c}}$ the span of the walks is limited by the finite size of the largest clusters, $\langle r^2(t) \rangle \sim \xi(p)^2$ ($t \to \infty$). These three regimes are illustrated in Fig. 6.1.

6.2 Two ensembles

Diffusion in percolation clusters can be interpreted in different ways. On the one hand, one could restrict the walks to the incipient infinite cluster at $p = p_{\text{c}}$. In this case finite-sized clusters are irrelevant and the problem is similar to that of diffusion in the Sierpinski gasket (Chapter 5). Research into this problem enriches our knowledge of the critical properties of the infinite cluster at criticality. One goal, for example, is to find the *dynamical* exponents d_w and $\tilde{\mu}$ of the percolation phase transition. (Notice that critical exponents of the percolation transition are defined in terms of the incipient infinite cluster alone.)

On the other hand, one could study diffusion in *all* the clusters of the system, unrestrictedly. This would make sense for many practical applications. Suppose for example that the percolation clusters represent islands of a conducting material within a nonconducting matrix. Diffusion of the charge carriers in the conductor phase would surely be described by considering the unrestricted ensemble of all clusters. We expect different results than those for the critical infinite cluster, because the span of the walks would now be bounded by the size of the finite clusters. We shall denote quantities associated with the ensemble of all clusters with a prime, to distinguish them from the case restricted to the incipient infinite cluster. Thus, for example, the diffusion exponent of walks in the ensemble of all clusters will be denoted by d'_w.

We now present a simple argument for the relation between d_w and d'_w. In the critical infinite cluster,

$$\langle r^2(t) \rangle \sim t^{2/d_w}. \tag{6.1}$$

The same is true for walks in finite clusters, as long as the mean displacement of the walk is smaller than the span of the cluster. The boundaries of the cluster are then never sensed by the walker.

The ensemble of all clusters can be formed at any p. Let us momentarily limit the discussion to p just below p_c, so that there is no infinite cluster. (Computer simulations are usually carried out in this regime.) The average squared displacement of a walker in s-site clusters is

$$r_s^2(t) \sim \begin{cases} t^{2/d_w} & (t^{2/d_w} < L_s^2), \\ L_s^2 & (t^{2/d_w} > L_s^2), \end{cases} \tag{6.2}$$

where $L_s \sim s^{1/d_f}$ is the average span of s-clusters. The first line corresponds to the case of Eq. (6.1). The second line accounts for walkers that have reached the edge of an s-cluster. Their average square displacement cannot increase beyond the finite size of the cluster. Put in a different way, if the time t is fixed, a different behavior would be observed for clusters smaller or larger than s_*;

$$s_*^{1/d_f} \sim t^{1/d_w}.$$

The probability that a walker is in an s-cluster is proportional to $sn_s \sim s^{1-\tau}$ (Section 2.4), so the average square displacement of diffusion in all clusters is given by

$$\langle r^2(t) \rangle_{p=p_c} \sim \sum_{s=1}^{\infty} s^{1-\tau} r_s(t)^2 \sim \sum_{s=1}^{s_*} s^{1-\tau} s^{2/d_f} + \sum_{s_*}^{\infty} s^{1-\tau} t^{2/d_w}$$

$$\sim s_*^{2-\tau+2/d_f} + s_*^{2-\tau} t^{2/d_w}. \tag{6.3}$$

We see that the sum splits into two parts that scale in the same way, and

$$\langle r^2(t) \rangle \sim t^{2/d_w'} \sim t^{2/d_w + (2-\tau)d_f/d_w},$$

or

$$\frac{d_w}{d_w'} = 1 + \frac{1}{2} d_f(2 - \tau) = 1 - \frac{\beta}{2\nu}. \tag{6.4}$$

In the last equality we have made use of the scaling relation $\tau = (2d\nu - \beta)/(d\nu - \beta)$. The fact that $d_w' > d_w$ is a consequence of the more compact walks that occur within small clusters of the unrestricted ensemble.

6.3 Scaling analysis

The results of the previous section apply for percolation at the critical threshold, $p = p_c$. We now develop an alternative scaling approach for the diffusion constant, $D(t, p) = \langle r^2(t) \rangle / t$ (and $D' = \langle r^2 \rangle'/t$), which is also valid off-criticality (Havlin *et al.*, 1983b).

Consider first the unrestricted ensemble of all clusters. In the conducting phase of $p > p_c$ diffusion is dominated by the infinite cluster. The infinite cluster is homogeneous on scales larger than $\xi(p)$, and hence in the long-time asymptotic limit one obtains regular diffusion, with a time-independent diffusion constant D'. On the other hand, D' is related through the Einstein relation to the dc conductivity σ (Eq. (3.21)). At large length scales the density of charge carriers is constant (since the infinite cluster is homogeneous). Hence,

$$D'(t \to \infty, p) \sim \sigma(p) \sim (p - p_c)^{\mu}, \qquad (p > p_c), \tag{6.5}$$

where we have used $\xi \sim |p - p_c|^{-\nu}$ and $\mu = \tilde{\mu}\nu$.

At criticality the diffusion coefficient is not a function of p (since $p = p_c$ is fixed) and its dependence on time follows from $\langle r^2(t) \rangle \sim t^{2/d_w'}$, as discussed in the previous section. Thus,

$$D'(t, p_c) \sim t^{(2-d_w')/d_w'}. \tag{6.6}$$

In the nonconducting phase, $p < p_c$, the dominant contribution comes from the largest clusters, of size $\xi \sim |p - p_c|^{-\nu}$. At long times $r_s^2(t) \sim \xi^2$, but the probability of being in any of the large clusters is proportional to $|p - p_c|^\beta$. Hence,

$$D'(t \to \infty, p) \sim t^{-1}(p_c - p)^{-2\nu+\beta}, \qquad (p < p_c). \tag{6.7}$$

Equations (6.5)–(6.7) may be combined in the scaling form:

$$D'(t, p) = t^{2/d_w'-1} f\left(\epsilon t^{(d_w'-2)/\mu d_w'}\right), \tag{6.8}$$

where $\epsilon = (p - p_c)/p_c$ measures the distance from criticality. The various limits discussed above determine the asymptotic behavior of the scaling function $f(x)$:

$$f(x) \sim \begin{cases} x^\mu & \text{as } x \to \infty, \\ \text{constant} & \text{as } x \to 0, \\ (-x)^{-2\nu+\beta} & \text{as } x \to -\infty. \end{cases} \tag{6.9}$$

Consistency with the time behavior in Eq. (6.7) requires

$$d_w' = \frac{2 + \mu/\nu + \beta/\nu}{1 - \beta/2\nu}. \tag{6.10}$$

Recalling that $d_w = 2 - d + d_f + \tilde{\mu}$ (Eq. (5.6)) and that, for percolation, $d_f = d - \beta/\nu$ (Eq. (2.8)), we recover d_w/d_w' of Eq. (6.4).

Suppose that the percolation system is above criticality. At early times, as long as $\langle r^2 \rangle \ll \xi^2$, diffusion would behave as if the system were actually sub-critical. The time for crossover between the two regimes – from that of Eq. (6.7) to that of Eq. (6.5) – is reflected in the crossover behavior of the scaling function $f(x)$, which happens at some (constant) $x = x_{\text{cross}}$. We thus find

$$t_{\text{cross}}' \sim |p - p_c|^{-\mu d_w'/(d_w'-2)} = |p - p_c|^{\beta-2\nu-\mu}. \tag{6.11}$$

Very similar arguments can be made for the restricted ensemble, consisting of only the incipient infinite cluster. Remembering that the probability of being on the infinite cluster is proportional to $|p - p_c|^\beta$, one infers that

(a) for $p > p_c$ and $r(t) \gg \xi$, $D \sim (p - p_c)^{\mu-\beta}$;
(b) for $p = p_c$, $D \sim t^{2/d_w-1}$; and
(c) for $p < p_c$ and $t \to \infty$, $D \sim t^{-1}(p_c - p)^{-2\nu}$.

The scaling function of D that is consistent with these properties is

$$D(t, p) = t^{2/d_w-1} g\left(\epsilon t^{(d_w-2)/(\mu-\beta)d_w}\right), \tag{6.12}$$

where

$$g(x) \sim \begin{cases} x^{\mu-\beta} & \text{as } x \to \infty, \\ \text{constant} & \text{as } x \to 0, \\ (-x)^{-2\nu} & \text{as } x \to -\infty. \end{cases} \tag{6.13}$$

Table 6.1. *Dynamical exponents for percolation.*

d	d_w	d_s	$\tilde{\mu}$	d_w^{BB}
2	2.878 ± 0.001^a	1.318 ± 0.001^a	0.9826 ± 0.0008^a	2.62 ± 0.03^b
3	3.88 ± 0.03^c	1.32 ± 0.01^c	2.26 ± 0.04^d	3.09 ± 0.03^b
4	4.68 ± 0.08^e	1.30 ± 0.04^f	3.63 ± 0.03^g	
5	5.50 ± 0.06^e	1.34 ± 0.02^f	4.81 ± 0.04^g	
6	6	$\frac{4}{3}$	6	4

[a]Grassberger (1999a), based on d_w and Eq. (5.13). Series expansions for $d = 2$ yield $\xi = \tilde{\xi}/\nu = 1.32 \pm 0.02$ (Essam *et al.*, 1996). Results based on exact enumeration yield $d_w = 2.879 \pm 0.008$ (Elran *et al.*, 1999) (Fig. 6.9); [b]Porto *et al.* (1997b); [c]Elran *et al.* (1999); [d]Normand and Herrmann (1995); [e]based on the values of d_f and $\tilde{\mu}$, and on Eq. (5.6); [f]based on d_w and Eq. (5.13); and [g]Adler *et al.* (1990).

Again, consistency yields the relation

$$d_w = 2 + \frac{\mu - \beta}{\nu},\tag{6.14}$$

the same as our previous general result of Eq. (5.6) (with $d_f = d - \beta/\nu$). Analysis of the crossover time in the unrestricted ensemble leads to the surprising conclusion that $t_{cross} \sim t'_{cross}$.

The scaling approach of this section not only extends our understanding of diffusion in percolation to the off-critical regime, but also provides us with a powerful technique for measuring critical exponents. The idea is to collect numerical data (either from simulations or from exact enumeration) for different values of p, and for various time scales. The data are then plotted in the scaling form of Eq. (6.8). If one uses the right values of the critical exponents then the data sets collapse into the universal scaling function $f(x)$ (Fig. 6.2), but otherwise the data scatter over nonoverlapping curves (cf. Figs. 3.2 and 3.4). This technique of *data collapse* is a lot more sensitive than is direct measurement of critical exponents (for example, from the relation $\langle r^2(t) \rangle \sim t^{2/d_w}$). This is especially true for complex systems, in which the asymptotic behavior under consideration is likely to be obscured by crossover phenomena. Numerical estimates of the transport exponents in percolation are listed in Table 6.1.

6.4 The Alexander–Orbach conjecture

Alexander and Orbach (1982) (AO) have conjectured that the fracton dimension of the incipient infinite percolation cluster is

$$d_s = \tfrac{4}{3}$$

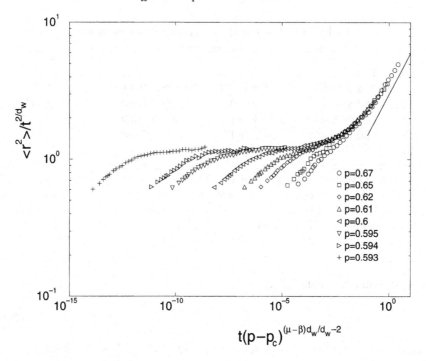

Fig. 6.2. Scaling of random walks in the incipient infinite percolation cluster. The data collapse illustrates the scaling of Eq. (6.12). At short times there exist strong corrections to scaling and scaling fails. After Elran *et al.* (1999).

in all dimensions $d > 1$. Indeed, the fracton dimension is exactly $\frac{4}{3}$ in $d \geq 6$, and is surprisingly close to $\frac{4}{3}$ in $2 \leq d < 6$ (Table 6.1). There have been numerous attempts to prove or disprove this remarkable result.

An interesting attempt at justifying the AO conjecture has been made by Rammal and Toulouse (1983). Noting that the fracton dimension characterizes the number of distinct sites visited by a walker,

$$S(t) \sim t^{d_s/2}$$

(see Sections 5.4 and 5.5), they suggest the evolution equation

$$\frac{dS(t)}{dt} \sim \frac{G(t)}{S(t)}, \qquad (6.15)$$

where $G(t)$ is the number of *growth* sites: the perimeter sites of $S(t)$, with accessible neighbors just outside of $S(t)$. The Rammal–Toulouse equation, Eq. (6.15), simply states that the rate of growth of the number of distinct sites visited by a walker is proportional to the fraction of sites from which $S(t)$ is still able to grow (Fig. 6.3).

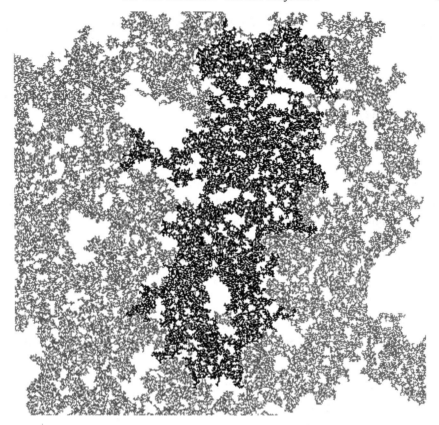

Fig. 6.3. Sites visited by diffusion. The sites visited by a random walker (black) on a percolation cluster at criticality (gray) are shown. The lattice size is 430×430, and the walk is 2×10^6 steps long. Notice that the cluster of sites visited by the walker is compact. Further growth is possible only from the perimeter sites of this cluster.

The AO conjecture is true provided that

(i) The Rammal–Toulouse equation is valid; and
(ii) $G(t)/S(t) \sim S(t)^{-1/2}$.

There is little doubt about (i), which is indeed supported by results of extensive numerical simulations. Assumption (ii) is equivalent to $G(t) \sim S(t)^{1/2} \sim L(t)^{d_f/2}$ ($L(t)$ is the average span of the walk) and is much harder to justify. Suppose for example that the diffusion front is spherical in shape; then one expects $G(t) \sim L(t)^{d_f-1}$, in violation of (ii). On the other hand, there is no compelling reason to believe that the diffusion front is spherical.

There is now a large body of evidence against the AO conjecture. Enormous computer simulations seem to rule out the conjecture for $d = 2$. It also violates an ϵ-expansion about $d = 6$. (This, however, is not definite proof: the conjecture could be wrong for small ϵ but nevertheless right for *integer* ϵ.) Even if the conjecture is wrong, it remains an amazingly good approximation. Its importance lies in the fact that it provides a simple relation between the *dynamical* critical exponent d_w and the *static* exponents $d_f = d - \beta/\nu$ of percolation. Note that the approach of Sections 6.2 and 6.3 fails to make a connection between the static and dynamic exponents.

6.5 Fractons

Consider now the vibrational modes of a percolation cluster. The density of states of fractons found for fractals, $g(\omega) \sim \omega^{d_s-1}$ ($d_s = 2d_f/d_w$), is also valid for the incipient infinite percolation cluster at criticality. Above criticality, the crossover behavior in the displacement of a random walker, $\langle r^2(t) \rangle \sim t^{2/d_w}$ ($t < t_{\text{cross}} \sim \xi^{d_w}$) and $\langle r^2(t) \rangle \sim t$ ($t > t_{\text{cross}}$), suggests the characteristic length scale

$$\Lambda(\omega) \sim \begin{cases} \omega^{-2/d_w} & \omega \gg \omega_\xi, \\ \omega^{-1} & \omega \ll \omega_\xi, \end{cases} \tag{6.16}$$

which may be identified with the wavelength of the vibrational modes. For vibrations, Eq. (5.10) implies the scaling relation $t \sim 1/\omega^2$, so the crossover frequency ω_ξ is

$$\omega_\xi \sim t_{\text{cross}}^{-1/2} \sim \xi^{-d_w/2} \sim (p - p_c)^{d_w \nu/2}. \tag{6.17}$$

This frequency separates two vibrational regimes: fractons, with a density of states $g(\omega) \sim \omega^{d_s-1}$, and phonons, with $g(\omega) \sim \omega^{d-1}$.

For percolation at $p > p_c$, it had been postulated that the vibrational modes are either *localized* fractons (their amplitude decays exponentially with distance), for $\omega > \omega_\xi$, or phonons, *extended* over the whole cluster, for $\omega < \omega_\xi$. This is referred to as the *localization–delocalization* transition. The transition is supposed to mirror the anomalous and regular regimes of diffusion in supercritical percolation. On the other hand, in $d = 2$ one might expect that the modes are always localized, in analogy to the electrons in the Anderson localization model. Kantelhardt *et al.* (1998) have shown that the latter is indeed the case, and the vibrational modes are always localized in $d = 2$, just like in the Anderson localization model. Thus, the crossover of Eq. (6.16) is about the wavelength's dependence on frequency but not about the spatial extent of the vibration modes. For $\omega < \omega_\xi$, the vibrational states can be regarded as *localized phonons*. In $d = 3$, there exists numerical evidence for a localization–delocalization transition, but the transition occurs at frequencies above the phonon–fracton crossover. Thus, there exists a regime of *extended fracton* states. A possible explanation of these facts is that the localization of electrons and vibrational excitations is caused by scattering and interference phenomena that have no effect on random walks.

6.6 The chemical distance metric

The chemical distance between two sites, ℓ, is the length of the shortest path through the percolation cluster which connects the two sites. It is a useful concept for the analysis of transport phenomena in percolation, and even for computer simulations of the clusters themselves – as in the Alexandrowicz–Leath algorithm, whereby clusters are constructed one shell at a time (Exercise 2.2).

The chemical distance ℓ between two sites and their Euclidean distance r are related through

$$r(\ell) \sim \ell^{\nu_\ell} \equiv \ell^{1/d_{\min}}, \tag{6.18}$$

where ν_ℓ is the *chemical length exponent* and $d_{\min}$ may be interpreted as the fractal dimension of the chemical path. The average number of sites in a cluster enclosed within shell ℓ is

$$S(\ell) \sim r(\ell)^{d_f} \sim \ell^{\nu_\ell d_f} \equiv \ell^{d_\ell}. \tag{6.19}$$

Here d_f is the fractal dimension of the cluster, and d_ℓ may be thought of as the fractal dimension in chemical space. The number of sites in the ℓth shell is $G(\ell) \sim dS/d\ell \sim \ell^{d_\ell - 1}$. Note that only one exponent is needed to characterize chemical space, since $d_\ell = \nu_\ell d_f$.

Consider now the scaling of $S(r)$, the number of sites in the incipient infinite cluster within a radius r. A sensible scaling form would be

$$S(r) \sim r^d P_\infty f(r/\xi) \sim r^d \xi^{-\beta/\nu} f(x); \qquad x = \frac{r}{\xi}, \tag{6.20}$$

where f has the asymptotic behavior

$$f(x) \sim \begin{cases} \text{constant} & x \gg 1, \\ x^{-\beta/\nu} & x \ll 1. \end{cases} \tag{6.21}$$

In the limit $x \gg 1$ the cluster is homogeneous, and, for $x \ll 1$ (when $p \to p_c$), there is no dependence on ξ, since it diverges. In the latter case $S(r) \sim r^{d-\beta/\nu}$, or $d_f = d - \beta/\nu$. This is in agreement with the cruder approach of Section 2.2.

In a very similar way, the number of cluster sites within a chemical distance ℓ may be written as

$$S(\ell) \sim \ell^{\nu_\ell d} \xi_\ell^{-\beta \nu_\ell/\nu} g(\ell/\xi_\ell), \tag{6.22}$$

where ξ_ℓ is the correlation length in chemical distance and g has the asymptotic behavior

$$g(x) \sim \begin{cases} \text{constant} & x \gg 1, \\ x^{-\beta \nu_\ell/\nu} & x \ll 1. \end{cases} \tag{6.23}$$

This is merely a transcription of Eqs. (6.20) and (6.21) into chemical distance variables. From the scaling form we get $d_\ell = \nu_\ell(d - \beta/\nu) = \nu_\ell d_f$, in agreement with (6.19).

One can also define dynamical exponents in chemical space. The fractal dimension of diffusion in chemical space is given by d_w^ℓ:

$$\ell \sim t^{1/d_w^\ell}. \tag{6.24}$$

Combining this with Eq. (6.18) and with $r \sim t^{1/d_w}$ we obtain

$$d_w^\ell = \nu_\ell d_w = d_w/d_{min}. \tag{6.25}$$

It then follows that *the fracton dimension in chemical space is the same as in regular space*, $d_s = 2d_f/d_w = 2d_\ell/d_w^\ell$.

A more detailed description of the relation between the regular Euclidean metric and the metric in chemical space is obtained through the conditional probability $P(r|\ell)$ – the probability that a chemical path of length ℓ possesses an end-to-end (Euclidean) distance r. The distribution $P(r|\ell)$ is useful, among other things, for the derivation of the probability density of diffusion in percolation clusters. In analogy with the theory of SAWs, it was suggested that

$$P(r|\ell) = \frac{A}{\ell^{\nu_\ell}}\left(\frac{r}{\ell^{\nu_\ell}}\right)^g \exp\left[-a\left(\frac{r}{\ell^{\nu_\ell}}\right)^u\right]. \tag{6.26}$$

Extensive numerical simulations of percolation in two dimensions support this notion, and also suggest that $u = 1/(1 - \nu_\ell)$ for $r \gg \ell^{\nu_\ell}$ (Fig. 6.4). $P(1|\ell)$ is the probability of having a chemical path of length ℓ between two neighboring sites. A Flory-type argument for g, similar to that of de Gennes (1979) for SAWs, yields

$$g = d_f + d_{min} - 1, \tag{6.27}$$

in agreement with the known exact value $g = 5$ in $d = 6$. Numerical simulations (Porto *et al.*, 1998; Dokholyan *et al.*, 1999) yield $g = 2.04 \pm 0.05$ for $d = 2$, and $g = 2.88 \pm 0.05$ for $d = 3$, in good agreement with the relation of Eq. (6.27). Recently, Ziff (1999), using a scaling relation of Dokholyan *et al.* (1999) and Grassberger (1999b), as well as the crossing-probability results of Cardy (1998), has found that, in two dimensions, $g = 49/24$ exactly.

The inverse relationship between chemical and Euclidean distances is $P(\ell|r)$ – the probability that two sites separated by a distance r are a chemical distance ℓ apart. It is related to $P(r|\ell)$ through

$$P(r|\ell)P(\ell) = P(\ell|r)P(r) = P(r, \ell), \tag{6.28}$$

where $P(r, \ell)$ is the joint probability that two sites are at chemical distance ℓ and

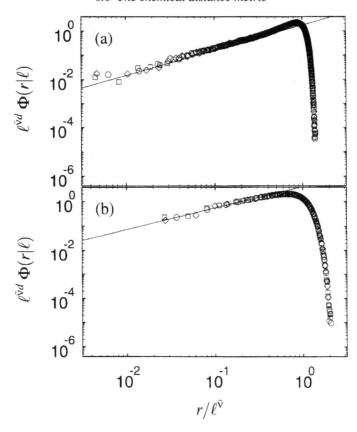

Fig. 6.4. Scaling of $P(r|\ell)$. Shown is $\ell^{\tilde{v}d}\Phi(r|\ell)$ versus $r/\ell^{\tilde{v}}$ for the incipient infinite cluster in the following cases: (a) $d = 2$, $\ell = 1000$ (circles), $\ell = 1400$ (diamonds), and $\ell = 1800$ (squares); and (b) $d = 3$, $\ell = 400$ (circles), $\ell = 600$ (diamonds), and $\ell = 800$ (squares). The plots are based on averages over more than 10^5 cluster configurations, for clusters grown up to a maximum chemical distance $\ell_{\max} = 2000$ on a square lattice ($d = 2$) and $\ell_{\max} = 1000$ on a simple-cubic lattice ($d = 3$). The straight lines represent fits for $\ell^{\tilde{v}d}\Phi(r|\ell) = f(x)$, ($x \equiv r/\ell^{\tilde{v}} \ll 1$), and have the slopes $g_1 = g - 1 = 1.04$ in (a), and $g_1 = g - 2 = 0.88$ in (b). In this figure $\ell^{\tilde{v}}\Phi(r|\ell) \equiv \ell^{v_\ell}P(r|\ell)$ of Eq. (6.26). After Porto *et al.* (1997b).

geometrical distance r apart. Here

$$P(r) = \int P(r, \ell)\, d\ell, \qquad P(\ell) = \int P(r, \ell)\, dr.$$

Since $P(\ell) \sim \ell^{d_\ell - 1}$ and $P(r) \sim r^{d_f - 1}$, we obtain

$$P(\ell|r) = \frac{B}{r^{d_{\min}}}\left(\frac{\ell}{r^{d_{\min}}}\right)^{-g_\ell} \exp\left[-b\left(\frac{\ell}{r^{d_{\min}}}\right)^{-u_\ell}\right], \qquad (6.29)$$

where $u_\ell = uv_\ell$ and $g_\ell = 1 + v_\ell(1 - d_f + g)$. If the Flory expression of Eq. (6.27) is correct, then $g_\ell = 2$ in all dimensions. On the other hand, unless $g_\ell > 2$ the average chemical distance between two sites, $\langle \ell \rangle$, diverges! Numerical simulations in $d = 2$ yield $g_\ell = 2.04 \pm 0.05$.

The concept of chemical paths is especially handy for obtaining bounds for the dynamical exponents in percolation. The problem of diffusion in loopless aggregates can be treated exactly, to a large extent (Chapter 7). In that case

$$d_w = d_f\left(1 + \frac{1}{d_\ell}\right), \qquad d_s = \frac{2d_\ell}{d_\ell + 1},$$

and the dynamics is completely controlled by d_ℓ and d_f. This is particularly true for percolation in $d \geq 6$, for which loops may be neglected. We then have $d_\ell = 2$, since the chemical paths behave essentially as regular random walks, and hence $d_s = \frac{4}{3}$ (in agreement with AO).

For percolation in $d < 6$ loops cannot be neglected. However, the resistance between two points on a cluster separated by a distance L, $R(L) \sim L^{\tilde{\zeta}}$ (Eq. (5.7)), is bounded as follows. On the one hand, R cannot exceed the resistance of the *shortest path* between the two points. This resistance is proportional to the length of the path, L^{1/ν_ℓ}. On the other hand, R is larger than the resistance of the red bonds connecting the two points. For percolation, the number of red bonds within a span L scales as $L^{1/\nu}$, hence

$$\frac{1}{\nu} \leq \tilde{\zeta} \leq \frac{1}{\nu_\ell}. \tag{6.30}$$

This may be combined with $d_w = \tilde{\zeta} + d_f$ and with $d_s = 2d_f/d_w$ to yield the corresponding bounds for d_w and d_s. The resulting bounds are narrower the higher the dimension. For percolation in $d = 6$, $\nu = \nu_\ell = 2$; the bounds coincide and one recovers the exactly known exponents.

Clearly, an expression for ν_ℓ as a function of the other static percolation exponents is highly desirable. Flory-type arguments have yielded

$$\nu_\ell = \frac{4}{d + 2}, \qquad d \leq 6,$$

and also

$$\nu_\ell = \frac{2}{d + 2 - d_f}, \qquad d \leq 6.$$

On the basis of numerical data available at the time, Havlin and Nossal (1984) conjectured the relation

$$d_f = \frac{1}{\nu} + \frac{1}{\nu_\ell}.$$

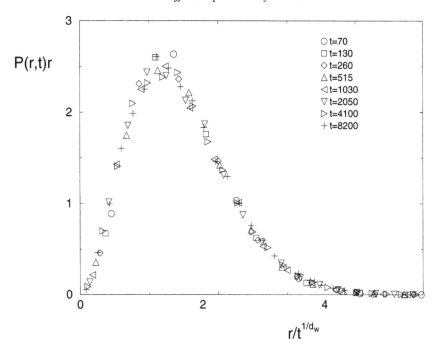

Fig. 6.5. Scaling of $P(r, t)$ for diffusion in percolation clusters. The scaling of $rP(r, t)$ with $r/t^{1/d_w}$, where $d_w = 2.87$, for diffusion in two-dimensional percolation clusters is shown for various values of t. The data collapse confirms Eq. (6.31).

These formulae are not exact and the question of whether ν_ℓ is a new, independent exponent is still open.

6.7 Diffusion probability densities

As stated earlier, probability densities carry a lot more information than do the critical exponents of anomalous diffusion. For diffusion in percolation clusters we expect the probability density to scale as in fractals (Section 5.5):

$$P_r(r, t) = \frac{1}{t^{1/d_w}} \phi\left(\frac{r}{t^{1/d_w}}\right).$$ (6.31)

Here $P_r(r, t)\, dr$ is the probability that, at time t, the walker is a distance between r and $r + dr$ from the origin. This is slightly different than $P(r, t)$ of Eq. (5.18) and requires a different normalization factor. Likewise, the probability density of diffusion in chemical space – for the walker being at chemical distance ℓ at time t – should scale as

$$P_\ell(\ell, t) = \frac{1}{t^{1/d_w^\ell}} \phi_\ell\left(\frac{\ell}{t^{1/d_w^\ell}}\right).$$ (6.32)

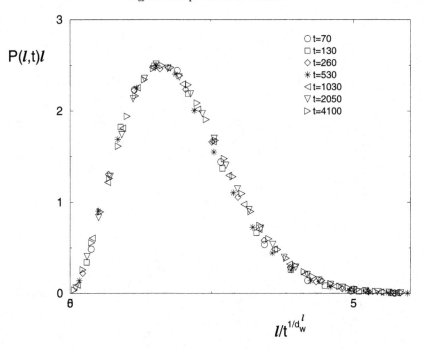

Fig. 6.6. Scaling of $P(\ell, t)$ for diffusion in percolation clusters: Shown is the scaling of $\ell P(\ell, t)$ with $\ell/t^{1/d_w^\ell}$, where $d_w^\ell = 2.54$, for diffusion in two-dimensional percolation clusters, at various values of t. The data collapse confirms Eq. (6.32).

Computer simulations are consistent with these scaling forms (Figs. 6.5 and 6.6). Moreover, simulations in $d = 2$ suggest that $\phi(x) \sim \exp(x^{-\delta})$, with $\delta = d_w/(d_w - 1)$; and $\phi_\ell(x) \sim \exp(x^{-\delta_\ell})$, with $\delta_\ell = d_w^\ell/(d_w^\ell - 1)$. These relations for δ and δ_ℓ are similar to what one finds for CTRW (Chapter 4), and also for diffusion in SAWs. Exact enumerations of walks in percolation clusters grown on the Cayley tree show that $\delta_\ell = \frac{3}{2}$ (Fig. 6.7), in accord with $\delta_\ell = d_w^\ell/(d_w^\ell - 1)$ and the exact result $d_w^\ell = 3$.

The diffusion probability densities in chemical space and in regular space are related through

$$P_r(r, t) = \int_0^\infty P(r|\ell) P_\ell(\ell, t) \, d\ell.$$

Using $P(r|\ell)$ from Eq. (6.26) we then find the condition

$$\delta = \frac{u\delta_\ell}{\delta_\ell + u\nu_\ell}. \tag{6.33}$$

Applying further $u = 1/(1 - \nu_\ell)$, this becomes

$$\nu_\ell = \frac{1 - 1/\delta}{1 - 1/\delta_\ell}. \tag{6.34}$$

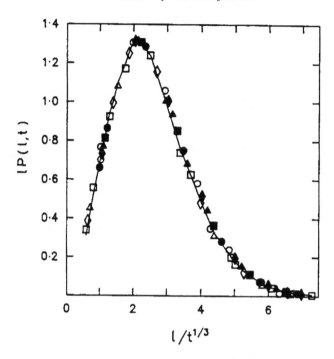

Fig. 6.7. $P(\ell, t)$ for diffusion in percolation clusters grown on the Cayley tree. The scaling of $P(\ell, t)$ with $\ell/t^{1/d_w^\ell}$ is shown for walks of time spans ranging from $t = 400$ to $t = 4000$. The data collapse suggests that in this case $d_w^\ell = 3$. After Havlin and ben-Avraham (1987).

This too seems well supported by the measured numerical values of δ, δ_ℓ, and ν_ℓ.

6.8 Conductivity and multifractals

The problem of conductivity may be studied at a more detailed level than the mere measurement of critical exponents. Suppose that each bond in a percolation cluster at criticality represents a $1 \, \Omega$ resistor, and that a voltage is applied between two sites separated by a distance L. Across each bond i there is a voltage drop V_i. The largest voltage drop V_{max} is across red bonds, since they carry all of the current. The distribution of voltage drops $\{V_i\}$ has a very long tail which cannot be fit by a Gaussian.

Let $v = V_i/V_{max}$ be the reduced voltage drop in bonds, and let $n(v)$ be the distribution of $\{v\}$. The k-moments of v are

$$v_k \equiv \langle v^k \rangle = \sum_v v^k n(v).$$

They define a hierarchy of exponents ζ_k through

$$\langle v^k \rangle \sim L^{\zeta_k}. \tag{6.35}$$

The ζ_k can be related to known percolation exponents. For example, consider the zeroth moment $v_0 = \sum_v n(v) \sim L^{\zeta_0}$. Since $v = 0$ across bonds belonging to dangling ends (there is no current flow in the dangling ends), $\sum_v n(v)$ simply counts the bonds of the backbone. Hence $v_0 \sim L^{d_f^{BB}}$, and $\zeta_0 = d_f^{BB}$ is identical with the fractal dimension of the backbone. Similarly, the second moment is proportional to the resistance of the cluster, $v_2 \sim L^{\tilde{\zeta}}$, or $\zeta_2 = \tilde{\zeta} = \tilde{\mu} + 2 - d$. When $k \to \infty$ only the red bonds contribute to v_∞, and $\zeta_\infty = 1/\nu$. The (-1)th moment yields the mean time for a tracer particle to move along a path of length ℓ (Lee *et al.*, 1999):

$$\langle v^{-1} \rangle = \frac{1}{\ell} \sum_{i=1}^{\ell} \frac{1}{v_i} \sim L^{\zeta_{-1}}.$$

The unusual distribution $n(v)$ has interesting consequences. When a probability density has a simple scaling form (as in Eq. (6.31), say) its moments are related to each other: $\zeta_k = \zeta_0 + \Delta k$. In this case, Δ is called the *gap exponent* of the distribution in question. It turns out that the distribution $n(v)$ cannot be put into scaling form and it has no gap exponent. Instead, one can write $n(v) \sim L^{f(\alpha)}$, where $\alpha \equiv \log v / \log L$ (Fig. 6.8). The function $f(\alpha)$ is called the *multifractal spectrum* and can be regarded as the "fractal dimension" of the subset of bonds having a voltage drop v. Knowledge of the multifractal spectrum is equivalent to knowing the infinite set of exponents, $\{\zeta_k\}$. Multifractal distributions have been observed in numerous phenomena, including the growth of diffusion-limited aggregates, the localization problem, percolation with nonlinear resistors, and diffusion in the presence of random fields, and they may be the rule rather than the exception.

In fact, there is numerical and analytical evidence that $P(r, t)$ and $P(\ell, t)$ too might be multifractal. It is found that the qth moment $\langle P^q(r, t) \rangle$ is of multifractal nature. Here, the average is taken over different sites at distance r from the start of the walks. The following relation was shown to hold (Bunde *et al.*, 1990):

$$\langle P^q(r, t) \rangle \sim \langle P(r, t) \rangle^{\tau(q)},$$

where $\tau(q) = q^\gamma$ and $\gamma = [1 + \delta(d_{\min} - 1)]^{-1}$. Since $\tau(q)$ is nonlinear in q, the various moments of $P(r, t)$ are described by an infinite number of exponents that represent the multifractal nature of $P(r, t)$.

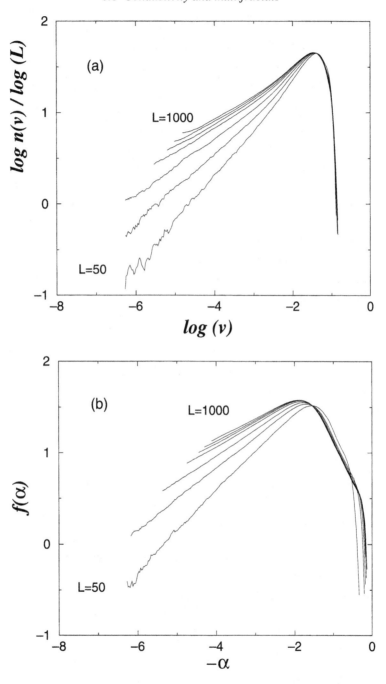

Fig. 6.8. Multifractal spectra $f(\alpha, L)$ for several values of L, namely $L =$ 50, 100, 200, 300, 400, and 500. Note the strong finite-size effect. After Barthelemy *et al.* (2000).

6.9 Numerical values of dynamical critical exponents

At present there is no complete analytical derivation of dynamical exponents for percolation, other than in $d = 6$. The resistance exponent $\zeta = \tilde{\zeta}\nu = \mu + (2 - d)\nu$ was calculated using a renormalization-group ϵ-expansion ($\epsilon = d - 6$) up to second order in ϵ:

$$\zeta = 1 + \frac{\epsilon}{42} + \frac{4\epsilon^2}{2987} + \mathcal{O}(\epsilon^3).$$

Further epsilon expansions, to order ϵ^4, including $d_{\min}$ and d_f^{BB}, have been presented by Janssen (1999). However, most of our knowledge about the values of percolation dynamical exponents comes from computer simulations.

In Fig. 6.9 we show numerical results for d_{w} obtained for percolation in $d = 2$. Notice that the exponent converges very slowly. This slow convergence, and the fact that simulation algorithms have to generate not only diffusion but also the random percolation substrate, make it difficult to obtain accurate results. The convergence is even slower in the case of d_{w}', when all clusters are considered.

An efficient method consists of simulating diffusion only along the backbone. Recall that the backbone includes only those parts of the cluster where there is a current flow when a voltage is applied. It follows that the conductivity of the backbone is the same as that of the full cluster, or $\tilde{\zeta} = d_{\mathrm{w}}^{\mathrm{BB}} - d_f^{\mathrm{BB}} = d_{\mathrm{w}} - d_f$. Thus, by measuring $d_{\mathrm{w}}^{\mathrm{BB}}$ and d_f^{BB} one can get estimates for $\tilde{\zeta}$.

6.10 Dynamical exponents in continuum percolation

In Chapter 2 we have stressed the universality of percolation theory and its insensitivity to the details of any particular model. This is not quite the case when it comes to dynamical exponents. Halperin *et al.* (1985) have studied the transport properties of a class of "Swiss-cheese" or random-void continuum models. These models consist of uniform substrates with randomly placed spherical holes. The static percolation exponents, such as ν and β, are the same as those for ordinary lattice percolation, in accordance with universality, but dynamical exponents can be quite different in the continuum models. It has been shown that the conductivity in the continuum model is the same as that in discrete percolation, provided that the conductivities of individual bonds, σ_i, are taken from the distribution

$$P(\sigma) \sim \sigma^{-\alpha}, \qquad \alpha < 1, \qquad 0 < \sigma \leq 1. \tag{6.36}$$

For the random-void model the mapping is achieved with $\alpha = (2d - 5)/(2d - 3)$. Similar remarks are true regarding diffusion, if we replace σ_i by the transition rates for movement between nearest sites.

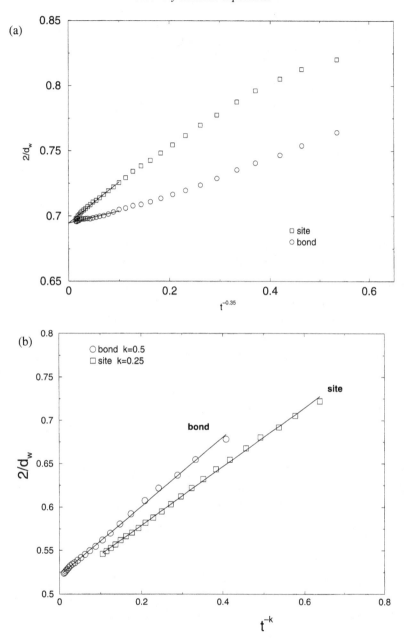

Fig. 6.9. (a) The diffusion exponent in two-dimensional percolation. The mean-square displacement of 10^6-step walks obtained from walks over $180\,000$ configurations of bond ($\square$) and site ($\circ$) percolation clusters yields slowly converging exponents. The data suggest $d_w = 2.879 \pm 0.008$. (b) The diffusion exponent in three-dimensional percolation. The mean-square displacement of 5×10^5-step walks obtained from walks over $180\,000$ configurations of bond ($\square$) and site ($\circ$) percolation clusters at criticality. The data suggest $d_w = 3.88 \pm 0.04$. After Elran *et al.* (1999).

Conductivity with the distribution (6.36) has been studied by several groups. The details are quite involved and beyond the scope of this book. The main achievements are bounds for the dynamical exponents:

$$1 + (d-2)\nu + \frac{\alpha}{1-\alpha} \le \mu(\alpha) \le \mu + \frac{\alpha}{1-\alpha}, \qquad 0 \le \alpha \le 1,$$

and

$$d_f + \frac{1}{(1-\alpha)\nu} \le d_w(\alpha) \le d_w + \frac{\alpha}{(1-\alpha)\nu}, \qquad 0 \le \alpha \le 1.$$

6.11 Exercises

1. Justify the limiting behavior leading to Eqs. (6.12) and (6.13). Obtain t_{cross} and show that it equals t'_{cross}. Show that in either case the crossover time is simply the typical time that a random walker requires to span the percolation correlation length $\xi \sim |p - p_c|^{-\nu}$.

2. Obtain a relation between the fracton dimension d'_s, for the ensemble of all percolation clusters, and d_s of the incipient infinite cluster alone. (Hint: consider the number of distinct sites visited by random walkers in both ensembles.) Write down scaling forms for the probability of being at the origin, $P_0(t, p)$ in both ensembles.

3. Given a set of linear polymers (SAWs) of length N, distributed according to $n(N) \sim N^{-\tau}$, find the relation between d'_w, the diffusion exponent of a walker randomly "dropped" on the polymers, and d_w for a single infinitely long polymer. Find also the corresponding fracton dimensions.

4. In order to test the Rammal–Toulouse equation, perform simulations of random walks in the Sierpinski gasket and check out the scaling of growth sites, $G(t)$. Compare simulations with the Rammal–Toulouse prediction $G(t) \sim L(t)^{2d_f - d_w}$. Notice the difference between $2d_f - d_w = \log(\frac{9}{5})/\log 2$ and the dimension of a spherical cut of the Sierpinski gasket, $d_f - 1 = \log(\frac{3}{2})/\log 2$.

5. Simulate percolation clusters with the Alexandrowicz–Leath algorithm (be sure to test sites for occupancy only once!) for various values of p. Check out the scaling form of Eq. (6.20): plotting $S(r)/r^d |p - p_c|^\beta$ against $r|p - p_c|^\nu$ should result in one curve, $f(x)$ (data collapse). Check also the scaling in chemical space of Eq. (6.22).

6. Use the bounds on resistance in percolation clusters to derive bounds also for d_w and d_s.

7. Prove the relation (6.33), by performing the necessary integration with a saddle-point approximation.

8. Show that, when a probability density has a scaling form, such as for example that in (6.31), the moments are characterized by a single gap exponent.

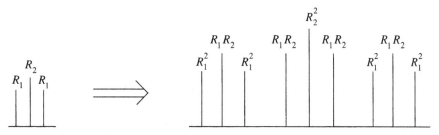

Fig. 6.10. A multifractal chain of resistors. An iteration f' consists of combining in series the scaled resistances of the previous generation, f; $f' = R_1 f + R_2 f + R_1 f$. Only one iteration is shown. The lengths of the vertical bars symbolize the magnitudes of the resistances.

9. Consider a multifractal chain of resistors: the genus consists of three consecutive resistors of resistances R_1, R_2, and R_1. In each iteration the construction is tripled; the resistance in the first section are multiplied by a factor R_1, those in the second section by R_2, and those in the third section by R_1 (Fig. 6.10). Assume that the chain consists of n such iterations and that a current I runs through the chain. Compute the exponents ζ_k and show that a gap exponent exists only asymptotically, for $k \to \infty$. (Answer: $\zeta_k = \log(2r_1^k + r_2^k)/\log 3$, where $r_{1,2} \equiv R_{1,2}/\max(R_1, R_2)$.)

6.12 Open challenges

1. The biggest open challenge is tying together the static and dynamical exponents of percolation. The Alexander–Orbach conjecture comes close to this goal, but we now know that it is not exact, other than for $d \geq 6$. Analytical results for $d < 6$ are of first priority! Improved numerical estimates of the diffusion exponents d_{w}, $d_{\mathrm{w}}^{\mathrm{BB}}$, and d_{s}, and the conductivity exponents $\tilde{\zeta}$ or $\tilde{\mu}$, as well as the devising of new numerical algorithms for their computation, are of great importance too.

2. An important question is that of whether the shape exponent δ or δ_ℓ can be expressed in terms of other critical exponents. Results of a study (Bunde *et al.*, 1990) suggest that the distribution $P(r, t)$ is multifractal, in which case the scaling form of Eq. (6.31) cannot be valid, and the very definition of the exponent δ is problematic! One possibility is that the scaling form of $P(r, t)$ and the relation $\delta = d_{\mathrm{w}}/(d_{\mathrm{w}} - 1)$ are only approximations that improve with increasing dimension – when loops may be neglected – and become exact at $d \geq 6$ (Barthelemy *et al.*, 2000).

3. The question of critical exponents of SAWs in percolation clusters is still open

and even controversial (Barat and Chakrabarti, 1995; Roman *et al.*, 1998). This problem may be relevant to polymers in porous systems.

4. The form of $P(r, t)$ for $r \ll t^{1/d_w}$ seems to be $\exp(-ar^{d_w}/t)$ (O'Shaugnessy and Procaccia, 1985a; 1985b; Havlin *et al.*, 1991). This must be analytically studied, and needs further numerical confirmation.

5. Numerical tests for g_ℓ are still needed, in particular for $d > 2$. In $d = 2$, one expects $g_\ell = 1 + 55/(48d_{min}) = 2.0134$ – based on the exact result $g = 49/24$ of Ziff (1999), and on the best available estimate for d_{min}, 1.1307 (Grassberger, 1992b).

6. Batrouni *et al.* (1988), and Aharony *et al.* (1993) have noticed that there exist strong logarithmic finite-size effects on the left-hand side of the multifractal spectrum (cf. Fig. 6.8). If the slope approaches zero when the system size $L \to \infty$, as suggested recently by Barthelemy *et al.* (2000), then the multifractal features found for small currents are merely a finite-size effect. This question is attracting much current interest and further studies are needed to clarify it.

6.13 Further reading

- Localization of optical excitations in fractal aggregates of silver colloidal nanoparticles has been studied experimentally by Safonov *et al.* (1998). Anomalous diffusion has been studied experimentally with inelastic-neutron-scattering techniques by Ikeda *et al.* (1997).

- Recent experimental applications of diffusion in percolation clusters: a summary of various experimental results is given by Balberg (1998a; 1998b); conductivity of lithium ions in perovskite-type oxides (Inaguma and Itoh, 1996); graphite–boron nitride under compression (Wu and McLachlan, 1997); conductivity of nonionic water-in-oil microemulsions (Weigert *et al.*, 1997); conduction in porous silicon (Hamilton *et al.*, 1998); insulator–superconductor transition and the quantum Hall transition (Shimshoni *et al.*, 1998). Applications in biology to diffusion in the disordered environment of the cell: Ogston *et al.* (1973), Peyrelasse and Boned (1990), Saxton (1994), Jones and Luby-Phelps (1996), and Seksek *et al.* (1997).

- Conductivity and diffusion in two ensembles and scaling: de Gennes (1976a), Mitescu *et al.* (1979), Webman (1981), ben-Avraham and Havlin (1982), Havlin *et al.* (1983b), Gefen *et al.* (1983b), Stauffer (1979), Kopelman (1976), Vicsek (1981), and Harris *et al.* (1987). The strict limit of diffusion in percolation, at which the random-walk step may be much smaller than the lattice step, was considered by Koplik *et al.* (1988).

- The Alexander–Orbach conjecture and the relation between static and dynamical exponents: Alexander (1983), Leyvraz and Stanley (1983), Harris *et al.* (1984),

Aharony and Stauffer (1984), Aharony *et al.* (1987), Havlin (1984a), Daoud (1983), Hong *et al.* (1984), Zabolitzky (1984), Herrmann *et al.* (1984a), Lobb and Frank (1984), Rammal *et al.* (1984a), Stanley and Coniglio (1984), Stanley *et al.* (1984), Essam and Bhatti (1985), de Gennes (1976b), Skal and Shklovskii (1975), and Grassberger (1999a).

- Fractons: Alexander *et al.* (1993). A review for fractons in percolation networks is presented by Nakayama *et al.* (1994). See also Kantelhardt and Bunde (1997). Vibration dynamics of cluster–cluster aggregations: Terao *et al.* (1998).
- Chemical distance and applications: Havlin and Nossal (1984), Havlin *et al.* (1985b; 1985c), Middlemiss *et al.* (1980), Alexandrowicz (1980), Pike and Stanley (1981), Hong and Stanley (1983a; 1983b), Vannimenus *et al.* (1984), Herrmann *et al.* (1984b), Ritzenberg and Cohen (1984), Stanley (1984), Edwards and Kerstein (1985), Margolina (1985), Banavar and Willemson (1984), O'Shaughnessy and Procaccia (1985a; 1985b), and Janssen (1985).
- The relation between ν_ℓ and d_f: Havlin (1984b), Family and Coniglio (1984), Roux (1985), Grassberger (1985), and Cardy and Grassberger (1985).
- Diffusion and conductivity exponents: Havlin and ben-Avraham (1983), Pandey and Stauffer (1983), Majid *et al.* (1984), Bug *et al.* (1986), Harris and Lubensky (1984), and Wang and Lubensky (1986).
- Conductivity and multifractals: Sen *et al.* (1985), Feng *et al.* (1987), Elan *et al.* (1984), Lubensky and Tremblay (1986), Balberg (1987), Kogut and Straley (1979), Straley (1982), Gawlinski and Stanley (1981), Ben-Mizrahi and Bergman (1981), Harris (1987), Benguigui (1986), Bunde *et al.* (1986b), Lobb and Forrester (1987), and Petersen *et al.* (1989). Multifractality of $P(r, t)$: Bunde and Havlin (1996) and Drager and Bunde (1996).
- Random resistor networks and multifractals: Mandelbrot (1974), de Arcangelis *et al.* (1985; 1986a; 1987), Meakin *et al.* (1985; 1986), Stanley and Meakin (1988), Halsey *et al.* (1986), Amitrano *et al.* (1986), Machta *et al.* (1986), Weiss and Havlin (1987), Rammal *et al.* (1985), Castellani and Peliti (1986), and Blumenfeld and Aharony (1985).

7

Diffusion in loopless structures

Nature abounds with types of structures for which loops may be neglected. The simplest example are perhaps linear polymers – modeled by self-avoiding walks – but also branched polymers (modeled by lattice animals), DLA aggregates, trees and tree-like structures, river systems, networks of blood vessels, and percolation clusters (in $d \geq 6$) are common examples.

Diffusion in loopless structures is a lot simpler than that in other disordered substrates, for which loops cannot be neglected, and it therefore yields itself to a more rigorous analysis. Chiefly, an *exact* relation between dynamical exponents (the walk dimension and spectral dimension) and structural exponents (the fractal dimension and chemical length exponent) may be derived.

Diffusion in combs is a reasonable model for diffusion in some random substrates: the delay of a random walker caused by dangling ends and bottlenecks may be well mimicked by the time spent in the teeth of a comb. This case can be successfully analyzed with a CTRW and other techniques.

7.1 Loopless fractals

A large class of fractals are tree-like in structure. They are characterized by the absence of loops (or loops are so scarce that they may be neglected). In Figs. 7.1 and 7.2 we show examples of deterministic loopless fractals. For the study of transport properties it is useful to define their *backbone*, or *skeleton*. It consists of the union of all shortest (chemical) paths connecting the root of the tree with the peripheral sites.

The fractal dimension d_f describes as usual the scaling of the mass (the number of nodes, or segments) with the linear size r,

$$M(r) \sim r^{d_f}. \tag{7.1}$$

Fig. 7.1. A loopless tree. This tree is similar to the Sierpinski gasket, and has the same fractal dimension, but loops are absent.

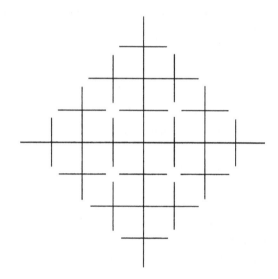

Fig. 7.2. A loopless tree. An example of a space-covering loopless tree with $d_f = 2$ (the dimension of the substrate), shown to three generations.

Similarly, d_f^{BB} is the fractal dimension of the backbone:

$$M_{BB}(r) \sim r^{d_f^{BB}}, \tag{7.2}$$

where M_{BB} is the mass (number) of backbone sites. Because of the absence of loops, tree-like fractals are more naturally analyzed in chemical space. The fractal dimensions in chemical space are given by

$$M(\ell) \sim \ell^{d_\ell}, \qquad M_{BB}(\ell) \sim \ell^{d_\ell^{BB}}. \tag{7.3}$$

The relation between regular and chemical space exponents follows from $r \sim \ell^{\nu_\ell}$

(Eq. (6.18)):

$$d_\ell = v_\ell d_f, \qquad d_\ell^{BB} = v_\ell d_f^{BB}. \tag{7.4}$$

The deterministic trees of Figs. 7.1 and 7.2 are particularly simple examples, with $d_f^{BB} = 1$ and $v_\ell = 1$.

Polynomial trees are a basic kind of loopless fractals, defined by their growth rate. The number of nodes in the ℓth chemical shell is

$$B(\ell) = B_0 \ell^{d_\ell - 1}. \tag{7.5}$$

The growth is completely random and uncorrelated, that is, the $B(\ell)$ nodes branch off from any of the nodes in the $(\ell - 1)$th shell with equal probabilities. Because of this Markovian growth it is possible to write down an exact recursion equation for $B_{BB}(\ell)$, the number of nodes in the ℓth shell of the backbone. One then finds an interesting relation between the backbone mass and the total mass of polynomial trees:

$$d_\ell^{BB} = \begin{cases} 1 & d_\ell < 2, \\ d_\ell - 1 & d_\ell \geq 2. \end{cases} \tag{7.6}$$

An important variation occurs when polynomial trees are grown on a d-dimensional lattice. Clearly, the growth process is no longer Markovian and d_ℓ is bounded by the lattice dimensionality d. In $d = 2$, a situation similar to that of Eq. (7.6) is observed, whereby the transition to nonlinear backbones happens at about $d_\ell \approx 1.65$ (Fig. 7.3).

A different class of trees is obtained from the *Eden growth model*. The Eden model was originally designed to describe the growth of tumors in healthy tissue. At each stage a healthy cell neighboring at least one tumor cell is randomly selected and it converts into a tumor cell. The resulting clusters are compact but may develop fractal outlines. If potential tumor cells are selected among healthy cells neighboring *only one* tumor cell, loops never form and one produces *Eden trees* (Fig. 7.4). The trees are still compact, with $d_f = 2$ (in $d = 2$), but are otherwise complex structures, with $d_f^{BB} \approx 1.3$ and $d_{min} \approx 1.2$ (see also Open challenge 3).

As a last example, consider percolation on Cayley trees (Section 2.4). This is again a class of loopless fractals whose structural properties can be analyzed exactly, just like for polynomial trees. One finds out that, at criticality ($p = p_c$), the trees are fractal, with $d_\ell = 2$ and $d_\ell^{BB} = 1$. When the trees are embedded in regular space, $v_\ell = \frac{1}{2}$, because the branches are essentially uncorrelated and behave as random walks. Percolation in Cayley trees is an idealized model for percolation clusters in $d \geq 6$. In this case loops are very rare and they may be neglected. A similar situation (in which loops may be neglected) occurs with lattice animals and diffusion-limited aggregates (Sections 7.3 and 7.4). Havlin *et al.* (1985a) have studied a stochastic model of tree growth on the Cayley tree, which is similar to

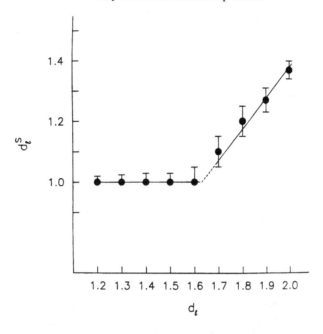

Fig. 7.3. The backbone mass of polynomial tree. The critical exponent describing the mass of the backbone (or skeleton) $d_\ell^{BB} \equiv d_\ell^S$ undergoes a sharp transition as trees become denser, around $d_\ell = 1.6$. After Havlin *et al.* (1985e).

percolation and to the case of polynomial trees, but with tunable dimensions. Other examples of loopless fractals include loopless percolation (Tzschichholtz *et al.*, 1989) and branched polymers (Lucena *et al.*, 1994; Porto *et al.*, 1996).

7.2 The relation between transport and structural exponents

In the special case of loopless random aggregates the exponents of diffusion and conductivity are easily obtained from the structural properties of the substrate (Havlin *et al.*, 1984a). Recall that the resistance between two sites separated by a distance L scales as

$$R \sim L^{\tilde{\zeta}}.$$

Since there are no loops, the shortest (chemical) path is the *only* path connecting any two sites. Hence, the resistance is proportional to the chemical distance between the two sites:

$$R \sim \ell \sim L^{d_f/d_\ell},$$

or

$$\tilde{\zeta} = d_f/d_\ell. \qquad (7.7)$$

Eden clusters   Eden trees

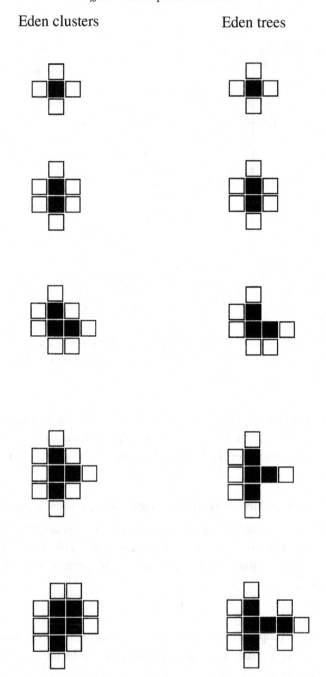

Fig. 7.4. Eden clusters and Eden trees. Both kinds of trees are generated one chemical shell at a time. At each time step nearest neighbors of the sites in the previous shell (empty squares) are occupied at random, with probability p. For Eden trees, sites that might lead to the closing of a loop are *not* considered (right).

Using the Einstein relation, $d_w = \tilde{\zeta} + d_f$, we then have

$$d_w = d_f\left(1 + \frac{1}{d_\ell}\right). \tag{7.8}$$

Thus, the transport exponents $\tilde{\zeta}$ and d_w are determined by the structural exponents d_f and d_ℓ.

The diffusion exponent in chemical space $d_w^\ell = \nu_\ell d_w = (d_\ell/d_f)d_w$ is

$$d_w^\ell = d_\ell + 1, \tag{7.9}$$

and the spectral dimension $d_s = 2d_f/d_w = 2d_\ell/d_w^\ell$ is

$$d_s = \frac{2d_\ell}{d_\ell + 1}. \tag{7.10}$$

Note that both d_s and d_w^ℓ depend only on d_ℓ, but not on d_f. In spite of the simplicity of Eq. (7.9), diffusion in trees in chemical space is generically anomalous, with $d_w^\ell \geq 2$ (since $d_\ell \geq 1$). This reflects the delay of walkers in side branches.

The number of growing sites in the diffusion front $G(\ell)$ in a loopless fractal can be calculated by means of the Rammal–Toulouse equation, $dS/dt \sim G/S$. By substituting for the number of distinct sites $S \sim t^{d_s/2} \sim t^{d_\ell/d_w^\ell} \sim \ell^{d_\ell}$, we find that the front of the walk scales as

$$G(\ell) \sim \ell^{d_\ell - 1}. \tag{7.11}$$

That is, the diffusion front is a spherical cut of the tree in chemical space, but not necessarily in regular space. This makes the chemical distance a natural length parameter for the study of diffusion in loopless aggregates.

If d_ℓ and d_f are known, one can use the results of this section to compute the dynamical exponents. Conversely, when the static exponents are not known, the study of diffusion yields insights into the structural properties of the aggregates. Such is the situation with lattice animals and diffusion-limited aggregation.

7.3 Diffusion in lattice animals

Lattice animals are unusually large percolation clusters ($L \gg \xi(p)$) obtained below the critical threshold, $p < p_c$. Such clusters are self-similar at all length scales up to L, but have a fractal dimension different than $d_f = d - \beta/\nu$ of regular percolation clusters. Loops are extremely rare, a fact that makes lattice animals a commonly accepted model for branched polymers.

The study of lattice animals is complicated by the fact that they are very rare. The probability of producing a cluster of span L decreases exponentially fast, as e^{-L}. Nevertheless, their numerical study is possible thanks to "enrichment" algorithms.

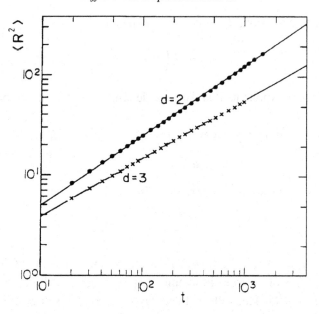

Fig. 7.5. Random walks on lattice animals. The mean-square displacement measured from computer simulations yields $d_w = 2.78 \pm 0.08$ ($d = 2$) and $d_w = 3.37 \pm 0.10$ ($d = 3$). After Havlin *et al.* (1984a).

Some exact results are also known. Lattice animals generated on Cayley trees (or in $d \geq 8$) have $d_f = 4$ and $d_\ell = 2$; $d_f = 12/5$ in $d = 4$; and $d_f = 2$ for lattice animals in $d = 3$. Other cases have been studied through numerical simulations including diffusion.

In Table 7.1 we summarize the various results known for the critical exponents of lattice animals and other loopless fractals. A Flory-type argument (Havlin, 1984b) which yields

$$\nu_\ell = \frac{5}{d+2},$$

predicts the correct upper critical dimension of $d = 8$ but otherwise does rather poorly for $d < 8$. Much of the data presented in the table has been derived from studies of diffusion. In Fig. 7.5 we present typical data, obtained from exact enumeration of walks in lattice animals generated by the enrichment method.

7.4 Diffusion in DLAs

As a second example of almost loopless fractals consider *diffusion-limited aggre-gates* (DLAs). The diffusion-limited-aggregation (DLA) model was introduced by Witten and Sander (1981). In its simplest version, a seed particle is fixed at the

Table 7.1. *Static and dynamic exponents of known loopless fractals.*

	d_f		d_{min}	
	$d = 2$	$d = 3$	$d = 2$	$d = 3$
Lattice animals	1.53 ± 0.05^a	2^b	1.16 ± 0.05^c	1.36 ± 0.04^c
DLAs	1.71 ± 0.07^d	2.50 ± 0.08^d	1.00 ± 0.02^e	1.02 ± 0.03^e
Eden trees	2	3	1.22 ± 0.02^f	1.32 ± 0.02^f

	d_s		d_w	
	$d = 2$	$d = 3$	$d = 2$	$d = 3$
Lattice animals	1.12 ± 0.06^g	1.18 ± 0.04^g	2.78 ± 0.08^c	3.37 ± 0.10^c
DLAs	1.20 ± 0.10^c	1.30 ± 0.10^c	2.56 ± 0.10^c	3.33 ± 0.25^c
Eden trees	1.42 ± 0.04^h	1.56 ± 0.06^h	2.82 ± 0.06^h	3.85 ± 0.15^h

[a] Ball and Lee (1996); [b]exact (Parisi and Sourlas, 1981), numerical simulations yield $d_f = 1.96 \pm 0.06$, Ball and Lee (1996); [c]Havlin *et al.* (1984c); [d]Vicsek (1991); [e]Meakin *et al.* (1984); and [f]Nakanishi and Herrmann (1993). For Eden trees in $d = 2$ Manna and Dhar (1996) claim the exact result $d_s = \frac{5}{4}$ and $d_f^{BB} = \frac{4}{3}$ (see Open challenge 3); [g]Havlin and ben-Avraham (1987); and [h]Reis (1996b).

origin and particles released from a large enclosing circle adhere to it upon contact. The particles move in random-walk fashion until they reach a site adjacent to the growing cluster, and then stick to it irreversibly. The aggregates formed in this way are tree-like fractals (an example is shown in Fig. 7.6), in which large-sized loops are extremely rare. DLA serves as a model of numerous natural phenomena, including sputter-deposited thin films of $NbGe_2$ (Orbach, 1986), copper aggregates in cathodes of electrolytic systems (Brady and Ball, 1984), electrostatic discharge (Niemeyer and Pinnekamp, 1982; Niemeyer *et al.*, 1984), and viscous fingering in immiscible fluids (Patterson, 1984; Jensen *et al.*, 1994).

DLA has been studied extensively, but in spite of some remarkable theoretical achievements the exact values of the critical exponents remain undetermined. The best available numerical values are summarized in Table 7.1. An intriguing possibility is that $\nu_\ell = 1$ in all dimensions, as suggested by the data in Table 7.1. This is in contrast to the situation with lattice animals, cluster–cluster aggregates, and percolation clusters, for which $\nu_\ell \to \frac{1}{2}$ as d approaches the upper critical dimension of the model. It supports the idea that DLA lacks an upper critical dimension. It also means that dynamic properties of DLAs are determined by d_f alone: Eqs. (7.8) and (7.10) would become

$$d_w = d_f + 1, \qquad d_s = 2d_f/(d_f + 1).$$

Fig. 7.6. An example of a DLA cluster consisting of 50 000 sites. Courtesy of Dr Stefan Schwarzer.

7.5 Diffusion in combs with infinitely long teeth

Diffusion in comb-like structures (Figs. 7.7–7.9) serves as a model for diffusion in more complicated fractal substrates, such as percolation clusters. The backbone of the comb (i.e., the x-axis in Figs. 7.7–7.9) is analogous to the quasilinear structure of the backbone of the clusters. A carefully chosen distribution of the teeth lengths may successfully reproduce the delaying effects of dangling ends and of the backbone irregularities in percolation clusters.

Let us first analyze a comb with teeth of equal lengths (Fig. 7.7). A random walker will behave differently along the horizontal and vertical directions. The motion along the x-axis may be regarded as a CTRW: the walker spends a certain time in a tooth before advancing to an adjacent tooth. Since the teeth are all equal the distribution of the delay times (waiting times) at each junction is also equal.

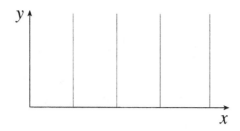

Fig. 7.7. A comb with teeth of finite, equal length.

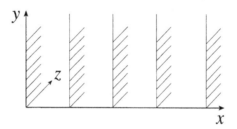

Fig. 7.8. A hierarchical comb in three dimensions. From the backbone (the x-axis), there emanate two-dimensional combs that lie in the y–z plane.

If the length of the teeth is finite, $L < \infty$, there is a finite characteristic waiting time $\tau \sim L$. Diffusion along the x-axis is then regular. The delay in the teeth merely dilates time by a factor of L; $\langle x^2 \rangle \sim t/L$. A more interesting situation occurs when $L \to \infty$, when the characteristic waiting time also diverges. Then the waiting-time distribution is proportional to the probability that a random walker (in $d = 1$) returns to the origin for the first time at time t:

$$\psi(t) \sim t^{-3/2}. \tag{7.12}$$

This is the same as Eq. (4.3), with $\gamma = \frac{1}{2}$, and results in anomalous diffusion along the backbone (Eq. (4.4)):

$$\langle x^2 \rangle \sim t^{2/d_{\mathrm{w}}}, \qquad d_{\mathrm{w}} = 4. \tag{7.13}$$

The probability of being at site x at time t, for $x \gg t^{1/4}$, is

$$P_x(x, t) \sim \frac{1}{t^{1/4}} \exp\left[-a\left(\frac{x}{t^{1/4}}\right)^{4/3}\right]. \tag{7.14}$$

Diffusion along the y-axis is normal, $\langle y^2 \rangle \sim t$. (The time delay in the x-axis when walkers reach a junction is typically of order unity.) The probability density of being at site y at time t is simply Gaussian,

$$P_y(y, t) \sim \frac{1}{t^{1/2}} \exp\left(-\frac{by^2}{t}\right). \tag{7.15}$$

Ball *et al.* (1987) find the interesting result that, in the long-time asymptotic limit, the *joint* probability of being at (x, y) at time t is

$$P(x, y, t) \sim \frac{1}{t^{3/4}} \exp\left[-a\left(\frac{x}{t^{1/4}}\right)^{4/3} - \frac{by^2}{t}\right], \tag{7.16}$$

as if the horizontal and vertical motions were uncorrelated. Similarly, the number of distinct sites visited by a walker is

$$\langle S(t) \rangle \sim \langle S^x(t) \rangle \langle S^y(t) \rangle \sim t^{d_s/2} \sim t^{d_s^x/2} t^{d_s^y/2}, \tag{7.17}$$

where the superscripts x and y refer to the backbone and the teeth of the comb, respectively. Thus,

$$\frac{d_s}{2} = \frac{d_s^x}{2} + \frac{d_s^y}{2} = \frac{d_f^x}{d_w^x} + \frac{d_f^y}{d_w^y} = \frac{1}{4} + \frac{1}{2} = \frac{3}{4}. \tag{7.18}$$

For the hierarchical, three-dimensional comb of Fig. 7.8 the results are

$$\langle x^2 \rangle \sim t^{1/4}, \qquad \langle y^2 \rangle \sim t^{1/2}, \qquad \langle z^2 \rangle \sim t, \tag{7.19}$$

and

$$P(x, y, z, t) \sim \frac{1}{t^{7/8}} \exp\left[-a\left(\frac{x}{t^{1/8}}\right)^{8/7} - b\left(\frac{y}{t^{1/4}}\right)^{4/3} - \frac{cz^2}{t}\right]. \tag{7.20}$$

Several authors have suggested a different form of the probability distribution function of Eqs. (7.16) and (7.20) (Open challenge 5).

7.6 Diffusion in combs with varying teeth lengths

Consider the more general case of combs with varying teeth lengths (Fig. 7.9). Suppose that the length of each tooth is independent of the others, and that it is characterized by the distribution $\varphi(L)$. In particular, we are interested in the distribution:

$$\varphi(L) = \gamma L^{-(1+\gamma)}, \qquad \gamma > 0, \quad L \geq 1. \tag{7.21}$$

When $\gamma < 1$ one obtains anomalous diffusion along the backbone.

We will first tackle this problem in an approximate fashion, neglecting the correlation between a tooth's length and its position along the backbone. That is to say, we assume that, when a random walker returns to a previously visited tooth, it may encounter a tooth of *different* length (but the "new" length is still drawn from

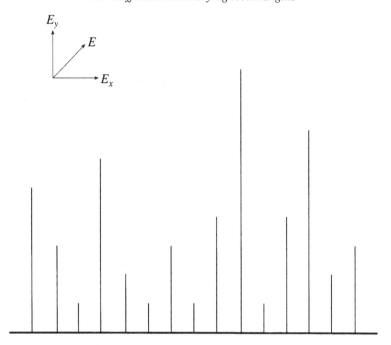

Fig. 7.9. A comb with varying teeth lengths. E, E_x, and E_y indicate possible directions of a bias field, considered in Chapter 9.

the probability $\varphi(L)$). The conclusions from this "mean-field" approach agree with a more rigorous derivation and with results from simulations.

With the above approximation, motion along the backbone can be viewed as a CTRW with a waiting-time probability

$$\psi(t) = \langle \psi_L(t) \rangle_L = \int_1^\infty \psi_L(t)\varphi(L)\,dL. \tag{7.22}$$

$\psi_L(t)$ is the distribution of waiting times for a tooth of length L, and, because of the mean-field *Ansatz*, this must be averaged over the distribution of lengths. Although $\psi_L(t)$ may be obtained exactly, the approximation

$$\psi_L(t) \sim \begin{cases} t^{-3/2} & (t < L^2), \\ 0 & (t > L^2), \end{cases} \tag{7.23}$$

suffices for our purpose. The idea is that, since diffusion along the teeth is regular, for times $t \ll L^2$ the tooth's edge is not sensed by the walker and ψ_L agrees with that of infinitely long teeth (Eq. (7.12)). On the other hand, once one has reached the edge of a tooth, the characteristic waiting time ceases to grow, hence the probability for longer waiting times vanishes. By combining Eqs. (7.22) and (7.23) we get $\psi(t) \sim t^{-(3/2+\gamma/2)}$, and CTRW theory then predicts anomalous diffusion with

$$d_w = \begin{cases} 4/(1+\gamma) & (0 < \gamma < 1), \\ 2 & (\gamma \geq 1). \end{cases} \tag{7.24}$$

We now present a more rigorous approach, that does not require the mean-field assumption. The delays in a *specific* tooth are characterized by an average waiting time $\tau(L) \sim L$. Given the distribution of L, Eq. (7.21), the distribution of effective *transition rates*, $w(L) \sim 1/\tau(L)$, is $\phi(w) = \varphi(L)/(dw/dL) \sim w^{\gamma-1}$. Diffusion along the backbone satisfies

$$\frac{t}{\langle x^2 \rangle} \sim \int_{w_{min}}^{1} \frac{\phi(w)}{w} \, dw \sim w_{min}{}^{\gamma-1}, \qquad \gamma < 1, \tag{7.25}$$

which follows from expressing the inverse of the effective diffusion coefficient D^{-1} in two different ways (see also Chapter 8). w_{min} is the smallest transition rate encountered within the span x of the walk. It is obtained from the smaller of two extremal lengths along the y-axis:

$$L_{max} = \min(x^{1/\gamma}, x^{d_w/2}). \tag{7.26}$$

The first is the length of the largest tooth likely to be found within the interval x. The second length is the maximum span of the walk along the y-direction in a comb with infinitely long teeth ($y_{max}^2 \sim t \sim x^{d_w}$). Clearly, if the teeth are long enough it is this second length which limits the vertical span of the walks. It can be seen *a posteriori* that this is indeed the case. On rewriting Eq. (7.25) in powers of x we obtain $t/\langle x^2 \rangle \sim x^{d_w-2} \sim w_{min}{}^{\gamma-1} \sim 1/L_{max}^{\gamma-1} \sim x^{-d_w(\gamma-1)/2}$, or

$$d_w = 2 - \tfrac{1}{2}d_w(\gamma - 1),$$

from which one recovers Eq. (7.24).

It is now clear why the mean-field *Ansatz* works. Our second analysis suggests that a walker rarely reaches the teeth edges, hence there is little harm in ascribing a new length to a tooth with each new visit. Both of our approaches involve approximations, but the predictions agree very well with exact enumeration of walks in simulated combs (Fig. 7.10).

7.7 Exercises

1. Derive recursion relations for the number of nodes in the ℓth shell of the backbone of polynomial trees (Havlin *et al.*, 1985e).
2. A loopless tree is grown with the rule that the probability of gaining or losing one unit mass (a node) per unit time is $p_\pm(M) = (A/2)M^{-\gamma}$. If the tree loses all its mass no further growth is possible and it becomes extinct. Derive a differential equation describing this process and find the distribution of M as

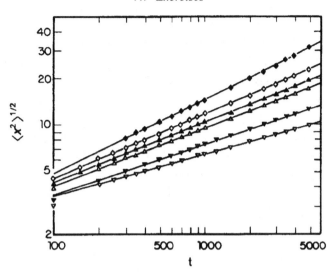

Fig. 7.10. Simulation of diffusion in combs with power-law distributions of teeth lengths. Exact-enumeration results for the displacement along the backbone of combs with $\gamma = 1, \frac{2}{3}, \frac{1}{2}, \frac{2}{5}, \frac{1}{5}$, and $\frac{1}{10}$ are shown (top to bottom). The solid lines represent the slopes predicted by Eq. (7.24). After Havlin *et al.* (1987b).

a function of time (assuming that growth starts at $t = 0$ from a single node). Compare this with polynomial trees.

3. Show that, in chemical space, the Rammal–Toulouse equation becomes $G(\ell) \sim \ell^{2d_\ell - d_w^\ell}$. Thus, the front of diffusion in chemical space is a spherical cut only when $d_w^\ell = d_\ell + 1$, as in loopless fractals. (See Aharony and Stauffer (1984), and Havlin (1984a).)

4. *The delay time in a tooth.* Consider a one-dimensional chain with sites $n = 0, 1, \ldots, L$. A random walk started off at $n = n_0$ is terminated when it reaches $n = 0$. The edge at $n = L$ is reflecting. Write down recursion relations for the time for the first passage to $n = 0$ and obtain the solution for $n_0 = 1$. (Hint: a more general case, with bias, is considered in Section 9.1.)

5. Prove Eq. (7.19). Consider a generalization of the hierarchical comb of Fig. 7.8 to d dimensions, and show that $d_w^x = 2^d$.

6. Consider a "brush" whose backbone is the two-dimensional x–y plane and whose teeth are parallel to the z-axis and distributed according to $\varphi(L) \sim L^{-(1+\gamma)}$. Compute the mean-square displacement of RWs along the backbone, $\langle r^2 \rangle = \langle x^2 \rangle + \langle y^2 \rangle$, using a self-consistent scaling theory. What is the probability of being at the origin, and what is the fracton dimension?

7. Diffusion in percolation clusters in the Cayley tree (or in $d > 6$ dimensions) may be modeled by diffusion in the backbone of a comb. The mass of

the dangling ends is distributed according to $p(M) \sim M^{1-\tau}$, similar to the distribution of clusters. Derive from this the known result $d_w^\ell = 3$.

8. Check out the self-consistent approach at the end of Section 7.6 and show that the reasoning in the text applies only for $0 < \gamma < 1$ (Eq. (7.24)). Extend the argument to $\gamma \geq 1$.

9. Compute the fracton dimension of the comb of Eq. (7.21) using two different approaches: (a) the Einstein relation ($\tilde{\zeta} = 1$, for the backbone), and (b) the self-consistent approach of the end of Section 7.6. (Hint: for (b), compute the number of distinct sites $S(t)$ visited by a walker, and recall that the probability of being at the origin is $P_0(t) \sim 1/S(t) \sim t^{-d_s/2}$.) (Answer: $d_s = (3 - \gamma)/2$ for $0 < \gamma < 1$, and $d_s = 1$ for $\gamma \geq 1$.)

7.8 Open challenges

1. Polynomial trees are grown according to Eq. (7.5). In two dimensions, a transition from linear to nonlinear backbones seems to take place at $d_\ell \approx 1.65$. For $d = 6$, a similar transition occurs at about $d_\ell = 2$. It would be of interest to find out whether such a crossover occurs in 3, 4, and 5 dimensions, and, if it does, at which values of d_ℓ.

2. A related unresolved question regarding polynomial trees concerns the values of $d_{\min}$ corresponding to various values of d_ℓ and different dimensions. The results for $d = 2$ suggest that $d_{\min} = 1$. If indeed $d_{\min} = 1$, it would be of interest to calculate the roughness exponent of paths in these trees.

3. Manna and Dhar (1996) relate the dynamic exponent z of the KPZ model to the critical exponents of Eden trees: $d_f^{BB} = 1 + (d - 1)(1 - 1/z)$, and $d_{\min} = 2(z + d - 1)/(2z + d - 1)$, with interesting consequences for $d = 2$, where $z = \frac{3}{2}$ exactly. Earlier work of Cieplak *et al.* (1996) had established that, in $d = 2$, the roughness exponent of the backbone paths (which measures how fast their width scales with length; $w \sim L^\alpha$), is $\alpha = \frac{2}{3}$, in agreement with $1/z$ of the KPZ model. Numerical work is necessary to confirm Manna and Dhar's hypothesis, until a rigorous proof becomes available.

4. The question of whether the Alexander–Orbach relation $d_s = 2d_f/d_w$ is valid for any fractal is controversial. See, e.g., the studies on diffusion on DLAs and Eden trees of Jacobs *et al.* (1994), Nakanishi and Herrmann (1993), and Reis (1996b). Reis (1996b) suggests that a finite-sized logarithmic correction to $S(t)$ could resolve the problem.

5. Banavar and Willemson (1984), Ohtsuki and Keyes (1984), and O'Shaughnessy and Procaccia (1985a) have suggested alternative forms to $P(r, t)$ of Eqs. (7.16) and (7.20). The issue is not yet resolved.

6. The relation between the distribution of dangling ends in the infinite percolation cluster and the nature of anomalous diffusion poses a great challenge. Some steps towards a better understanding were taken by Huber *et al.* (1995) and Porto *et al.* (1999) in studying the mass distribution of dangling ends. A good approximation is $P(M) \sim M^{-(1+k)}$, with $k = d_f^{BB}/d_f$, which is exact for $d = 6$ (see also Exercise 7).

7.9 Further reading

- Lattice animals: Zimm and Stockmayer (1949), Lubensky and Isaacson (1981), Parisi and Sourlas (1981), Dhar (1983), Martin (1972), Redner (1982), and Djordjevic *et al.* (1984). Diffusion in lattice animals: Havlin *et al.* (1984a) and Wilke *et al.* (1984).
- DLA: Meakin (1983a; 1983b; 1998), Meakin *et al.* (1984), Family and Landau (1984), Herrmann (1986), Witten and Sander (1983), Witten and Cates (1986), Ball and Witten (1984), Nittman *et al.* (1985), and Nittman and Stanley (1986). Diffusion in DLAs: Meakin and Stanley (1983).
- Combs: Shlesinger (1974), Barma and Dhar (1983), White and Barma (1984), Blumen *et al.* (1984), Weiss and Havlin (1986), Goldhirsch and Gefen (1987), Havlin *et al.* (1987b), Dean and Jansons (1993), Pottier (1995), and Revathi *et al.* (1996).
- Diffusion in *infinitely ramified* loopless fractals: Dhar and Ramaswamy (1985) and Havlin *et al.* (1984c; 1985d; 1985e).

8

Disordered transition rates

Anomalous diffusion arises even in regular lattices, when the distribution of transition times is disordered. We consider diffusion with random barriers and wells in one and higher dimensions. The case of barriers can be mapped to an equivalent conduction problem, and wells can successfully be analyzed with the powerful Zwanzig formalism. The more general case of disordered transition rates (other than barriers or wells) displays very rich behavior but is harder to analyze. We discuss it in one dimension only.

8.1 Types of disorder

In order to classify important types of disorder it is useful to look at the generic master equation

$$\frac{\partial P(r, t)}{\partial t} = \sum_{r'} w_{r,r'} P(r', t) - \sum_{r'} w_{r',r} P(r, t). \tag{8.1}$$

$P(r, t)$ is the probability distribution for the walker being at r at time t, and $w_{x,y}$ is the rate of transition from y to x. The order of the subscripts is important: $w_{x,y}$ and $w_{y,x}$ are not necessarily equal. Throughout most of this book we have restricted ourselves to nearest-neighbor random walks, for which $w_{x,y} = 0$ unless x and y are nearest neighbor sites. In one dimension, for example, the master equation for nearest-neighbor walks would be

$$\frac{\partial P(x, t)}{\partial t} = w_{x,x-1} P(x-1, t) + w_{x,x+1} P(x+1, t) - (w_{x-1,x} + w_{x+1,x}) P(x, t). \tag{8.2}$$

When all the transition rates are equal this master equation corresponds to symmetric, or unbiased, random walks or simply regular diffusion. We are interested in the case in which the $\{w_{x,y}\}$ are drawn from a random distribution. The first distinction to be made is that between *strong* and *weak* disorder. Fractal

114

substrates may be understood as an instance of strong disorder, for which the transition rates assume the extreme value of zero for transitions into sites that do not belong to the fractal. We know that this results in anomalous diffusion. We will now see that anomalous diffusion may occur in cases of much weaker disorder, simply as a result of inhomogeneities in the transition rates in regular Euclidean lattices.

Weak disorder may be grossly divided into three kinds: cases with (a) random barriers, (b) random wells, and (c) no particular symmetry. The first two types are important cases of symmetry, namely

$$w_{x,x+1} = w_{x+1,x} \qquad \text{for barriers,} \qquad (8.3a)$$

$$w_{x-1,x} = w_{x+1,x} \qquad \text{for wells.} \qquad (8.3b)$$

For simplicity, we specialize to the case of nearest-neighbor walks in one dimension (Fig. 8.1). The symmetry conditions assure that walkers experience no bias at all length scales larger than the lattice spacing. In cases of no symmetry, however, bias plays an important role. These cases will be examined in Section 8.6.

In one-dimensional systems there is little difference between barriers and wells. The distinction becomes important in higher dimensions, since barriers may then be avoided by simply walking around them (a possibility that does not exist in $d = 1$) and are of less consequence than are wells. We shall investigate the effect of the crossover from one dimension to two dimensions by examining diffusion in strips of increasing widths.

A last classification is made according to the type of distribution of the transition rates. The possibilities here are endless, but we will limit our discussion to power-law distributions. Our choice is guided by the fact that such distributions are frequently encountered in specific physical models, and they lead to anomalous diffusion. Some selected applications are the following.

(i) The temperature dependence of the conductivity exponent observed in the one-dimensional superionic conductor hollandite can be explained by a model with power-law rates (Bernasconi *et al.*, 1979).

(ii) Continuum random systems such as the Swiss-cheese percolation model can be mapped onto lattices with a power-law distribution of bond conductivities (Halperin *et al.*, 1985).

(iii) Anomalous relaxation in spin glasses can be interpreted in terms of stochastic motion (in phase space) with a power-law distribution of transition rates (Sompolinsky, 1981; Teitel and Domany, 1985).

(iv) Diffusion along the backbone of random aggregates (percolation, DLA) can be modeled with power-law transition rates, because of the power-law mass

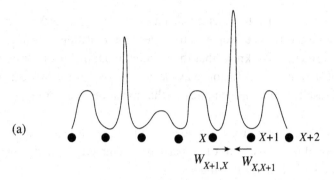

(a)

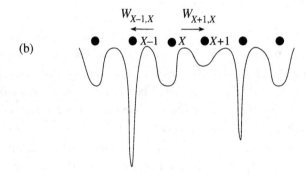

(b)

Fig. 8.1. Barriers and wells: the transition frequencies in barriers (a) and wells (b) satisfy different symmetry relations (Eq. (8.3)).

distribution of dangling ends (Dyson, 1953; Bunde *et al.*, 1986a; Havlin *et al.*, 1986b; Porto *et al.*, 1997b).

(v) Transitions between sites in hierarchical structures such as trees can be interpreted to follow a power-law distribution.

An example of (v) is shown in Fig. 8.2. The hierarchical structure in part (a) is constructed iteratively. The genus is a barrier of height $R^0 = 1$, and the nth iteration is obtained by doubling the existing structure and adding a barrier of height R^n between the two replicas. It is easy to see that the barrier heights follow a power-law distribution. On the other hand, the structure is closely related to the tree in part (b). The heights of the barriers represent the depths to which one must reach in order to move from one "leaf" to the next. Such rules characterize *ultrametric* spaces – a concept that has been found useful in many physics applications (Rammal *et al.*, 1986).

(a)

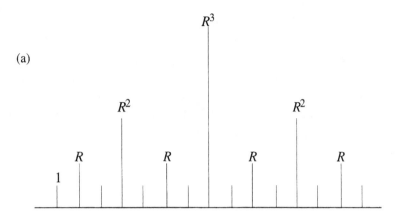

(b)

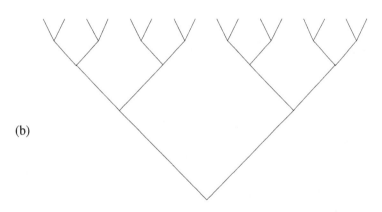

Fig. 8.2. Hierarchical transition rates. (a) Transition rates are represented by resistors in this hierarchical structure. The chain is obtained from the genus – a barrier of height $R^0 = 1$ – and the nth generation is obtained by duplicating the existing structure and adding a barrier of height R^n between the two replicas. The hierarchical nature of the chain is seen explicitly in the equivalent tree structure of part (b).

8.2 The power-law distribution of transition rates

Imagine a one-dimensional chain with bond conductivities $\{\sigma_i\}$ distributed in a power-law fashion:

$$P(\sigma)\,d\sigma = (1-\alpha)\sigma^{-\alpha}\,d\sigma, \quad \alpha < 1, \quad 0 \le \sigma \le 1. \tag{8.4}$$

The conductance Σ (the inverse of the total resistance) of a segment of length L is

$$\Sigma^{-1} = \sum_{i=1}^{L} 1/\sigma_i. \tag{8.5}$$

For $L \gg 1$ we may pass to the continuum limit

$$\Sigma^{-1} \approx L \int_{\sigma_{\min}}^{1} \frac{1}{\sigma} P(\sigma) \, d\sigma = \frac{1-\alpha}{\alpha} L(\sigma_{\min}^{-\alpha} - 1). \tag{8.6}$$

To estimate $\sigma_{\min}$, we choose a random variable u ($0 \le u \le 1$) distributed uniformly so that $P(\sigma) \, d\sigma = du$, i.e., $u \sim \sigma^{-\alpha+1}$. The expected minimum value of u is $u_{\min} = 1/L$ (since there are L conductivities), hence $\sigma_{\min} \sim L^{-1/(1-\alpha)}$ and the resistance scales as

$$R = \Sigma^{-1} \sim \begin{cases} L & (\alpha < 0), \\ L \ln L & (\alpha = 0), \\ L^{1/(1-\alpha)} & (\alpha > 0). \end{cases} \tag{8.7}$$

Thus, we find for the resistance exponent $\tilde{\zeta}$, $(R \sim L^{\tilde{\zeta}})$:

$$\tilde{\zeta} = \begin{cases} 1 & (\alpha \le 0), \\ (1-\alpha)^{-1} & (\alpha > 0). \end{cases} \tag{8.8}$$

The anomalous diffusion exponent can now be computed through the Einstein relation, Eq. (5.6), $d_{\mathrm{w}} = \tilde{\zeta} + d_{\mathrm{f}} = \tilde{\zeta} + 1$:

$$d_{\mathrm{w}} = \begin{cases} 2 & (\alpha \le 0), \\ \dfrac{2-\alpha}{1-\alpha} & (\alpha > 0). \end{cases} \tag{8.9}$$

An alternative derivation due to Zwanzig may be obtained from the distribution of transition rates $\{w_i\}$. These are proportional to the bond conductivities, so $P(w)$ is the same as $P(\sigma)$ of Eq. (8.4). The inverse diffusion coefficient may be written in two different ways:

$$D^{-1} = \frac{t}{\langle x^2 \rangle} = \frac{1}{N} \sum_{i=1}^{N} \frac{1}{w_i}. \tag{8.10}$$

N is the number of distinct sites visited by the walker in time t, and the sum on the RHS equals $\langle 1/w \rangle$. This way of expressing D^{-1} parallels Eq. (8.5). Passing again to the continuum limit, carrying out the integration, and rewriting everything in powers of L; namely $\langle x^2 \rangle \sim L^2$, $N \sim L$, and $t \sim L^{d_{\mathrm{w}}}$, one recovers Eq. (8.9).

8.3 The power-law distribution of potential barriers and wells

Consider a one-dimensional system with potential barriers or wells $\{V_i\}$ distributed according to

$$\psi(V) = (\gamma V_0^{\gamma}) V^{-(1+\gamma)}, \quad V_0 \le V < \infty, \quad \gamma > 0. \tag{8.11}$$

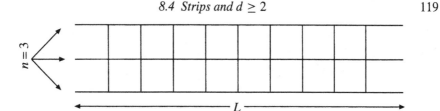

Fig. 8.3. An $n \times \infty$ strip of resistors. In this example $n = 3$. The length L is assumed to be infinite.

If we associate with each well (or barrier) a transition rate $w_i \sim 1/V_i$, then the distribution of transition rates is similar to Eq. (8.4), with $\alpha = 1 - \gamma$, and the results of Section 8.2 apply.

A more realistic relation between the transition rates and the potentials would be given by the Boltzmann factors

$$w_i \sim \exp(-\beta V_i),$$

where $\beta \equiv 1/(k_\mathrm{B} T)$. The distribution of transition rates is then

$$P(w) = \psi(V)\frac{dV}{dw} = \frac{\gamma |\ln w_0|^\gamma}{w|\ln w|^{1+\gamma}}; \qquad w_0 = e^{-\beta V_0}. \qquad (8.12)$$

We then use the Zwanzig formalism of Eq. (8.10):

$$D^{-1} = \gamma |\ln w_0|^\gamma \int_{w_\mathrm{min}}^{w_0} \frac{dw}{w^2 |\ln w|^{1+\gamma}} \sim w_\mathrm{min}^{-1}|\ln w_\mathrm{min}|^{-(1+\gamma)},$$

and compute $w_\mathrm{min} \sim \exp(-L^{1/\gamma})$, in a similar way to that of the previous section. Thus, in the long-time asymptotic limit the r.m.s. displacement grows very slowly,

$$x \sim (\ln t)^\gamma, \qquad 0 < \gamma < \infty. \qquad (8.13)$$

Such a logarithmic growth is also seen in the Sinai problem – a special case of random transition rates without the symmetry of barriers or wells (Section 8.6).

8.4 Barriers and wells in strips ($n \times \infty$) and in $d \geq 2$

One-dimensional diffusion is conceptually different than that in higher dimensions, because the walker is forced to pass through each bond. This restriction is lifted already with the addition of one parallel chain. Generally, one can think of similar structures consisting of n cross-linked parallel chains, or ($n \times \infty$) strips (Fig. 8.3). The crossover between $d = 1$ and $d = 2$ can then be observed as n varies from 1 to ∞.

The case of wells is easiest, because then one may apply the Zwanzig formalism. If n is finite ($n < \infty$), there are essentially no changes from the case of $n = 1$. The number of distinct sites scales now as nL, instead of L, with trivial consequences. However, the cases of $n \to \infty$ ($d = 2$) and of $d > 2$ are interesting. Consider first a power-law distribution of wells. The Zwanzig *Ansatz* yields

$$D^{-1} \sim L^{d_\mathrm{w}-2} \sim w_\mathrm{min}^{-\alpha} \sim N^{\alpha/(1-\alpha)}.$$

The number of distinct sites visited, N, is simply $N \sim t \sim L^2$ ($d \geq 2$), so

$$d_\mathrm{w} = \frac{2}{1 - \alpha}, \qquad d \geq d_\mathrm{c} = 2. \qquad (8.14)$$

Note that this result is the same as that for CTRWs, for which the waiting time is always taken from the same distribution, regardless of the order of the sites visited (Exercise 5). This confirms the argument that *quenched* disorder is not relevant in $d \geq 2$, due to the fact that in our case a walker rarely returns to the same site. For the Boltzmann distribution of wells of Eq. (8.12) a similar treatment yields

$$x \sim (\ln t)^{\gamma/2}, \qquad d \geq d_\mathrm{c} = 2. \qquad (8.15)$$

The more difficult case of barriers may be analyzed through the analogy to a network of conductivities (distributed in the same way as the barriers). A bound for ($n \times \infty$) strips is obtained by assuming that the vertical bonds have infinite conductivity. The problem then reduces to that of a linear chain for which each conductivity $\bar{\sigma}$ is the equivalent of n conductivities connected in parallel, $\bar{\sigma} = \sum_{i=1}^{n} \sigma_i$. The distribution of $\bar{\sigma}$ can be computed from that of the $\{\sigma_i\}$ and finally one follows the approach of Eq. (8.5). In this way it was found that, for a power-law distribution of transition rates, there is a phase transition, depending on the width of the strip n and the power exponent α (Fig. 8.4):

$$\tilde{\zeta} = \frac{1}{n(1 - \alpha)}, \qquad \alpha > \alpha_\mathrm{c} = 1 - 1/n. \qquad (8.16)$$

The diffusion exponent is $d_\mathrm{w} = 1 + \tilde{\zeta}$. For $\alpha < \alpha_\mathrm{c}$ conduction and diffusion are normal. The behavior at $d \geq d_\mathrm{c} = 2$ may be obtained from the limit $n \to \infty$ and also through other methods. Diffusion and conduction are then normal.

For Boltzmann-distributed barriers (Eq. (8.12)), the same technique of equivalent conductivities $\bar{\sigma}$ shows that there exists a transition about $\gamma_\mathrm{c} = 1/[2(d - 1)]$:

$$x \sim \begin{cases} (\ln t)^{\gamma/[1-\gamma(d-1)]} & \gamma < \gamma_\mathrm{c}, \\ t^{1/2} & \gamma > \gamma_\mathrm{c}. \end{cases} \qquad (8.17)$$

In contrast to all previous cases, here there is no upper critical dimension, but the anomalous region shrinks with increasing dimensionality. A summary of the various cases discussed so far is presented in Table 8.1.

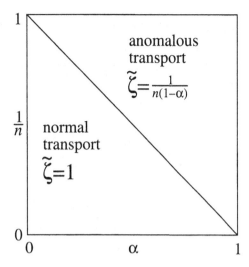

Fig. 8.4. The phase diagram for $n \times \infty$ strips as in Fig. 8.3, consisting of vertical bonds that are perfect conductors, and horizontal conductances distributed in power-law fashion (Eq. (8.4)). Two phases arise, characterized by different values of the exponent $\tilde{\zeta}$.

8.5 Barriers and wells in fractals

Random distributions of barriers in fractals may be replaced by the equivalent conductivity problem. In general, this results in very hard problems and there exist few rigorous results. For power-law distributions of barriers (conductivities) in percolation clusters some useful bounds have been derived (Section 6.10).

The problem of wells is, as usual, simpler. In this case the Zwanzig approach may be generalized by assuming that $D^{-1} \sim L^{d_w - d_w^0}$, where d_w^0 is the walk dimension in the fractal when the transition probabilities are all equal. A distinction between fractals with fracton dimensionalities d_s greater than or less than two must be made. If $d_s = 2d_f/d_w^0 < 2$ then diffusion is recurrent and the number of distinct sites visited scales as $N \sim L^{d_f}$. In the opposite case of $d_s > 2$ the walks are no longer compact and $N \sim L^{d_w^0}$.

For a power-law distribution of transition rates we get

$$
d_w(\alpha) = \begin{cases} d_w^0 + \dfrac{d_f \alpha}{1 - \alpha} & d_s \le 2, \\[2mm] \dfrac{d_w^0}{1 - \alpha} & d_s > 2, \end{cases} \tag{8.18}
$$

which is a generalization of Eqs. (8.9) and (8.14). The results for $d_s < 2$ have been verified numerically for the Sierpinski gasket and for percolation clusters.

Table 8.1. *A summary of results. Cases A and B correspond to the distributions* $P_A(w) \sim w^{-\alpha}$ *and* $P_B(w) \sim (w \ln w)^{-(1+\gamma)}$.

	Wells	
	A	B
$d = 1$	$\langle x^2 \rangle \sim t^{2/d_w}$ $d_w = (2-\alpha)/(1-\alpha)$ $d_s = 2(d_w - 1)/d_w$	$\langle x^2 \rangle \sim (\ln t)^{2\gamma}$
n-strip	Same as $d = 1$	Same as $d = 1$
$d \geq 2$	$d_w = 2/(1-\alpha)$	$\langle r^2 \rangle \sim (\ln t)^{\gamma}$

	Barriers	
	A	B
$d = 1$	$\langle x^2 \rangle \sim t^{2/d_w}$ $d_w = (2-\alpha)/(1-\alpha)$ $d_s = 2/d_w$	$\langle x^2 \rangle \sim (\ln t)^{2\gamma}$
n-strip	$d_w = \begin{cases} 2 & \alpha < \alpha_c = 1 - 1/n \\ 1 + 1/[n(1-\alpha)] & \\ & \alpha > \alpha_c \end{cases}$	Same as $d = 1$
$d \geq 2$	$d_w = 2$	$\langle r^2 \rangle = \begin{cases} t & \gamma > \gamma_c = 1/[2(d-1)] \\ (\ln t)^{2\gamma/[1-\gamma(d-1)]} & \\ & \gamma < \gamma_c \end{cases}$

8.6 Random transition rates in one dimension

Imagine now a one-dimensional chain with a random distribution of transition rates, and without the special symmetry of barriers or wells. For concreteness, we shall consider the case in which at each site i there is a probability p_i of stepping to the right (and a probability $q_i = 1 - p_i$ of stepping to the left). The $\{p_i\}$ are independent of each other, and drawn from the distribution $\phi(p)$. Diffusion in such an environment may be regarded as Brownian motion under the influence of the effective potential

$$V_n = \sum^{n} \ln\left(\frac{q_i}{p_i}\right). \tag{8.19}$$

The case of $\langle \ln q \rangle = \langle \ln p \rangle$ is known as the Sinai problem. The potential is then constant on the average, just like for barriers or wells, and one might expect similar dynamics. This is not the case, however. Even when the Sinai condition holds, there may develop strong biases along finite chain segments. Suppose for

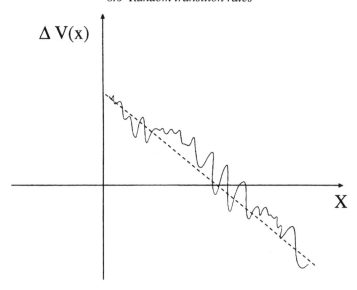

Fig. 8.5. A schematic representation of the effective potential of Eq. (8.22), in a chain of random transition rates. The dashed line indicates the average bias (the first term on the r.h.s. of Eq. (8.22)). Fluctuations around that average grow as $x^{1/2}$ (the second term). After Bouchaud and Georges (1990).

example that $\phi(p)$ is the dichotomous distribution

$$\phi(p) = \tfrac{1}{2}\delta(p - p_0) + \tfrac{1}{2}\delta[p - (1 - p_0)], \qquad p_0 > \tfrac{1}{2}. \qquad (8.20)$$

Although $\langle \ln q \rangle = \langle \ln p \rangle = \ln[p_0(1 - p_0)]/2$, the potential within a segment of length x would still grow (or decrease) by as much as $\Delta V(x) \sim x^{1/2} \ln[p_0/(1 - p_0)]$, because of the typical binomial fluctuations. To overcome this potential a particle would take a time of the order of $t \sim \exp[\Delta V(x)]$, hence

$$x \sim (\ln t)^2. \qquad (8.21)$$

In contrast, the symmetry of barriers and wells ensures that $\Delta V(x)$ is zero at all length scales.

Generally, the effective potential is (expanding (8.19))

$$\Delta V(x) \sim x \left\langle \left(\ln \frac{q}{p} \right) \right\rangle \pm x^{1/2} \left\langle\!\!\left\langle \left[\ln \left(\frac{q}{p} \right) \right]^2 \right\rangle\!\!\right\rangle^{1/2}, \qquad (8.22)$$

where $\langle\!\langle (\cdot)^2 \rangle\!\rangle \equiv \langle (\cdot)^2 \rangle - \langle \cdot \rangle^2$. When the Sinai condition is violated ($\langle \ln q \rangle < \langle \ln p \rangle$, say) there is an effective drift, expressed by the linear term in x. The actual kinetics is then the result of the competition between the average drift and the superposed growing fluctuations (Fig. 8.5). It has been shown that, for μ such that

$$\left\langle \left(\frac{q}{p}\right)^{\mu}\right\rangle = 1, \tag{8.23}$$

the displacement grows as

$$x \sim \begin{cases} t^{\mu} & \mu < 1, \\ vt \pm t^{1/\mu} & 1 < \mu < 2, \\ vt \pm t^{1/2} & \mu > 2, \end{cases} \tag{8.24}$$

where $v \sim -\langle \ln(q/p)\rangle$. That is, there is a phase transition between linear drift and a slower *creeping* motion, as a function of the distribution. The Sinai problem is an extreme instance of the creeping phase, corresponding to the limit $\mu \to 0$.

As a specific example, consider the distribution

$$\phi(p) = c\delta(p - p_0) + (1 - c)\delta[p - (1 - p_0)], \qquad c \geq p \geq \tfrac{1}{2},$$

for which $\mu = \ln[(1 - c)/c]/\ln[(1 - p_0)/p_0]$. In the limit $1 - c \equiv \epsilon \ll 1$ and $1 - p_0 \equiv \eta \ll 1$ this becomes $\mu \approx \ln \epsilon / \ln \eta$. We can understand the above result from the following argument. Most of the bonds in the chain are strongly biased to the right. A walker would therefore advance rapidly until it meets a sequence of n bonds that are all biased to the left. It will then take the walker a time of the order of $t \sim (1/\eta)^n$ to overcome this barrier. The left-biased bonds are Poisson-distributed, therefore within a span x their longest typical sequence is given by $1/x \approx \epsilon^n$, or $n \approx -\ln x/\ln \epsilon$. This yields $t \sim 1/(\eta)^n \sim x^{\ln \eta / \ln \epsilon}$, in agreement with Eqs. (8.23) and (8.24).

8.7 Exercises

1. Find the distribution of barriers in Fig. 8.2. Compute the resistance exponent of the backbone, assuming that barriers of height R^n represent conductivities $\sigma = 1/R^n$. (Answer: $\tilde{\zeta} = 1$ if $R < 2$, and $\tilde{\zeta} = \log R/\log 2$ if $R \geq 2$.)

2. In the multifractal structure of Fig. 8.2, the transition rates are inversely proportional to the barriers' heights, $w = 1/R^n$. Compute d_w, using the Zwanzig formalism, and assuming for simplicity that the walker starts at the left edge of the structure. (Answer: $d_w = 1 + \log(2 + R)/\log 3$.)

3. In the previous exercise the limit $R \to 0$ corresponds to random walks in the Cantor set, where the walker takes a unit time to step between nearest sites regardless of their distance. Obtain the *super-diffusive* result of $d_w = 1 + \log 2/\log 3$ directly from this picture. Explain why there is no super-diffusive motion in the hierarchical structure of Exercise 1.

4. Consider an exponential distribution of potential wells, $\psi(V) = (1/V_0)\exp(-V/V_0)$, with a Boltzmann distribution of transition rates, $w \sim \exp(-\beta V)$. Show that there is a phase transition about $T_c = V_0/k_B$. For $T > T_c$ diffusion is normal, but below the transition temperature $d_w = 1 + T_c/T$, for $d = 1$; and $d_w = 2T_c/T$, for $d \geq 2$.

5. Complete the details in the derivation of Eq. (8.14). Obtain the same result by considering a CTRW with waiting-time distribution $\psi(t) \sim t^{-(1+\alpha)}$.

6. Consider random walks in strips of $n \times \infty$ where the nodes are wells with a power-law distribution of transition rates, according to Eq. (8.4). One expects that for short times diffusion is as in a two-dimensional disordered lattice, with d_w of Eq. (8.14), but for long times it behaves as in one-dimension, with d_w of Eq. (8.9). Calculate the crossover time $t_\times(n)$ between these two regimes. (Answer: $t_\times \sim n^{(1-\alpha)/2}$.)

7. Assume that the mass distribution of dangling ends in percolation at criticality is $P(M) \sim M^{-(1+k)}$ (Porto *et al.*, 1999). Assume also that the time τ spent in a dead end is proportional to its mass, $\tau \sim M$. Show that $d_w = d_w^{BB} + [(1+k)/k]d_f^{BB} = d_w^{BB} + d_f - d_f^{BB}$.

8. Complete the details in the derivation of Eq. (8.17).

8.8 Open challenges

1. The form of $P(r, t)$ for diffusion in the presence of random wells and random barriers is not known analytically, and, to the best of our knowledge, neither has it been studied numerically. An intriguing open question is that of whether the form of $P(r, t)$ in $d = 1$ is different than the one suggested by the CTRW model of Section 4.2.

2. The transport properties of the case in which both particles and barriers (or wells) diffuse has not been studied. Diffusion of the disordered transition rates is a reasonable model of many annealing processes.

3. Consider a distribution of transition rates $\phi(w) \sim 1/w$ in $d = 2$. In this case the time for transport between two given sites is dominated by the smallest w along a connecting path, since $\sum 1/w_i \sim 1/w_{min}$. The optimal path is therefore the one for which $1/w_{min}$ is minimal. This is the problem of optimal paths in strong disorder (Cieplak *et al.*, 1996; Porto *et al.*, 1997a); its length scales as $\ell \sim r^{1.24}$ in $d = 2$, and as $\ell \sim r^{1.4}$ in $d = 3$. It is an open question how transport properties are related to the optimal path.

8.9 Further reading

- Bernasconi *et al.* (1978), Alexander (1981), Alexander *et al.* (1981), Haus and Kehr (1987), Robillard and Tremblay (1986), Kogut and Straley (1979), Straley (1982), Stephen and Kariotis (1982), Bouchaud and Georges (1990), Dentener and Ernst (1984), Havlin *et al.* (1986c; 1987a; 1987b), Havlin and Weissman (1986), Huberman and Kerszberg (1985), Kutasov *et al.* (1986), Grossmann *et al.* (1985), Blumen *et al.* (1986a), Sinai (1982), Pandey (1986), Winter *et al.* (1984), Derrida and Pomeau (1982), Machta (1985), Scher and Montroll (1975), Scher and Lax (1973), Klafter and Silbey (1980), Kivelson (1980), and Harder *et al.* (1987).

9

Biased anomalous diffusion

An important aspect of transport physics is the response to external fields. In homogeneous systems, constant external fields give rise to constant currents which increase with the strength of the biasing field. In contrast, it has been pointed out by Böttger and Bryskin (1982) that the drift velocity in disordered media would respond nonmonotonically to the biasing field; increasing, initially, when the bias is weak, but ultimately decreasing when the bias grows stronger, as the walkers get mired inside dangling ends and backbends of the disordered substrate.

The bias field E is modeled by giving random walkers a higher probability P_+ of moving along the direction of the field, and a lower probability P_- of moving against it,

$$P_\pm = A(1 \pm E),$$

where A is an appropriate normalization factor. The motion of walkers in a direction perpendicular to the field is not affected by it. Alternatively, the bias could be represented by a similar asymmetry in the transition rates. Bias random walks in fractals have attracted recent interest because of the log-periodic behavior found for $\langle r^2 \rangle$ (Sornette, 1998).

One must distinguish between *Cartesian bias*, in which bias is along a fixed direction in space, and *topological bias*, in which bias is in chemical space – in the direction of increasing chemical distance (Fig. 9.1). An example of Cartesian bias is a disordered conductor placed between the charged plates of a condensor. Topological bias occurs in hydrodynamic flow through porous media when pressure is applied at one point, such as in the recovery of oil from rocks.

Bias in combs is simpler to understand than is that in fractals in general. We use it to model general aspects of biased anomalous diffusion, and to interpret the results of topological bias in the more physically relevant case of percolation clusters.

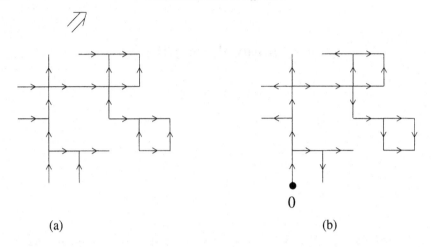

(a)   (b)

Fig. 9.1. Types of bias. A schematic representation of a random cluster in which a walker might be subject to (a) Cartesian bias, in the direction of the slanted arrow, or (b) topological bias, towards increasing chemical-shell numbers and away from the entry point O.

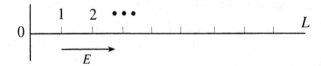

Fig. 9.2. Delay in a tooth. The bias field E pushes the walker away from site 0 (on the backbone), and towards the edge of the tooth, L. The first-passage times for leaving the tooth from site i, T_i, are related to each other as specified in Eq. (9.1).

We also discuss briefly transport behavior in fractals under the effect of sinusoidal bias fields (ac conductivity). The subject may be tackled through scaling analysis, linear-response theory (in the appropriate regime), or numerical simulations and experiments.

9.1 Delay in a tooth under bias

The first, basic problem to consider is the delay caused to a walker in a comb as it enters a tooth of length L, under the influence of a bias field E acting in parallel to the tooth (Fig. 9.2). This serves as a model for the delay caused to walkers in dangling ends and backbends of fractals and disordered structures.

Let the sites of the tooth be numbered $i = 0, 1, 2, \ldots, L$. Let T_i be the average time taken to leave the tooth (reach site 0) for the first time, given that the walker is at site i at time $t = 0$. These first-passage times satisfy the recursion relations (see also Section 5.2)

$$T_i = 1 + \tfrac{1}{2}(1 - E)T_{i-1} + \tfrac{1}{2}(1 + E)T_{i+1}, \qquad 0 < i < L. \tag{9.1}$$

For the endpoints we have the special equations

$$T_0 = 0, \qquad T_L = T_{L-1} + \frac{2}{1 - E}, \tag{9.2}$$

which correspond to absorbing and reflective boundary conditions, respectively. The second boundary condition is equivalent to $T_{L+1} = T_L$.

The general solution of difference equations like (9.1) is exponential in i, with an additional linear term that accounts for the inhomogeneous constant. Beginning with the latter we guess $T_i = ai$, and, on plugging this into (9.1), we find $a = -1/E$. The exponential terms come from the homogeneous equation $T_i' = \tfrac{1}{2}(1 - E)T_{i-1}' + \tfrac{1}{2}(1 + E)T_{i+1}'$. Here the guess $T_i' = \theta^i$ leads to $\theta_- = (1 - E)/(1 + E)$ and $\theta_+ = 1$. Therefore the general solution is

$$T_i = b\theta_-^i + c\theta_+^i - \frac{1}{E}i,$$

where b and c are constants found from the boundary conditions. We finally get

$$T_i = \frac{1 + E}{2E^2}\left(\frac{1 + E}{1 - E}\right)^L \left[1 - \left(\frac{1 - E}{1 + E}\right)^i\right] - \frac{i}{E}. \tag{9.3}$$

We can now compute several limits of interest. The time delay in a tooth of length L is $\tau(L) \sim T_1$, since a walker has to first enter the tooth in order to experience a delay. For long teeth we obtain (from (9.3)) the exponential behavior

$$\tau(L) \sim \left(\frac{1 + E}{1 - E}\right)^L \equiv e^{L/\Lambda}, \qquad L \gg 1. \tag{9.4}$$

In the limit of very strong bias, $E \to 1$, the time delay diverges. On the other hand, when $E \to 0$, Eq. (9.4) predicts $\tau \to 0$. This is, however, not the case. To obtain the correct answer when there is no bias we take the limit $E \to 0$ in Eq. (9.3). Thus,

$$\tau(L) \sim T_1 = 2L + 1 \sim L, \qquad E = 0, \tag{9.5}$$

a result that we have used in previous chapters.

9.2 Combs with exponential distributions of teeth lengths

Consider a random comb with a distribution of teeth lengths

$$\varphi(L) = |\ln \lambda|\lambda^L = \frac{1}{L_0}\exp\left(-\frac{L}{L_0}\right), \qquad 0 \le L \le \infty, \tag{9.6}$$

where $\lambda \equiv \exp(-1/L_0)$. This choice is relevant to biased diffusion in percolation clusters above criticality, for which the size of the dangling ends in the infinite cluster is also distributed in exponential fashion.

We first deal with biased anomalous diffusion when the bias field is along the y-direction, parallel to the teeth (Fig. 7.9). The comb may be regarded as a one-dimensional chain with characteristic waiting times determined by the length of the teeth, or with transition rates proportional to the inverse of these waiting times:

$$w(L) \sim \frac{1}{\tau(L)} \sim \left(\frac{1 - E_y}{1 + E_y}\right)^L = e^{-L/\Lambda}. \tag{9.7}$$

The distribution of the rates $\{w(L)\}$ is given by the distribution of teeth lengths, $\phi(w) = \varphi(L)\, dL/dw$, or, using (9.6) and (9.7),

$$\phi(w) \sim w^{-\alpha}, \qquad \alpha = 1 - \frac{\Lambda}{L_0}. \tag{9.8}$$

This is our problem of Section 8.2, and the anomalous-diffusion exponent along the backbone is (Eq. (8.9))

$$d_w = \begin{cases} 2 & \alpha < 0, \\ 2 + \alpha/(1 - \alpha) & 0 \le \alpha \le 1. \end{cases} \tag{9.9}$$

If there is a nonzero field component along the backbone, $E_x > 0$ (Fig. 7.9), then the time t that a walker takes to travel a distance x is proportional to the sum of waiting times along its path:

$$t \sim \sum_{i=1}^{x} \tau_i = \sum_{i=1}^{x} \frac{1}{w_i}. \tag{9.10}$$

Using the techniques of Chapter 8 and the given distribution (Eq. (9.8)), we find

$$t \sim x \int_{w_{\min}}^{1} \frac{1}{w} \phi(w)\, dw \sim \begin{cases} x & \alpha < 0, \\ x^{1+\alpha/(1-\alpha)} & 0 \le \alpha \le 1, \end{cases}$$

or

$$d_w = \begin{cases} 1 & \alpha < 0, \\ 1 + \alpha/(1 - \alpha) & 0 \le \alpha \le 1. \end{cases} \tag{9.11}$$

The results for general bias may be summarized as

$$d_w = \begin{cases} d_w^0 & E_y < E_y^c(\lambda), \\ d_w^0 - 1 + \dfrac{\ln[(1 + E_y)/(1 - E_y)]}{\ln(1/\lambda)} & E_y > E_y^c(\lambda), \end{cases} \tag{9.12a}$$

where d_w^0 is the diffusion exponent along the backbone in the absence of teeth,

$$d_w^0 = \begin{cases} 2 & E_x = 0, \\ 1 & E_x > 0. \end{cases} \tag{9.12b}$$

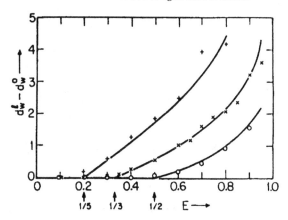

Fig. 9.3. The dependence of d_w on the bias field E_y. Data for the cases $\lambda = \frac{1}{3}$ (o) and $\frac{1}{2}$ ($\times$) with $E_x > 0$, and for $\lambda = \frac{2}{3}$ (+) with $E_x = 0$ are shown. The solid curves represent the theoretical result of Eqs. (9.12) and (9.13). After Bunde *et al.* (1986a).

Thus, there is a phase transition in the quality of diffusion along the backbone as a function of the transverse field. The critical bias field $E_y^c(\lambda)$ corresponds to $\alpha = 0$:

$$E_y^c(\lambda) = \frac{1 - \lambda}{1 + \lambda}. \tag{9.13}$$

For $E < E_y^c$ diffusion is normal, but when $E_y > E_y^c$ the longer delays of the walker in the teeth give rise to anomalous diffusion. Notice also the sharp transition in the average *drift velocity* of the walkers: this is finite for $E_y < E_y^c$, but vanishes when $E_y \geq E_y^c$.

Equations (9.12) and (9.13) were tested through exact enumeration of diffusion in computer-simulated combs. The numerical results agree very nicely with the theory (Fig. 9.3).

9.3 Combs with power-law distributions of teeth lengths

Consider now a comb with a power-law distribution of teeth lengths

$$\varphi(L) = \gamma L^{-(1+\gamma)}, \qquad 1 \leq L \leq \infty. \tag{9.14}$$

This choice is relevant to percolation clusters at criticality. In this case the mass of the dangling ends emanating from the quasilinear backbone is algebraically distributed.

For a bias in the *y*-direction, and with the distribution of teeth lengths of Eq. (9.14), we get

$$\phi(w) = \frac{\gamma \Lambda}{w (\ln w)^{1+\gamma}}. \tag{9.15}$$

This is our problem of Section 8.3, and the solution is (Eq. (8.13))

$$\langle x^2 \rangle \sim (\Lambda \ln t)^{2\gamma}, \qquad \Lambda = \left[\ln\left(\frac{1 + E_y}{1 - E_y}\right) \right]^{-1}. \tag{9.16}$$

When a field is applied also along the backbone, $E_x > 0$, the time t that a walker requires to cover a distance x is

$$t \sim \sum_{i=1}^{x} \tau_i \sim x \int_{w_{\min}}^{1} \frac{dw}{w^2 (\ln w)^{1+\gamma}}.$$

However, $w_{\min} \sim \exp(-x^{1/\gamma}/\Lambda)$ and therefore $\langle x^2 \rangle \sim (\Lambda \ln t)^{2\gamma}$, the same as when $E_x = 0$. That is, one obtains the same logarithmic time dependence for any direction of the bias field! The theory has been tested through computer simulations for various values of γ, yielding generally good agreement.

Interestingly, the results of this section hold almost without change even when the backbone of the comb is of a higher dimension. In fact, when the bias field has a nonzero component along the backbone the results do not change at all. If $E_x = 0$ and the backbone is d-dimensional, one finds

$$\langle r^2 \rangle \sim \begin{cases} [\Lambda(E) \ln t]^{2\gamma} & d = 1, \\ [\Lambda(E) \ln t]^{\gamma} & d \geq 2. \end{cases} \tag{9.17}$$

9.4 Topological bias in percolation clusters

Consider topological biased diffusion in percolation clusters. The bias is such that it always pushes the walkers away from the backbone and into dangling ends. Thus, the situation is similar to biased diffusion in combs, wherein the teeth of the comb play the same role as the dangling ends of percolation clusters, and the backbone of the comb is analogous to the backbone of the clusters.

Above criticality, $p > p_c$, the mass of the dangling ends is exponentially distributed: their typical length is finite, of order $\xi(p) \sim (p - p_c)^{-\nu}$. It is then reasonable that the time delays in dangling ends are distributed as the delays in combs with exponential distributions of teeth lengths. Accordingly, we expect a phase transition as a function of the bias field (Bunde *et al.*, 1990):

$$d_w^{\ell} = A(p) \ln\left(\frac{1 + E}{1 - E}\right), \qquad E > E_c(p), \tag{9.18}$$

and $d_w^{\ell} = 1$ for $E < E_c(p)$. Here $A(p)$ is analogous to $\ln(1/\lambda)^{-1}$ of Eq. (9.12a), it decreases monotonically with p and vanishes as $p \to p_c$. Similarly, $E_c(p)$ is zero at $p = p_c$ and increases monotonically with p, that is, the region of anomalous diffusion widens as $p \to p_c$. These results have been supported by exact enumeration in computer-simulated clusters above criticality.

For clusters at criticality $\xi \to \infty$, and the exponential distribution of the mass of the dangling ends gives way to a power-law distribution. In this case we expect

$$\langle \ell \rangle \sim [\Lambda(E) \ln t]^{\gamma}, \tag{9.19}$$

where $\Lambda(E) = 1/\ln[(1 + E)/(1 - E)]$, in analogy to Eq. (9.16). Numerical results confirm this situation (Fig. 9.4). The exponent γ is hard to compute, because of the difficulty in estimating delay times in dangling ends of complex fractal structure. As one can see from Fig. 9.4, the data for $d = 2$ suggest that $\gamma \approx 1$.

For percolation in Cayley trees (or in $d \geq 6$) the dangling ends are topologically linear and γ may be computed exactly. We argue that, at criticality, the mass of the dangling ends is roughly distributed as the mass of the clusters themselves,

$$P(S) \sim S^{-\tau+1}, \qquad \tau = \tfrac{5}{2}. \tag{9.20}$$

The chemical length ℓ of the dangling ends is related to their mass through $S \sim \ell^{d_\ell} \sim \ell^2$. Hence the distribution of ℓ is

$$\varphi(\ell) = P(S)\frac{dS}{d\ell} \sim \ell^{2(1-\tau)+1} = \ell^{-2}. \tag{9.21}$$

Comparing this with Eq. (9.14), we conclude that $\gamma = 1$. This is indeed supported by simulations of biased diffusion in percolation clusters grown on the Cayley tree. It is not yet clear whether $\gamma = 1$ exactly for percolation in $d < 6$.

9.5 Cartesian bias in percolation clusters

Cartesian bias in percolation clusters is a lot harder to analyze than topological bias. In this case, the analogy with combs breaks down. Walkers inside dangling ends are pushed by the biasing field as often towards the backbone as away from it. This makes it virtually impossible to map dangling ends onto the teeth of a comb with homogeneous bias.

Barma and Dhar (1983), and Dhar (1984) have argued that, above criticality, a phase transition in the drift velocity, similar to the one with topological bias, should be observed. It was also argued that this transition disappears if there exist repulsive interactions between the particles (Barma and Ramaswamy, 1994). However, early numerical work had failed to reveal such a transition (Pandey, 1984; Seifert and Suessenbach, 1984). The situation is complicated by the fact that the effective diffusion exponent for Cartesian bias in percolation clusters undergoes log-periodic oscillations (Stauffer and Sornette, 1998), making extrapolations to the long-time asymptotic limit difficult. These fluctuations apparently result from increasingly long waiting times inside a hierarchy of ever larger, but ever rarer, dangling ends.

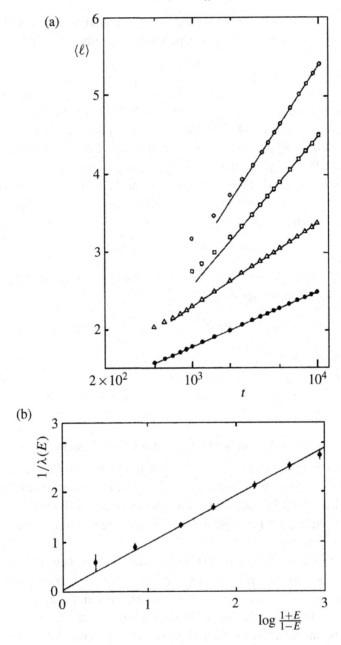

Fig. 9.4. Biased diffusion in percolation clusters in $d = 2$. (a) A plot of $\langle \ell \rangle$ versus t for various values of the bias fields: $E = 0.9$ ($\bullet$), 0.8 ($\triangle$), 0.7 ($\circ$), and 0.5 ($\square$). The solid lines are best linear fits, suggesting that $\gamma = 1$ in this case. (b) The dependence of Λ upon the bias E. The points represent the inverse of the slopes in part (a), and the solid line is $1/\Lambda(E) = \ln[(1 + E)/(1 - E)]$. After Havlin *et al.* (1986a).

Recent work of Dhar and Stauffer (1998) supports the hypothesis of there being a sharp transition in the drift velocity. They argue theoretically that, above the percolation threshold and for bias fields smaller than the critical one, $E = E_c - \epsilon$, the drift velocity tends to a constant proportional to ϵ; it vanishes as $1/\ln t$ for $E = E_c$, and as t^{-x} for $E > E_c$. The drift exponent x depends on the bias field, and vanishes as $E \to E_c$. They also propose a formula that interpolates among these various behaviors:

$$v_{\text{drift}} = \frac{K(\epsilon)x}{(t/t_0)^x - 1},$$

and which fits very nicely with their numerical simulations. If this formula is correct, then the convergence of the drift velocity to its long-time asymptotic value should occur after a time $\sim \exp(\text{constant}/\epsilon)$. Such a slow convergence is yet another reason for the lack of definite results.

For Cartesian bias at criticality, simulations suggest that there is a logarithmic time dependence (Pandey, 1984; Stauffer, 1985b), which is also supported by the heuristic arguments of Dhar and Stauffer (1998). However, a renormalization-group approach leads to the conclusion that $\langle r \rangle \sim t^y$ with $y < 1$ – which agrees with the logarithmic prediction only when y is strictly zero (Ohtsuki and Keyes, 1984).

9.6 Bias along the backbone

Topological bias along the backbone is a very important case of diffusion. This is the situation in hydrodynamic flow under a pressure gradient generated between two points (of the backbone), since the net flow into dangling ends is then zero and diffusion there is not biased.

Consider first as a model a random comb with a power-law distribution of teeth lengths (Eq. (9.14)) and with bias along the backbone only; $E_x > 0$ and $E_y = 0$. The average time delay in a tooth of length L is $\tau(L) \sim L$ (Eq. (9.5)) and hence the distribution of transition rates is

$$\phi(w) \sim w^{\gamma - 1}. \tag{9.22}$$

The total time spent in the backbone when a walker advances a distance x is

$$t \sim \sum_{i=1}^{x} \frac{1}{w_i} \sim x \int_{w_{\min}}^{1} \frac{1}{w} \phi(w)\, dw \sim \begin{cases} x w_{\min}^{\gamma - 1} & 0 < \gamma < 1, \\ x & \gamma \geq 1, \end{cases}$$

where $w_{\min}$ is found by the self-consistent method of Section 7.6. That is, $w_{\min} \sim 1/L_{\max}^{\gamma - 1}$, and $L_{\max}$ is the smaller of two characteristic lengths, namely the maximum length of a tooth in the y-direction, $x^{1/\gamma}$, and the maximum vertical span if the walk were to take place in a comb with infinitely long teeth, $x^{d_w/2}$. Assuming that the latter length is the smaller of the two (this is justified *a posteriori*), one derives

$$d_w = \begin{cases} 2/(1+\gamma) & 0 < \gamma < 1, \\ 1 & \gamma \geq 1. \end{cases} \tag{9.23}$$

It is interesting to compare this with the result in the absence of bias, Eq. (7.24). The phase transition as a function of the distribution of teeth lengths is similar, in either case.

A similar argument applies almost without change for diffusion along the backbone in percolation clusters grown in the Cayley tree. The transition rates in the dangling ends are proportional to $w \sim 1/S$, and, with the known distribution of S (at criticality), we have

$$\phi(w) \sim P(s)\frac{dS}{dw} \sim w^{\tau-3}.$$

Following the self-consistent argument to the end, we obtain

$$d_w^\ell = \frac{1}{1 - (3-\tau)d_\ell/d_w^\ell(0)}, \tag{9.24}$$

where $d_w^\ell(0)$ is the chemical-walk dimension in the absence of bias. Substitution of the known values of the various exponents ($d_w^\ell(0) = 3$, $d_\ell = 2$, and $\tau = \frac{5}{2}$) yields $d_w^\ell = \frac{3}{2}$, in agreement with the fact that $d_w^\ell/d_w^\ell(0) = \frac{1}{2}$ (cf. Eqs. (7.24) and (9.23)). The result has not been tested numerically. It would be interesting to see whether d_w^ℓ is close to $\frac{3}{2}$ also for topological bias along the backbone of percolation clusters in $d < 6$, in analogy to the situation for full topological bias.

9.7 Time-dependent bias

Until now we have discussed static bias fields. When the bias is time-dependent it can be represented by its Fourier decomposition as a superposition of sinusoidally varying fields, $E(\omega) \sim e^{i\omega}$. It is therefore sufficient to study the response to $E(\omega)$. This response is characterized by the ac conductivity $\sigma(\omega)$. The Einstein relation for the dc conductivity (Eq. (3.21)) may be generalized to

$$\sigma(\omega) = \frac{ne^2 D(\omega)}{k_B T}, \tag{9.25}$$

where $D(\omega)$ is given by the Wiener–Khinchine theorem

$$\text{Re } D(\omega) = \int_0^\infty \cos\omega \, \langle v(t)v(0)\rangle \, dt. \tag{9.26}$$

$D(\omega)$ is the spectrum of the autocorrelation function of the velocity of a walker (in the absence of external fields), at frequency $f = \omega/(2\pi)$.

The velocity autocorrelation function scales as $\langle v(t)v(0)\rangle \sim (r/t)^2 \sim t^{2/d_w-2}$ and therefore $D(\omega) \sim \omega^{(d_w-2)/d_w}$. The density of charge carriers scales as $n \sim r^{d_f-d} \sim t^{(d_f-d)/d_w} \sim \omega^{(d-d_f)/d_w}$, so

$$\sigma(\omega) \sim (-i\omega)^{\tilde{\mu}/d_w} = \exp\left(-i\frac{\tilde{\mu}}{d_w}\frac{\pi}{2}\right)\omega^{\tilde{\mu}/d_w}. \tag{9.27}$$

where we have used $d_w = 2 - d + d_f + \tilde{\mu}$ (Eq. (5.6)), and the phase is determined from a Kramers–Kronig relation. The dielectric constant $\varepsilon(\omega)$ may also be computed from the Kramers–Kronig relation:

$$\varepsilon(\omega) - 1 \sim \text{P.V.}\int \frac{d\omega'\,\sigma(\omega')}{\omega'(\omega'-\omega)},$$

where P.V. refers to the principal value of the integral. In the linear response regime, $\omega\tau \gg 1$,

$$\varepsilon(\omega) \sim \omega^{-1+\tilde{\mu}/d_w}. \tag{9.28}$$

An alternative derivation of (9.27) is due to the scaling *Ansatz*

$$\sigma(\omega) = \xi^{-\tilde{\mu}}s(\omega\tau), \qquad \tau = \xi^{d_w}, \tag{9.29}$$

proposed by Gefen *et al.* (1983b). The factor $\xi^{-\tilde{\mu}}$ accounts for the dc ($\omega \to 0$) behavior, and τ is a characteristic diffusion time of the system. The linear-response regime corresponds to the limit $\omega\tau \gg 1$. In this case $\sigma(\omega)$ should be independent of ξ, and $\sigma(\omega) \sim \omega^{\tilde{\mu}/d_w}$ follows.

The theory has had some interesting applications. The one-dimensional superconductor hollandite undergoes a mobility transition as a function of temperature. Conduction is normal for $T > T_c$, and anomalous below T_c. Bernasconi *et al.* (1979) successfully modeled this situation by means of an exponential distribution of barriers with a Boltzmann distribution of transition rates (see Exercise 8.4). Their model predicts $\sigma(\omega) \sim (-i\omega)^\nu$, with $\nu = (1 - T/T_c)/(1 + T/T_c)$ for $T < T_c$, and $\sigma(\omega)$ independent of ω for $T > T_c$.

Laibowitz and Gefen (1984) measured $\sigma(\omega)$ and $\varepsilon(\omega)$ for thin gold films near the percolation threshold. They found $\sigma(\omega)/\varepsilon(\omega) \sim \omega$, in agreement with the theory, but $\sigma(\omega) \sim \omega^{0.95}$ – in sharp contrast to $\sigma(\omega) \sim \omega^{\tilde{\mu}/d_w} \approx \omega^{0.34}$ of Eq. (9.27). The discrepancy has been attributed to the neglect of electron–electron interactions.

Harder *et al.* (1986) studied sinusoidal bias in percolation in $d = 2$ numerically. They found that, for large applied fields, the phase shift is still consistent with that predicted by linear-response theory, but that the amplitude of the response exhibits strong nonlinear effects when the frequency is small (Fig. 9.5).

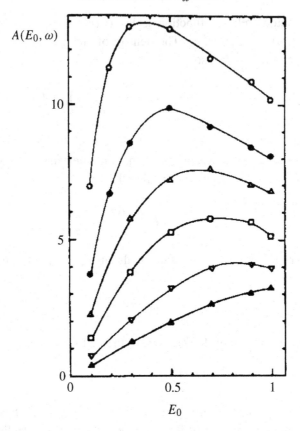

Fig. 9.5. Periodically biased diffusion in percolation clusters in $d = 2$. A plot of the amplitude of the response (span of the walks) $A(E_0, \omega)$ versus the amplitude of the bias field E_0, for various values of the bias frequency: $\omega = 0.002, 0.005, 0.01, 0.02, 0.05$, and 0.1 (top to bottom). After Harder *et al.* (1986).

9.8 Exercises

1. Show that, when the field is antiparallel to a tooth (i.e., directed towards the backbone), the delay time in the tooth remains finite even as $L \to \infty$.

2. Calculate the analogous equations to (9.9) and (9.12) for a *brush*: that is, a comb with a two-dimensional backbone.

3. A comb has a distribution of teeth lengths $\varphi(L) \sim L^{-(1+\gamma)}$. A bias field is applied along the backbone so that the probability of a walker stepping to the right is $p_+ = \frac{1}{3} + \epsilon$, to the left $p_- = \frac{1}{3} - \epsilon$, and towards the teeth $p = \frac{1}{3}$. Compute the mean-square displacement along the backbone, using a self-consistent scaling argument. What is the probability of being at the origin?

4. Derive Eq. (9.17) and show that it is valid irrespective of whether $E_x = 0$ or $E_x > 0$.

5. Complete the details in the derivation of (9.24).

6. Discuss the case of bias along the backbone for an exponential distribution of teeth lengths.

7. Confirm the results of Bernasconi *et al.* (1979), regarding hollandite. (Hint: recall Exercise 8.4, and use Eq. (9.27) and the fact that $\tilde{\mu} = d_w - 2$ for $d = 1$.)

9.9 Open challenges

1. Topological bias in percolation clusters. Is $\gamma = 1$ also in dimensions $d < 6$ (see the discussion following Eq. (9.19))? The problem could be, at the very least, studied numerically.

2. The same question is open for topological bias along the backbone of percolation clusters. In analogy to bias in whole clusters, one expects $d_w^\ell = \frac{3}{2}$ for all dimensions, the same as for $d = 6$, but we know of no numerical (or theoretical) results.

3. Cartesian bias in percolation clusters. The subject enjoys considerable current interest but remains mostly an open challenge. The work of Dhar and Stauffer (1998) is very revealing, but the theoretical arguments are not rigorous, and simulations are limited to $d = 3$. Much additional work is needed to clarify the situation, including the long-standing controversy about bias in percolation clusters at criticality.

4. The effect of backbends with Cartesian bias. The subject has only been studied numerically, but it might be possible to obtain exact results for finitely ramified fractals, for example with renormalization techniques similar to that of Chapter 5. An interesting question is that of whether backbends alone could bring about a transition in the drift velocity, or whether dangling ends are absolutely necessary.

5. It has been argued that the transition in the drift velocity under bias disappears when one introduces hard-core repulsions between the diffusing particles (Ramaswamy and Barma, 1987; Barma and Ramaswamy, 1994). More recently, Andrade *et al.* (1997a) studied fluid flow along the backbones of percolation clusters, with a bias generated by a pressure gradient applied between two points. They compute the velocities in different bonds, by solving the Navier–Stokes equation numerically, and find them to follow a power law. The transport of tracer particles in percolation remains largely an open question, even in the limit of very low Reynolds numbers.

9.10 Further reading

- Biased diffusion in fractals: Scher and Lax (1973), Ohtsuki (1982), Ohtsuki and Keyes (1984), Barma and Dhar (1983), Dhar (1984), White and Barma (1984), Pandey (1984), Seifert and Suessenbach (1984), Luck (1985), Gefen and Goldhirsch (1985), Goldhirsch and Gefen (1987), Chowdhury (1985), Stauffer (1985b), Barma and Ramaswamy (1986), Bunde *et al.* (1986a), Havlin *et al.* (1986a, 1986b), de Arcangelis *et al.* (1986b), Redner *et al.* (1987), Roux *et al.* (1986), Bug *et al.* (1986), Giona (1994), Sartoni and Stella (1997), Kirsch (1998), and Drager and Bunde (1999). An extensive analysis of first-passage times of diffusing particles in percolation, with and without bias, was given by Koplik *et al.* (1988).

- Potential applications of biased diffusion include chromatography – in which particles in gels are subjected to gravity or centrifugal forces (Fischer, 1969), and conduction in doped semiconductors in the presence of strong electric fields (Böttger and Bryskin, 1980; van der Meer *et al.*, 1982; van Lien and Shklovskii, 1981). Experiments on ac conductivity of fractals are reported in Bernasconi *et al.* (1979), and numerical simulations in Harder *et al.* (1986). See also Clerk *et al.* (1990).

- Log-periodic oscillations arising from a hierarchy of waiting times have been seen in many cases – aside from biased diffusion – including stock-market crashes (Sornette *et al.*, 1996b; Sornette and Johansen, 1997) and earthquakes (Saleur *et al.*, 1996). A recent review is given by Sornette (1998). Recent simulations were presented by Stauffer (1999).

10

Excluded-volume interactions

Until now we have considered systems involving noninteracting walkers. This is an enormous simplification that allows analysis of such problems in terms of a *single* walker. Reality, however, is more complex and interactions cannot always be neglected. In this chapter we consider an elementary type of hard-core repulsion, known as *excluded-volume interactions*: walkers, or particles, are not allowed to occupy the same site simultaneously. We describe the dramatic effects that this simple interaction has on diffusion.

A *self-avoiding walk* (SAW) is a random walk that does not intersect itself. SAWs are a useful model for linear polymer chains: the visited sites represent monomers, and self-avoidance accounts for the excluded-volume interactions between monomers. The study of polymers in random media finds applications in enhancing recovery of oil, gel electrophoresis, gel-permeation chromatography, etc. Flory's theory provides us with a beautiful, intuitive understanding of the anomalous properties of SAWs in regular Euclidean space, and it may be extended to percolation and fractals. The problem of SAWs in finitely ramified fractals can be solved exactly.

10.1 Tracer diffusion

Imagine a regular lattice of lattice spacing a with a density c of particles, i.e., each site is occupied with probability c. The particles perform nearest-neighbor random walks, with hopping rate Γ, and are subject to excluded-volume interactions: at most one particle may occupy a site at any given moment. Clearly, diffusion of the particles is hindered by these interactions. For example, in the limit $c = 1$, when all sites are occupied, motion is impossible and the system is frozen. The motion of a specific particle in the system is known as *tracer diffusion*. It is an elementary model for diffusion of particles under *short-range* repulsive interactions. The

141

emphasis on short range is important: long-range interactions, such as Coulombic repulsion between like charges, should be treated separately.

In dimensions $d \geq 2$, tracer diffusion may be described at a "mean-field" level. The rate at which a particle hops to a nearest-neighbor site is simply proportional to the *global* probability that the site is empty. Diffusion is then normal, with a diffusion coefficient $D \sim (1 - c)$. The problem is more interesting in one dimension (e.g., linear chains and channels). For short times, smaller than the typical time separating encounters between particles, tracer diffusion is still normal,

$$\langle x^2 \rangle = 2a^2 \Gamma (1 - c)t \qquad \text{for small } t, \tag{10.1}$$

but it is anomalous ($d_w = 4$) in the long-time asymptotic limit:

$$\langle x^2 \rangle = \frac{2(1 - c)}{c} a^2 \left(\frac{\Gamma t}{\pi} \right)^{1/2}, \qquad \text{as } t \to \infty. \tag{10.2}$$

An elegant derivation of (10.2) was provided by Alexander and Pincus (1978). They observe that the displacement of the lth particle from a uniform configuration, $u_l(t)$, is coupled to the local density fluctuation, $\rho_l(t)$;

$$\rho_l(t) \approx -c^2 \frac{du_l}{dl},$$

while the fluctuations themselves obey a regular diffusion equation. Equation (10.2) follows from these two facts.

A simple intuitive argument valid in the very-dense limit ($c \to 1$) was presented by van Beijeren *et al.* (1983). In this limit there are very few vacancies and they can be assumed to perform regular diffusion. One may further imagine that, when vacancies meet, they are allowed to bypass each other. The displacement of a tracer particle is given by $x(t) = n_{rl}(t) - n_{lr}(t)$, where n_{rl} (n_{lr}) is the number of vacancies that had crossed the tracer from right to left (left to right) by time t. Both quantities grow as $n_{rl} \sim n_{lr} \sim (1 - c)t^{1/2}$, just like in regular diffusion. Assuming a normal distribution with square-root fluctuations, we recover $\langle x^2 \rangle \sim \langle (n_{rl} - n_{lr})^2 \rangle \sim (1 - c)t^{1/2}$.

A different argument can be made for the opposite limit of very low density, $c \ll 1$. In this case, imagine that there are no excluded-volume interactions, other than between the tracer and the rest of the particles. The motion of the tracer particle is not affected by the labeling of all other particles and hence the excluded-volume interactions between them are irrelevant (Fig. 10.1). During a time interval t the tracer makes of the order of $ct^{1/2}$ collisions with the other particles, since they undergo regular diffusion. The average displacement of the tracer between collisions is of the order of $1/c$ – the average distance between particles. Assuming

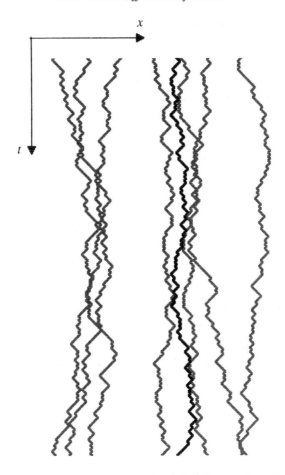

Fig. 10.1. Tracer diffusion. The tracer particle (black) undergoes excluded-volume inter-actions with the other particles (gray) and can never bypass them. From the perspective of the tracer it is irrelevant whether the remaining particles undergo hard excluded-volume interactions among themselves. This fact is used in the heuristic derivation of $\langle r^2 \rangle$ of the tracer in the low-density limit.

that these displacements may be positive or negative in a purely random fashion, we get $\langle x^2 \rangle \sim (1/c)^2 c t^{1/2} \sim (1/c) t^{1/2}$, in agreement with (10.2).

10.2 Tracer diffusion in fractals

Tracer diffusion is anomalous in $d = 1$, so it is potentially interesting in other confined geometries, such as fractals. The question has received some attention. Amitrano *et al.* (1985) studied tracer diffusion in DLA clusters. They find that the long-time behavior is well described by

$$\langle r^2 \rangle \sim [(1 - c) f_T(c) t]^{2/d_w}. \tag{10.3}$$

The function $f_T(c)$ depends on the correlations between consecutive jumps of the tracer particle. For c close to 1, $f_T \approx 1 - c$. The anomalous-diffusion exponent d_w in (10.3) is the same as that for the noninteracting case; that is, hard-core interactions do not change the fractal dimension of the walks in DLAs, at least not in the long-time asymptotic limit. This holds true for all concentrations, c. However, for high densities close to $c = 1$,

$$\langle r^2 \rangle = g\left(\frac{t}{\tau}\right), \qquad \tau = \frac{c^2}{(1 - c)^2}, \tag{10.4}$$

where

$$g(x) \sim \begin{cases} x^{1/2} & x \ll 1, \\ x^{2/d_w} & x \gg 1. \end{cases} \tag{10.5}$$

Thus, for high densities there exists a regime for which tracer diffusion in DLAs is exactly the same as that in one dimension. Amitrano *et al.* explained this result by invoking the fact that, for short times, the tracer particle sees a finite quasi-one-dimensional portion of the aggregate, therefore reproducing the behavior of Eq. (10.2). Similar results were found for tracer diffusion in percolation clusters.

10.3 Self-avoiding walks

A *self-avoiding walk* (SAW) is a random walk that does not intersect itself. In other words, the *equilibrium ensemble* of N-step SAWs is obtained by discarding the intersecting walks from the ensemble of N-step random walks (Exercises 2 and 3). In one dimension, the only possible SAW is a stretched line. In dimensions $d \geq 4$ the likelihood of a walk intercepting itself is negligible and SAWs are then similar to regular random walks. The problem becomes interesting in $d = 2$ and $d = 3$, in which the excluded-volume interactions (of the walk with itself) impose a nontrivial restriction. Examples of SAWs in $d = 2$ are shown in Fig. 10.2.

SAWs are a common model for linear polymer chains (de Gennes, 1979). The visited sites represent monomers whose excluded-volume interactions are modeled by the self-avoidance rule. SAWs are of interest in the theory of equilibrium phase transitions, because they are equivalent to the $n = 0$ limit of the n-vector model (Daoud *et al.*, 1975). However, SAWs are also studied for their own sake, as a fundamental problem in probability theory.

The end-to-end length of a SAW consisting of N steps is

$$\langle r_N^2 \rangle^{1/2} \sim N^{\nu_{SAW}}, \tag{10.6}$$

SAW　　　　　　　　RW

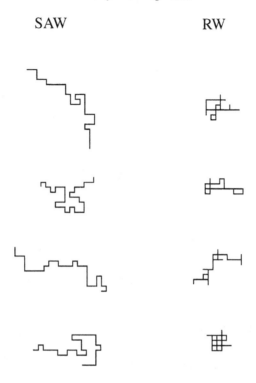

Fig. 10.2. Self-avoiding walks (SAWs). Examples of computer-generated 40-step SAWs (left) compared with 40-step RWs (right) are shown. Notice the swelled configuration of the SAWs, which is due to the excluded-volume interaction (self-avoidance).

where ν_{SAW} is the *end-to-end exponent*. The fractal dimension of the SAW is $d_{SAW} = 1/\nu_{SAW}$. The end-to-end exponent ranges from $\nu_{SAW} = \frac{1}{2}$, for SAWs in $d = 4$, to $\nu_{SAW} = 1$ in $d = 1$. Generally, SAWs are more extended than noninteracting walks and $d_{SAW} \leq d_w$.

Another quantity of interest is the total number of N-step SAWs, $\mathcal{N}(N)$. For noninteracting random walks $\mathcal{N}(N) = z^N$, where z is the coordination number of the lattice (the number of nearest-neighbor sites). For SAWs,

$$\mathcal{N}(N) \sim N^\gamma \mu^N. \tag{10.7}$$

μ may be understood as an effective coordination number, smaller than z – because of the excluded-volume interaction. The prefactor N^γ is less obvious, but can be predicted from the analogy to the $n = 0$-vector model, in which γ is the usual exponent for susceptibility.

Much of the research on SAWs is done through computer simulations. For small N, an exact enumeration of all possible SAWs may be performed. For larger values of N, one simulates a large number of random walks and discards

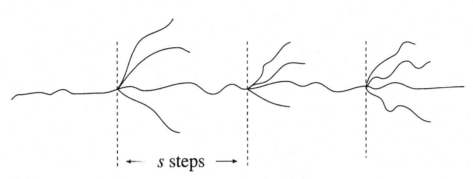

Fig. 10.3. The enrichment method. At each stage p attempts to extend the walk by an additional s-step section are made. Failed attempts (walks that intersect existing parts of the SAW, or themselves) are discarded. A clever choice of p and s increases the odds of producing long walks without compromising their conformational properties.

all of the walks that intersect themselves. The remaining walks are SAWs. However, the fraction of SAWs decreases rapidly with increasing N, rendering this straightforward approach impractical. The *enrichment* algorithm is a popular technique that overcomes the effects of attrition. In this method, an s-step SAW is first produced in the straightforward way. Then p attempts to extend this SAW by s additional steps are made. If none of these attempts is successful, the original walk is discarded. Otherwise, the procedure may be repeated to obtain $3s$-step SAWs, etc. (Fig. 10.3). The parameters s and p should ideally satisfy $p(\mu/z)^s = 1$, and can be chosen so as to greatly enhance the overall success rate.

10.4 Flory's theory

Flory (1971) has suggested a very elegant way to compute the end-to-end exponent of SAWs in d dimensions. The argument is based on a mean-field estimate of the free energy of an N-step SAW,

$$F = U - TS.$$

The entropy S is approximated by the entropy of an N-step random walk. Since $P(r, N) \sim \exp[-dr^2/(2N)]$, the entropy is

$$S(r) = k_{\mathrm{B}} \ln P(r, N) = S(0) - k_{\mathrm{B}} \frac{dr^2}{2N}.$$

Thus, the free energy of a noninteracting random walk is

$$F_{el} = F(0) + k_B T \frac{dr^2}{2N}.$$ (10.8)

This can be understood as an elastic energy, arising purely from entropic considerations: the random walk tends to be coiled in a state such that the density of accessible configurations is largest.

The internal energy U is a result of the excluded-volume interactions. The repulsion energy per unit volume is proportional to $k_B T v \rho^2$, where ρ is the local density of the SAW and v is a measure of the excluded volume per monomer in d dimensions. Hence, the total repulsion energy is

$$F_{rep} = U \sim k_B T v \langle \rho^2 \rangle r^d \sim k_B T v \frac{N^2}{r^d},$$ (10.9)

where we have used the mean-field approximation $\langle \rho^2 \rangle \approx \langle \rho \rangle^2 \sim (N/r^d)^2$.

The total free energy is

$$\frac{F}{k_B T} \sim v \frac{N^2}{r^d} + \frac{dr^2}{2N}.$$ (10.10)

Because of the opposite tendencies of F_{el} to grow with the end-to-end distance r, and of F_{rep} to decrease with increasing r, F has a well-defined minimum that determines the likeliest configurations of SAWs (Fig. 10.4). Minimizing Eq. (10.10) one finds $r \sim N^{3/(d+2)}$, or

$$\nu_{SAW} = \frac{3}{d+2},$$ (10.11)

which is the Flory formula for the end-to-end exponent.

Flory's formula works surprisingly well. For $d = 1$ and $d = 4$ it reproduces the exact results of $\nu_{SAW} = 1$ and $\nu_{SAW} = \frac{1}{2}$, respectively. For $d = 2$ Nienhuis (1982) has argued that $\nu_{SAW} = \frac{3}{4}$, as predicted by Flory, by mapping SAWs onto the exactly solvable hard-hexagon model. For $d = 3$ Flory's prediction of $\nu_{SAW} = \frac{3}{5}$ is close to the best numerical estimates: $\nu_{SAW} = 0.5877 \pm 0.0006$ (Li *et al.*, 1995), and $\nu_{SAW} = 0.5882 \pm 0.0009$ (Eizenberg and Klafter, 1996).

Why is Flory's theory so successful? On the one hand, the repulsion energy is clearly overestimated, because of the neglect of correlations. On the other hand, the entropic contribution to the free energy is underestimated, because the number of possible configurations $\mathcal{N}$ is much smaller for SAWs than it is for RWs. These two sources of error tend to cancel each other out, but it remains unclear why this happens so remarkably well.

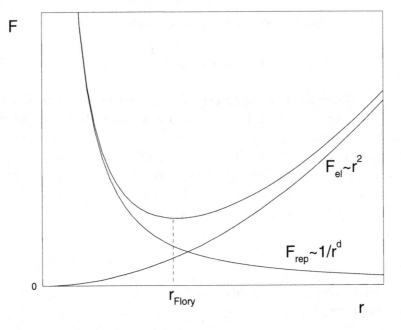

Fig. 10.4. The Flory argument. The free energy of SAWs is made up of two parts (the elastic energy, F_{el}, and the energy of repulsion, F_{rep}), which contribute opposite trends. The end-to-end distance of the SAW, r_{Flory}, is the compromise distance which keeps the free energy at a minimum.

10.5 SAWs in fractals

SAWs in fractals are interesting for three main reasons: (a) they serve as a model of polymers in porous media – a problem relevant to numerous applications such as enhancing recovery of oil, gel electrophoresis, and gel-permeation chromatography; (b) their conformational properties are determined by the backbone of the fractal rather than by its dangling ends, so that SAWs complement the information gained from noninteracting random walks; and (c) their study may shed further light on Flory's theory and the reasons for its success.

In finitely ramified fractals ν_{SAW} may be computed from an exact real-space renormalization-group procedure. The idea is based on $G_N(L)$, the total number of N-step SAWs between two points of the fractal, separated by a distance L. The generating function of G_N,

$$G(x, L) = \sum_{N=1}^{\infty} x^N G_N(L),$$

satisfies a scaling relation of the form $G(x', L/\lambda) = f[G(x, L)]$, where λ is the rescaling factor of space. This scaling equation has a fixed point at $x' = x = x_*$.

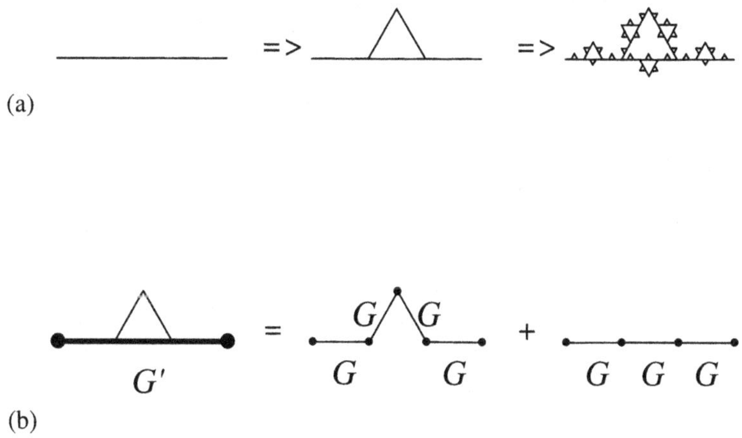

(a)

(b)

Fig. 10.5. Real-space renormalization of SAWs. (a) The branching Koch curve. (b) Renormalization of the generation function G upon magnification of the branching Koch curve by a factor of three.

x is known as the *fugacity* of the SAW, and the fixed point occurs at the critical value $x_* = 1/\mu$ (cf. Eq. (10.7)). In the "thermodynamic" limit of $L \to \infty$, $G(x, L) \sim \exp(-L/\xi(x))$, where $\xi(x) \sim (x_* - x)^{-\nu_{\text{SAW}}}$ is a correlation length of the order of the end-to-end length of the SAW. Thus,

$$\nu_{\text{SAW}} = \ln \lambda \Big/ \ln\left(\left.\frac{dG'}{dG}\right|_{G_*}\right). \tag{10.12}$$

As an example, consider the branching Koch curve of Fig. 10.5a. From Fig. 10.5b we see that

$$G(x', L/\lambda) = G^3(x, L) + G^4(x, L),$$

where the initial value of the recursion is $G(x) = x$, and $\lambda = 3$. One finds the fixed point $x_* \approx 0.755$ ($\mu \approx 1.324$) and hence, using (10.12), $\nu_{\text{SAW}} \approx 0.891$.

The analysis of SAWs in the Sierpinski gasket is a bit more complicated. With endpoints at A and B, one must distinguish between SAWs that pass through the third vertex C and those which do not (Fig. 10.6). Inspection yields the recursion relations

$$G'_1 = G_1^2 + G_1^3 + 2G_1 G_2 + G_2^2 + 2G_1^2 G_2,$$
$$G'_2 = G_1^2 G_2 + 2G_1 G_2^2,$$

with initial values $G_1 = x$, $G_2 = x^2$, and $\lambda = 2$. This flows to the fixed point $G_{1*} = x_* = (\sqrt{5} - 1)/2$, $G_{2*} = 0$, and $\nu_{\text{SAW}} = \ln 2/\ln(2G_{1*} + 3G_{1*}^2) \approx 0.798$.

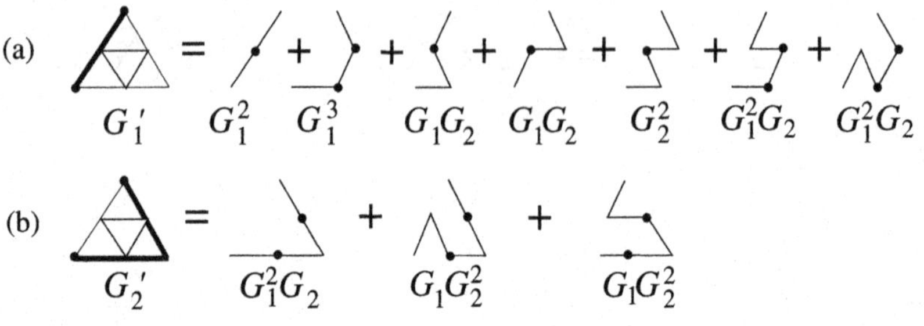

(a)

$$G_1' \qquad G_1^2 \qquad G_1^3 \qquad G_1 G_2 \qquad G_1 G_2 \qquad G_2^2 \qquad G_1^2 G_2 \qquad G_1^2 G_2$$

(b)

$$G_2' \qquad G_1^2 G_2 \qquad G_1 G_2^2 \qquad G_1 G_2^2$$

Fig. 10.6. Renormalization of SAWs in the Sierpinski gasket. Two kinds of SAWs need to be considered: those that go from one vertex to another without touching the third vertex (a), and those that do pass through the third vertex (b).

SAWs in the Sierpinski gasket in three dimensions were analyzed in a similar way, yielding $\nu_{SAW} = 0.729$ (Rammal *et al.*, 1984b).

Rammal *et al.* (1984b) also tried to generalize the Flory formula for SAWs in fractals, suggesting

$$\nu_{SAW} = \frac{1}{d_f^{BB}} \frac{3 d_s^{BB}}{d_s^{BB} + 2}, \tag{10.13}$$

in such a way that it depends only on properties of the backbone. This formula seems to underestimate the true value of ν_{SAW}. For example, for the Sierpinski gasket in $d = 2$ and 3 dimensions it predicts $\nu_{SAW} = 0.768$ and 0.654, respectively.

An alternative derivation yields a different formula, as follows. Restricting our attention to the backbone of the fractal, and working in chemical space, the probability density of random walks is $P(\ell, t) \sim \exp[-(\ell/t^{1/d_w^\ell})^{\delta_\ell}]$ (Eq. (6.32)). The Flory free energy is therefore, barring unimportant prefactors;

$$F \sim \frac{N^2}{\ell^{d_\ell}} + \left(\frac{\ell}{N^{1/d_w^\ell}}\right)^{\delta_\ell}.$$

Here the fractal dimension in chemical space, d_ℓ, and the dimension of random walks, d_w^ℓ, refer to the *backbone* of the fractal in question. Minimizing this free energy, we find

$$\nu_{SAW} = \nu_\ell \frac{2 + \delta_\ell/d_w^\ell}{\delta_\ell + d_\ell}, \tag{10.14}$$

where we have transformed back to Euclidean space, using $r \sim \ell^{\nu_\ell} \sim N^{\nu_{SAW}}$. This formula requires the value of δ_ℓ, which is known only from numerical studies. On the other hand, when ν_{SAW} is known, the formula can be inverted to obtain estimates of δ_ℓ! For the Sierpinski gasket in two and three dimensions one obtains $\delta_\ell = 1.955$ and $\delta_\ell = 1.584$, respectively. In order to resolve the question of the accuracy of the Flory approximation, better estimates of δ_ℓ (in the case of finitely ramified deterministic fractals) and also of d_w^ℓ, d_ℓ, and ν_ℓ (for random fractals) would be needed.

The question of SAWs in percolation clusters has attracted considerable interest. Kremer (1981) finds numerically that, for $p > p_c$, ν_{SAW} is the same as that in regular Euclidean space. At $p = p_c$ the value of ν_{SAW} changes, but Harris (1983) claims that, when averages include *all* clusters, the exponent is the same as that in regular space. On the other hand, Derrida (1982) has argued that ν_{SAW} is affected even by the presence of weak disorder, and that its value would be different than that in normal Euclidean space at all p. There is now a growing consensus that it is only at the percolation threshold that the critical exponents differ from the case of a regular lattice (Nakanishi, 1994; Barat and Chakrabarti, 1995). Indeed, in $d = 2$, for example, the results of $\nu_{SAW} = 0.77 \pm 0.02$ (Roman *et al.*, 1995), $\nu_{SAW} = 0.77 \pm 0.01$ (Woo and Lee, 1991; Vanderzande and Komoda, 1992) $\nu_{SAW} = 0.775 \pm 0.005$ (Rintoul *et al.*, 1994), $\nu_{SAW} = 0.78 \pm 0.01$ (Nakanishi and Moon, 1992), $\nu_{SAW} = 0.783 \pm 0.003$ (Grassberger, 1993), and $\nu_{SAW} = 0.786 \pm 0.010$ (Roman *et al.*, 1998), are all slightly higher than $\nu_{SAW} = \frac{3}{4}$ of SAWs in regular two-dimensional space.

10.6 Exercises

1. Simulate tracer diffusion in a 1000-site ring, for various particle densities, and test Eq. (10.2).
2. Obtain 100-step SAWs (in the square lattice, $d = 2$) in two different ways: (a) by direct simulation of walks in the equilibrium ensemble, and (b) by the enrichment method, with $s = 20$. Analyze their fractal dimensions. Compute μ from your simulations in method (a).
3. Consider the ensemble of growing SAWs (GSAWs). The walks are generated randomly, and each time that an attempt to extend the walk by one step results in an intercept, one simply tries again. Simulate 100-step GSAWs and show by comparison that they differ from SAWs in the equilibrium ensemble. Note that GSAWs correspond to the limit of $s = 1$ and $p \to \infty$ of the enrichment method. Note also that GSAWs in $d = 2$ have a finite probability of terminating, when all the sites adjacent to the growing tip have already been visited.
4. Obtain the result $\langle r_N^2 \rangle \sim N$, for a regular, noninteracting random walk, by minimizing the appropriate free-energy expression.
5. Compute d_f and d_s for the branching Koch curve and compare the exact ν_{SAW} in the text with the Flory prediction, Eq. (10.13). Compare this also with the Flory formula, Eq. (10.14), assuming that $\delta_\ell = d_w/(d_w - 1)$. Repeat this for the Sierpinski gasket.
6. Show that Eqs. (10.13) and (10.14) reduce to the original Flory formula,

Eq. (10.11), when the fractal substrate is replaced by regular space. Show also that both equations predict a crossover to mean-field behavior when the fracton dimension of the substrate is $d_s = 4$. (Hint: in the mean-field limit SAWs behave like random walks; $\nu_{SAW} = 1/d_w$.)

7. Under what condition do Eqs. (10.13) and (10.14) agree? (Answer: $\delta_\ell = \nu_\ell d_w = d_w^\ell$.)

8. Equation (10.14) uses chemical-space exponents, rather than regular-space exponents. Show that in this way one gets the exact result for percolation in $d = 6$.

10.7 Open challenges

1. Exact results have been derived for tracer diffusion in one dimension, but tracer diffusion in fractals has been studied only numerically. It should be possible to obtain exact results, analogous to the ones in one dimension, at least for tracer diffusion in finitely ramified fractals.

2. Tracer diffusion in fractals in the limit of high particle concentrations is fairly well understood, particularly in the long-time asymptotic limit, but additional work is required to shed light on the remaining regimes of low and intermediate concentrations, and short and intermediate times.

3. Concerning tracer diffusion in percolation clusters above the percolation threshold, $p > p_c$, there exists the intriguing possibility of a crossover between "anomalous" tracer diffusion – the type observed in fractals – and "regular" tracer diffusion – the type observed in regular space – depending on the concentration of walkers, c. Such an effect would parallel the transition between one-dimensional behavior and fractal behavior of tracer diffusion in DLAs, which was discovered by Amitrano *et al.* (1985).

4. Tracer diffusion under *bias* has been considered in one dimension (van Beijeren and Kutner, 1988), and also in random networks (Ramaswamy and Barma, 1987; Barma and Ramaswamy, 1994). The problem is relevant to the anomalous kinetics of diffusion-limited reactions of particles with hard-core interactions (Janowsky, 1995a; 1995b; Ispolatov *et al.*, 1995; ben-Avraham *et al.*, 1995; Lee 1997), and needs further elucidation.

5. Bunde *et al.* (1985; 1986c) have studied a model of tracer diffusion on a line, involving two kinds of particles and randomly placed sinks. Only one type of particles can be trapped, while the other type is unaffected by the traps. This serves as a model for slow release of a drug in a porous matrix; however, the study has never been generalized to the more relevant case of fractal substrates. Such a generalization would be important to the pharmaceutical industry.

6. When is it advantageous to use the Flory approximation of Eq. (10.13) for SAWs in fractals, and when that of Eq. (10.14)? Can one suggest yet a better approximation? Kumar *et al.* (1990a) have studied this question in connection with a class of exactly solvable finitely ramified fractals.

7. SAWs in three-dimensional percolation at criticality are of great interest, since they may help clarify the question of whether ν_{SAW} is larger than that of SAWs in regular space.

8. Attracting random walks, in which the walker is attracted to previously visited sites, have been studied neither on fractals nor for percolation. In regular space it seems that there might be a critical level of attraction above which there is a collapse (Ordemann *et al.*, 2000).

10.8 Further reading

- Early interest in tracer diffusion arose in the field of biology, as a model for diffusion of proteins through very narrow pores in membranes. That constraint was known as "single-filing conditions" (Harris, 1960; Lea, 1963; Heckmann, 1972). Later on, the problem was taken over by mathematicians, and solid-state physicists, who regarded it as a model for conduction in superionic conductors. See Harris (1965), Spitzer (1970), Arratia (1983), Levitt (1973), Sankey and Fedders (1977), Richards (1977), Fedders (1978), Kehr and Binder (1987), van Beijeren *et al.* (1983), and van Beijeren and Kehr (1986). More recent simulations have been performed by Pandey (1992). Single-file diffusion has been recently observed in zeolites (Hahn *et al.*, 1996).

- SAWs as models for polymers, and applications: Flory (1949), Fisher (1966), Domb (1969), de Gennes (1979), Dullien (1979), Andrews (1986), Doi and Edwards (1986), des Cloizeaux and Jannink (1990), and Baumgärtner (1995). Eizenberg and Klafter (1996) present a useful summary of numerical and theoretical results for ν_{SAW} in three dimensions.

- SAWs in fractals, percolation and porous media: Klein and Seitz (1984), Kim and Kahng (1985), Chakrabarti and Kertesz (1981), Roy and Chakrabarti (1987), Lam and Zhang (1984), Kumar *et al.* (1990a; 1990b), Hattori *et al.* (1993), Nakanishi (1994), Milosevic and Zivic (1991; 1993), Narasimhan (1996), Lee and Woo (1995), Lee (1996), Hovi and Aharony (1997b), Perondi *et al.* (1997), Roman *et al.* (1998), and Zivic *et al.* (1998). Directed SAWs were studied too, both in deterministic fractals (Reis and Riera, 1995) and in percolation clusters (Dudek, 1995).

- A recent application of tracer diffusion in percolation clusters to the study of the anomalous conductivity of aliovalently doped fluorite-related oxides has been presented by Meyer *et al.* (1997).

Part three

Diffusion-limited reactions

Diffusion-limited reaction processes are those for which the transport time (the typical time until reactants meet) is much larger than the reaction time (the typical time until reactants react, when they are constrained to be within their reaction-range distance). The transport properties of the reactants largely determine the kinetics of diffusion-limited reactions. One then naturally wonders how the (often anomalous) diffusion of particles discussed so far may affect such processes. This, and the need to account for the effects of fluctuations in the concentration of the reactants at all length scales, as well as other sources of fluctuations, make the study of diffusion-limited reactions notoriously difficult. The topic is discussed in the next four chapters.

In Chapter 11, we begin with the far simpler case of reaction-limited processes. In their case the system may be assumed to be homogeneous at all times: the transport mechanism and fluctuations play no significant role. The kinetics of reaction-limited processes is well understood, since they may be successfully analyzed by means of classical rate equations. We also touch upon the important subject of reaction–diffusion equations, but only at the mean-field level, without the addition of noise terms.

Chapter 12 discusses the Smoluchowski model for binary reactions, and trapping. It is instructive to see how diffusion-limited processes depart from their reaction-limited counterpart, even for such elementary reaction schemes.

In the absence of a systematic, comprehensive approach to the study of diffusion-limited processes, the preferred strategy has been to focus on the simplest conceivable processes, hoping that their understanding would shed light on more complex situations. In Chapter 13 we review some of the models which have attracted the most attention. The processes were chosen so as to highlight the effect of the kinetics of transport, and those of various sources of fluctuations.

The evolution of domains of reactants is ubiquitous in reaction–diffusion systems involving more than one species. In such cases, reactions take place only at the boundary between domains, and the kinetics in these "reaction fronts" is a fundamental problem. The diffusion-limited reaction $A + B \rightarrow C$, with initially separated reactants (A and B), treated in Chapter 14 is a prototype model of such situations.

11

Classical models of reactions

When a reaction takes place the reacting entities, or at least some of them, are fundamentally altered into "products". In this sense, reactions differ from the purely elastic interactions of Chapter 10. Reaction kinetics are strongly influenced by the transport properties of the reactants, and in several cases anomalous diffusion plays a central role. We shall elucidate this problem in the following chapters. Here we describe the different kinetic regimes of reaction processes. In the reaction-controlled limit, reactions may be understood at the level of classical rate equations. We take advantage of the simplicity of this limit to introduce the concept of kinetic phase transitions. Reaction–diffusion equations improve upon the simple-minded approach of classical rate equations, and they are discussed briefly.

11.1 The limiting behavior of reaction processes

Imagine a sealed reactor tank in which particles, possibly of different species, undergo a reaction process. Assume that the particles are within an inert, undisturbed fluid medium, and hence that their motion is Brownian, due to the thermal bombardment of the medium's molecules. We are interested in the reaction kinetics, i.e., in the time dependences of the concentrations of reactants and products, $c(t)$. Here we have in mind *global* concentrations, that is,

$$c(t) = \frac{N(t)}{V},$$
(11.1)

where $N(t)$ is the total number of particles (of a particular species) at time t, and V is the reactor's volume. More detailed questions about the *local* distribution of particles could be asked, but let us postpone those until later.

The reaction kinetics would then be governed by two characteristic time scales: the *diffusion time* and the *reaction time* (Fig. 11.1). The diffusion time, τ_{diff},

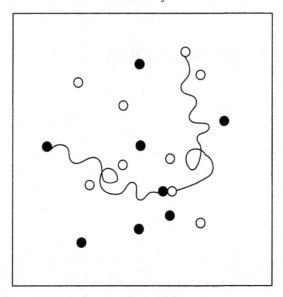

Fig. 11.1. Characteristic time scales. The diffusion time is the typical time that it takes for reactants (● and ○) to encounter each other through the diffusion process (shown in the figure). The reaction time is the time necessary for a reaction between reactants that are held in proximity to each other, within reaction range, to take place (not depicted).

is the typical time that it takes for two arbitrary particles in the reactor tank to meet each other. The reaction time, τ_{reac}, is the typical time for two particles to react when they are held in mutual proximity; that is, within the reaction-range distance.

When $\tau_{\text{diff}} \ll \tau_{\text{reac}}$ the process is slowed down by the relatively large reaction time and is thus known as a *reaction-limited* or *reaction-controlled* process. In this limit particles may come within reaction range of each other numerous times before they actually react. In other words, a particle may sample large volumes of the reactor tank before reacting and hence it effectively responds to the *global* concentration of the other particles. The problem is then relatively simple and may be approached by means of classical rate equations.

In the opposite limit, when $\tau_{\text{diff}} \gg \tau_{\text{reac}}$, one speaks of *diffusion-limited* or *diffusion-controlled* processes. In this case, particles most likely react upon their first encounter with a neighbor and hence the kinetics is largely influenced by *local* fluctuations in the concentration of the reactants. One can no longer rely on classical rate equations, which employ global concentrations, and a different approach is necessary. Moreover, the kinetics is then sensitive to the peculiarities of the diffusion process, which may be anomalous if the medium is fractal or of low dimensionality.

Stirring is a very efficient method of shortening the characteristic diffusion time and smoothing out local fluctuations in concentration. Both effects tend to speed up the overall reaction process, and stirring is therefore widely used in the chemical industry and in other practical applications. The result is that many reactions of interest may be well understood at the level of classical rate equations. However, many other physical systems can only be described as diffusion-limited processes. Examples include electron–hole recombination in semiconductors, fusion of molecular excitations in doped polymers, reaction of electronic excitations in doped matrices, reactions taking place in viscous solvents, heterogeneous catalysis (when the surface diffusion is slow compared with the reaction step), and spin dynamics in the Ising model in the limit of $T \to 0$.

Our goal is to discuss reaction kinetics in general, rather than limiting ourselves to a specific class of processes such as chemical reactions. We shall therefore attempt to use as general a terminology as possible – speaking of "species" rather than reactants – but we shall occasionally mention "molecules", "particles", etc., merely to ease our task.

11.2 Classical rate equations

A generic reaction may involve n different species $X_1, X_2, \ldots, X_n$, which convert to each other according to a reaction scheme;

$$\nu_1 X_1 + \nu_2 X_2 + \cdots + \nu_n X_n \longrightarrow \mu_1 X_1 + \mu_2 X_2 + \cdots + \mu_n X_n. \qquad (11.2)$$

For example, the chemical reaction $C_2H_6 + H_2 \to 2CH_4$ involves the three species $X_1 = C_2H_6$, $X_2 = H_2$, and $X_3 = CH_4$. The ν's and μ's of Eq. (11.2) are known as the *stoichiometric coefficients* representing the process in question. This defines them only up to a common factor, but the ambiguity may be removed by letting them equal the actual number of particles involved in individual reaction events. Thus, in our example the stoichiometric coefficients are $\nu_1 = \nu_2 = 1$, $\nu_3 = 0$, and $\mu_1 = \mu_2 = 0$ and $\mu_3 = 2$.

In the reaction-limited case, the rate of process (11.2), i.e., the number of reaction events per unit volume per unit time, is

$$k_f \prod_{j=1}^{n} c_j^{\nu_j}, \qquad (11.3)$$

where c_j is the (global) concentration of the jth species. The product expresses the probability that particles participating in a reaction event are within reaction range of each other. The *reaction constant*, k_f, denotes the rate of reactions, given that the required particles are in the desired configuration. Essentially, $k_f \sim 1/\tau_{\text{reac}}$.

Of course, some conditions must be satisfied in order to justify the rate (11.3). Chiefly, the system must be homogeneous so that the number of particles at any given point is well represented by the global concentrations. This is achieved in the limit $\tau_{\text{diff}} \ll \tau_{\text{reac}}$, for then a particle samples large volumes of space before reacting. Indeed, in writing (11.3) one assumes that $\tau_{\text{diff}} = 0$, neglecting the diffusion time altogether. Another requirement may be that the temperature remain stationary and constant throughout the reactor, so that k_f is truly constant – a condition that can be reasonably approached in most practical situations. We shall assume that this and other necessary conditions are satisfied, and that (11.3) holds as long as the process is reaction-limited.

The change in concentration of the ith species is then given by the *classical rate equation*:

$$\left[\frac{dc_i(t)}{dt}\right]_{\text{forward}} = (\mu_i - \nu_i)k_f \prod_{j=1}^{n} c_j^{\nu_j}. \tag{11.4a}$$

In many situations the reverse reaction of (11.2), $\sum_j \mu_j X_j \to \sum_j \nu_j X_j$, may also take place. The change in $c_i(t)$ due to the reverse process is

$$\left[\frac{dc_i(t)}{dt}\right]_{\text{reverse}} = (\nu_i - \mu_i)k_r \prod_{j=1}^{n} c_j^{\mu_j}, \tag{11.4b}$$

assuming that it too is reaction-limited. The total change in c_i is simply given by the sum of these two processes:

$$\frac{dc_i(t)}{dt} = (\mu_i - \nu_i)k_f \prod_{j=1}^{n} c_j^{\nu_j} - (\nu_i - \mu_i)k_r \prod_{j=1}^{n} c_j^{\mu_j}. \tag{11.4c}$$

With both processes (forward and reverse) taking place, the system will generally reach an equilibrium state that can be characterized by setting the LHS of Eq. (11.4c) to zero:

$$\frac{k_f}{k_r} = \prod_{j=1}^{n} c_j^{\mu_j - \nu_j}. \tag{11.5}$$

This is known as the *law of mass action*. Since in thermodynamic equilibrium the concentrations $\{c_j\}$ can be determined independently, from the system's partition function, the law of mass action predicts a nontrivial relation between the forward and the reverse reaction constants.

11.3 Kinetic phase transitions

As an example of applications of classical rate equations, consider the system

$$A + X \underset{k_{-1}}{\overset{k_1}{\rightleftharpoons}} A + 2X, \qquad X \xrightarrow{k_2} B, \tag{11.6}$$

where the subscripts and superscripts indicate the rate constants associated with the different processes. Our interest is in the species X. A can be thought of as an agent fostering the self-catalysis of X, since it is unaffected by the reaction process, and its concentration c_A remains constant. Likewise, we assume that the product B is an inert species of no consequence to the process. The system is known as *Schlögl's first model* (Schlögl, 1972).

Assuming that the process (11.6) is reaction-limited, the rate equation satisfied by c – the concentration of species X – is

$$\frac{dc(t)}{dt} = (k_1 c_A - k_2)c - k_{-1} c_A c^2.$$

Rescaling time by $k_{-1}c_A$ and defining $(k_1 c_A - k_2)/k_{-1} c_A \equiv r$, this is recast into the more succinct form

$$\frac{dc(t)}{dt} = rc - c^2. \tag{11.7}$$

The rate r may be tuned to different values (positive and negative), for example by varying c_A.

The solution of Eq. (11.7) is

$$c(t) = \frac{r c_0 e^{rt}}{r + c_0(e^{rt} - 1)}, \tag{11.8}$$

where $c_0 = c(t = 0)$ is the initial concentration of species X. Taking the limit $t \to \infty$, we see that, depending on the value of r, the system evolves into one of two stationary states:

$$c(t) \underset{t \to \infty}{\to} c_s = \begin{cases} 0 & r < 0, \\ r & r > 0. \end{cases} \tag{11.9}$$

The state $c_s = 0$ describes a situation in which there are no more X particles and therefore all of the reactions (11.6) stop. The other state is an active steady state in which reactions take place but the concentration of X remains constant (and finite). A different way to obtain the stationary behavior is by plotting dc/dt versus c. One can then determine the stable and unstable states of the model by inspection (Fig. 11.2).

The situation described above is an example of a *dynamical*, or *kinetic*, phase transition. If one characterizes the state of the system by its long-time asymptotic behavior (truly, a kinetic property), one sees that the system undergoes a sharp

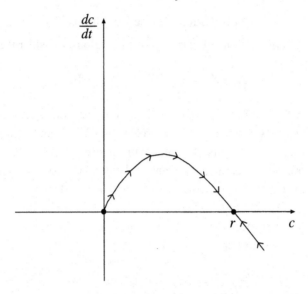

Fig. 11.2. Stability analysis of Eq. (11.7). The evolution of c may be inferred from the sign of dc/dt (arrows). In this way one can see that the steady-state solution of $c_s = r$ is stable, whereas that of $c_s = 0$ is unstable.

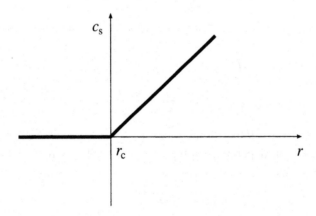

Fig. 11.3. Dynamical phase transition. The steady-state concentration c_s undergoes a sharp transition (from zero to finite) as r crosses the critical value $r_c = 0$.

transition as r increases from negative to positive values (Fig. 11.3). Thus, c_s plays the same role as that of an order parameter in equilibrium phase transitions: it is analogous to the magnetization in a ferromagnet. Likewise, r is the critical field, akin to temperature in ferromagnetic transitions. One may then define critical exponents, similar to those of equilibrium phase transitions. For example, the order-parameter critical exponent, β, is given by

$$c_s(r) - c_s(r_c) \sim (r - r_c)^\beta, \tag{11.10}$$

where $r_c = 0$ is the critical value of r at the transition point. The foregoing analysis predicts that, for Schlögl's first model, $\beta = 1$ (Eq. (11.9)).

An interesting question concerns the extent to which our results are affected by the assumption that the process is reaction-limited. For equilibrium phase transitions we know that the transition is ultimately dominated by fluctuations in the system. It is then reasonable that the nature of dynamical phase transitions would be dramatically altered when the underlying processes are diffusion-limited, and sensitive to fluctuations. This is indeed the case: for example, the critical exponents assume different values than the ones derived from classical rate equations, and they depend on the dimensionality of space!

11.4 Reaction–diffusion equations

Diffusion-limited conditions affect the kinetics of all dynamic processes, to some extent, even those which do not undergo kinetic phase transitions. In this case, the need to account for fluctuations makes the problem a lot more complex and interesting. We shall devote the remainder of this book to it.

One obvious way to account (partially) for fluctuations is by replacing the global concentrations $c(t)$ by local density functions. Thus, let $\rho_i(x, t)$ represent the *local* density of species X_i at x at time t. Imagine that we subdivide space into cells of volume v. The local densities are simply $\rho_i(x) = \langle n_i \rangle / v$, where n_i is the number of X_i particles in the cell centered around x. The number of particles in a cell changes due to reactions inside the cell, and due to diffusion of particles between cells. The diffusion time *within* a cell depends on the cell's size, so the cells can be made small enough that reactions inside them could be treated at the level of classical rate equations. In that case the concentrations satisfy the *reaction–diffusion* equation:

$$\frac{\partial \rho_i(x, t)}{\partial t} = D_i \nabla^2 \rho_i(x, t) + \mathcal{R}[\{\rho_j(x, t)\}]. \tag{11.11}$$

Here $\mathcal{R}$ is a functional of $\{\rho_j\}$ of the same form as in the corresponding classical rate equation. It accounts for reactions inside the cells. The remaining term describes diffusion of particles between cells, allowing for the possibility that diffusion of different species is characterized by different diffusion coefficients.

As an example consider the single-species coalescence process

$$A + A \rightarrow A. \tag{11.12}$$

This surprisingly simple reaction scheme is a reasonable model for such complex systems as coagulation of droplets in aerosols (Friedlander, 1977) and exciton fusion in optical pumps (Digonnet *et al.*, 1994). Its reaction–diffusion equation would be

$$\frac{\partial \rho_A(x, t)}{\partial t} = D\nabla^2 \rho_A(x, t) - \kappa \rho_A(x, t)^2. \tag{11.13}$$

Notice the similarity of the reaction term to the classical rate equation $dc_A/dt = -kc_A^2$. In fact, we can obtain the classical rate equation by simply ignoring fluctuations and substituting $\rho_A(x, t) = c_A(t)$ in (11.13). For this reason, the limit of classical rate equations is also referred to as "mean-field".

A minor difficulty with reaction–diffusion equations is the determination of the effective diffusion coefficients and reaction constants. One expects, however, that the main features of the system's kinetics are independent of their precise values. A more serious difficulty stems from the discrete nature of the reactants. The density ρ represents the average number of particles in a cell (over many different realizations of the process), $\langle n \rangle$, but it neglects fluctuations from this average, $\langle n^2 \rangle - \langle n \rangle^2$. If n is large these fluctuations *are* negligible. Thus, the cells ought to be small enough to justify classical rates, but large enough to obliterate fluctuations in the number of particles. This elusive compromise cannot always be achieved. In fact, Eq. (11.13) fails to predict the kinetics of the diffusion-limited coalescence process precisely for this reason.

Because of the inability of reaction–diffusion equations to deal with discrete fluctuations, they are viewed as a mean-field approach. A very exciting possibility is that of adding stochastic noise terms. Depending on the nature of the noise, reaction–diffusion equations reproduce different types of anomalous behavior and kinetic phase transitions. The study of reaction–diffusion equations with noise has blossomed in recent years (see Marro and Dickman (1999), and references therein).

11.5 Exercises

1. Justify the rate (11.3), on probabilistic grounds.
2. Derive a classical rate equation for the reaction-limited one-species coalescence process, $A + A \rightarrow A$, when there is a steady homogeneous input of A particles at rate R per unit volume per unit time. Work out the kinetics and the long-time asymptotic behavior.
3. *Conservation laws.* Consider the reaction $A + B \rightarrow C$, in the reaction-limited regime. Write down the classical rate equations for this system and show that $c_A(t) - c_B(t)$ and $c_A(t) + c_C(t)$ are constant in time. Use this to derive the kinetics of A. Such conservation laws are generally effective at reducing the complexity of dynamic processes involving multiple species. (Answer: $c_A(t) = (\Delta c_0 e^{k\Delta t})/(\Delta - c_0 - c_0 e^{k\Delta t})$, where $\Delta = c_A(0) - c_B(0)$ and $c_0 = c_A(0)$.)

4. Show that, in a normal bi-molecular reaction, an increase in concentration of either reactant results in a proportional increase of the reaction rate, whereas for catalysis an increase in the concentration of the catalyst increases the reaction rate up to a point, and then the rate saturates. A. Brown and V. Henri used this effect to prove the catalytic nature of the action of invertase – an enzyme that cleaves sucrose into glucose and fructose (De Turck *et al.*, 1994). (Hint: represent catalysis as a two-stage process: (i) $X + C \overset{k_1}{\rightarrow} C^*$, and (ii) $C^* \overset{k_2}{\rightarrow} C + Y$. C is the catalyst, which converts X into the (inert) product(s) Y. C^* is an intermediate state of X bound to the catalyst C.)

5. *Schlögl's second model.* Consider the system

$$A + 2X \underset{k_{-1}}{\overset{k_1}{\rightleftharpoons}} A + 3X, \qquad X \underset{k_{-2}}{\overset{k_2}{\rightleftharpoons}} B,$$

where A and B are in excess, so their concentrations may be deemed constant. Analyze the steady state and show that there is a *discontinuous* jump in the concentration of X when $r \equiv k_2 - k_{-2}c_B$ is increased beyond the critical value $r_c = -k_1^2 c_A / 4k_{-1}$.

6. As a simple-minded diffusion-limited realization of Schlögl's first model, simulate the *contact process* on a one-dimensional lattice: sites are either empty or occupied by a particle X. Occupied sites become empty at rate 1, and particles give birth to an additional particle at an adjacent site (to their right or left, with equal probability) at rate η. Birth into an already occupied site leaves that site unchanged. Conduct enough simulations to convince yourself that there exists a critical rate $\eta = \eta_c$ such that particles become extinct if $\eta < \eta_c$, but there is an active steady state if $\eta > \eta_c$. Plot the stationary concentration of particles as a function of η and compare it with Fig. 11.3.

7. *Discrete fluctuations.* To understand fluctuations in the number of particles, consider production of particles at constant rate k; $dc/dt = k$. The master equation for the probability of having n particles at time t, $p_n(t)$, is

$$\frac{\partial p_n(t)}{\partial t} = kp_{n-1} - kp_n, \qquad n = 1, 2, \ldots$$

and $\dot{p}_0 = -kp_0$. Show that, for an initially empty system, $p_n(t = 0) = \delta_{n,0}$, the solution is

$$p_n(t) = \frac{(kt)^n}{n!} e^{-kt}.$$

Thus, the average number of particles is $\langle n \rangle = kt$, and the fluctuations are $\langle n^2 \rangle - \langle n \rangle^2 = kt$. This demonstrates that the relative fluctuations are small only when $\langle n \rangle$ is large. Notice that $\langle n \rangle$ satisfies the same classical rate equation as c does. (Prove it!)

11.6 Open challenges

There exists a complete theory for the kinetics of reaction-limited processes (i.e., classical rate equations) and in this sense there remain no open problems. However, stirring – the chief mechanism for reaching this classical regime – is notoriously hard to model and remains poorly understood. An interesting approach using Lévy flights is reviewed by Zumofen *et al.* (1997). See also Reigada *et al.* (1997).

11.7 Further reading

- Classical reaction kinetics: Benson (1960) and Laidler (1965).
- General texts on diffusion-limited reactions: Oppenheim *et al.* (1977), van Kampen (1981), Rice (1985), and Ligget (1985). See also references in the following chapters.
- Kinetic phase transitions: a comprehensive text is that of Marro and Dickman (1999).

12

Trapping

The Smoluchowski model and the trapping problem are among the simplest models for diffusion-limited reaction processes. They illustrate the important roles played by the specific dynamics of diffusion (regular or anomalous) and by fluctuations in the concentration of particles in a clear, intuitive way.

12.1 Smoluchowski's model and the trapping problem

Smoluchowski's model (von Smoluchowski, 1917) consists of a sphere of radius a centered about the origin and surrounded by a sea of Brownian particles. The sphere acts as an ideal trap: any particle that hits the sphere is immediately absorbed and removed from the system. The total flux of particles into the trap serves as a simple-minded model for the rate of diffusion-limited binary reactions. The radius of the trap may be regarded as the reaction range, and the kinetics is completely determined by the diffusion process, since no time scale is associated with the immediate trapping events.

In the most basic version of the model the trap is static and the density of particles is taken to be initially uniform, $\rho(r, t = 0) = \rho_0$. The density evolves according to

$$\frac{\partial}{\partial t}\rho(r, t) = D\nabla^2\rho, \qquad r > a,$$
$$\rho(r, t)|_{r=a} = 0.$$

(12.1)

The flux into the trap is then $\gamma_d a^{d-1} D(d\rho/dr)_{r=a}$, where γ_d is the surface area of the d-dimensional unit sphere.

A more realistic model of binary reactions would deal with a finite concentration of traps, instead of with a single trap; the traps could be mobile, possibly diffusing with a different diffusion constant than that of the particles; and trapping need not necessarily be immediate. With each additional generalization the model becomes more realistic, though less tractable.

Consider first the case of a finite concentration of static, perfect traps. Assume that we have a lattice whose sites may be occupied by traps, with probability c. Walkers perform regular diffusion on the trap-free sites and get removed from the system upon hitting a trap. Since the density of free sites is $1 - c$, the *survival probability* – the probability that a walker is still in the system after time t – is

$$P_s(t) = \langle (1 - c)^{S(t)} \rangle, \tag{12.2}$$

where $S(t)$ is the number of distinct sites visited by a random walker during the time period t (Appendix B). The survival probability determines the particle concentration, $\rho(t) = \rho_0 P_s(t)$.

Equation (12.2) is exact, but not entirely useful, since in most cases we do not know the full distribution of $S(t)$. A more practical form may be derived from the expansion

$$P_s(t) = \langle e^{-\alpha S(t)} \rangle = \sum_{n=0}^{\infty} \frac{(-\alpha)^n}{n!} \langle S(t)^n \rangle, \tag{12.3}$$

where $\alpha = -\ln(1 - c)$. By keeping just the first-order term, one obtains the Rosenstock approximation (Rosenstock, 1969):

$$P_s(t) \approx (1 - c)^{\langle S(t) \rangle}. \tag{12.4}$$

Blumen *et al.* (1983), have studied the effect of terms of higher order, Eq. (12.3), extensively. The analysis is complicated by the fact that we have full knowledge of the behavior of only the first moment, $\langle S(t) \rangle$. The survival probability predicted by the Rosenstock approximation is well suited for low concentrations of traps in the early-time regime, and is generally smaller than the exact result.

12.2 Long-time survival probabilities

To obtain the long-time asymptotic result, consider first the problem of trapping in one dimension. If the traps are perfect, the problem decomposes into a collection of disconnected intervals. The particles diffuse freely within each interval, and get trapped at its edges. The density of particles within an interval of length L is given by

$$\frac{\partial}{\partial t} \rho(x, t) = D \frac{\partial^2}{\partial x^2} \rho, \qquad 0 \le x \le L,$$
$$\rho(0, t) = \rho(L, t) = 0; \qquad \rho(x, 0) = \rho_0. \tag{12.5}$$

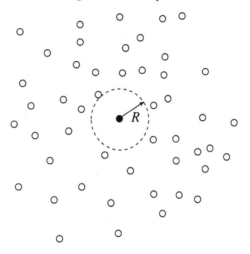

Fig. 12.1. A walker (•) surrounded by traps (○). The traps may effectively be replaced by an absorbing circular boundary of radius R – the distance from the walker to the nearest trap. This boundary places a more severe constraint on the walker than do the traps, and its survival probability would then be smaller.

Using the method of separation of variables, we find the long-time asymptotic behavior

$$\rho \sim \frac{4\rho_0}{\pi} e^{-\pi^2 Dt/L^2} \sin\left(\frac{\pi x}{L}\right), \tag{12.6}$$

corresponding to an average density within the interval of $\rho_L = (1/L)\int_0^L \rho(x)\,dx = (8\rho_0/\pi^2)\exp(-\pi^2 Dt/L^2)$. However, if the traps are distributed randomly with uniform density c, the prevalence of intervals of length L is ce^{-cL}. Therefore, the overall density of particles is

$$\rho = \int_0^\infty c^2 L e^{-cL} \rho_L\,dL \sim \exp\left(-\tfrac{3}{2}(2\pi^2 c^2 Dt)^{1/3}\right), \qquad t \to \infty. \tag{12.7}$$

The integral is evaluated by a saddle-point approximation. Remarkably, the decay of the concentration is slower than purely exponential. This results from the unusually long survival times of particles in very large but *very rare* trap-free intervals. It is crucial for the derivation that the upper integration boundary is infinity. For systems with a finite length the result is modified (Exercise 4).

Similar considerations apply in higher dimensions. Assume that a walker is surrounded by a perfect absorbing spherical boundary of radius R, equal to the walker's distance to the nearest trap (Fig. 12.1). The survival probability of the walker within this artificial boundary serves as a lower bound to the actual

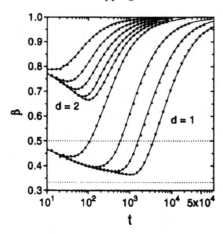

Fig. 12.2. A plot of β, the local exponent of t in Eq. (12.9), in finite systems with N traps, in two and three dimensions. For infinite systems and $t \to \infty$ it is expected that $\beta = d/(d+2)$. The horizontal dashed lines represent these asymptotic limits (for $d = 2$ and 3). After Bunde *et al.* (1997).

probability of surviving (without the boundary). The analogue of Eq. (12.5) is now

$$\frac{\partial}{\partial t}\rho(r, t) = D\frac{1}{r^{d-1}}\frac{\partial}{\partial r}\left(r^{d-1}\frac{\partial\rho}{\partial r}\right), \qquad 0 \leq r \leq R,$$

$$\rho(R, t) = 0; \qquad \rho(r, 0) = \rho_0.$$

(12.8)

Thus, at long times $\rho_R \sim \exp(-\eta_d Dt/R^2)$, where η_d is a dimensionless number related to the lowest eigenvalue of (12.8). If the traps are randomly distributed with density c, the prevalence of trap-free regions of radius R is proportional to $\exp[-(\gamma_d/d)R^d]$. Therefore, the survival probability is bounded by

$$P_s \geq \int_0^\infty \exp\left(-\eta_d\frac{Dt}{R^2} - \frac{\gamma_d}{d}R^d\right) dR \sim \exp[-a_d(c^{2/d}Dt)^{d/(d+2)}], \qquad t \to \infty,$$

(12.9)

where we have again used a saddle-point approximation, and a_d is a dimensionless constant. This bound excludes the possibility of a pure exponential decay and therefore the kinetics of trapping cannot be captured by a classical reaction–diffusion equation, even in arbitrarily large dimensions. Using more rigorous arguments, Donsker and Varadhan (1979) have shown that the actual long-time behavior of P_s takes the same form as that of the bound in (12.9).

Finite-size effects due to the limited size of the system (which for fixed volume can be represented by the total number of traps, N) have been studied by Bunde *et al.* (1997). They find that Eq. (12.9) is valid for $t \ll t_\times \approx [\gamma_d/(2\eta_d D)](d \ln N/\gamma_d)^{(d+2)/d}$, whereas for $t \gg t_\times$ the decay is dominated by the finite size of the system and is purely exponential (see Fig. 12.2). At the crossover time $t_{\times}$, the concentration of particles is $\rho(t) \sim \rho(0)N^{-d/2}$. Because of the logarithmic dependence of $t_{\times}$ upon N, such crossover behavior should be apparent even in macroscopic systems, and it should be taken into account in all numerical studies.

The Donsker–Varadhan behavior of Eq. (12.9) generalizes to the case in which particles and traps are confined to a fractal object, but with d replaced by the medium's fracton dimension, d_s (Exercise 6). Thus, from the trapping problem we learn how the kinetics of diffusion-limited reaction processes is influenced by two key ingredients: (a) spatial fluctuations in concentration, and (b) the physics of transport. The role of spatial fluctuations is illustrated by the importance of large trap-free regions, and the effect of anomalous transport physics is seen in the dependence of the survival probability upon d_s.

12.3 The distance to the nearest surviving particle

Since spatial fluctuations are important, it makes sense to be asking more detailed questions about the distribution of particles, rather than only about their global concentration. A commonly studied problem is that of the distance from the nearest surviving particle to a trap. A naive guess might place the nearest particle at distance $\sim\sqrt{Dt}$ away from the trap, but we will see that this is not usually the case.

Suppose that we have a single, static trap, centered about the origin. The density of surviving particles is given by the solution to Eq. (12.1). The probability that the infinitesimal volume dv is empty of particles is $1 - \rho(r, t)\,dv$, and therefore the probability that there are no particles up to a distance r from the origin is

$$P(r, t) = \exp\left(-\int_a^r \rho(u, t)\,\gamma_d u^{d-1}\,du\right). \tag{12.10}$$

The distribution of distances from the nearest particle to the trap is then $p(r, t) = -\partial P(r, t)/\partial r$, or

$$p(r, t) = \gamma_d r^{d-1} \rho(r, t) \exp\left(-\int_a^r \rho(u, t)\,\gamma_d u^{d-1}\,du\right). \tag{12.11}$$

The average minimal distance can then be computed from this distribution, $\langle r_{\min}\rangle = \int p(r, t)r\,dr$. However, the following simpler argument yields the correct scaling as well. The average number of particles up to a distance r is given by $\int_a^r \rho\,dr$, and $r_{\min}$ may then be estimated by demanding that there be only one particle (the nearest particle) within a distance r_{min}:

$$\int_a^{r_{min}} \rho(u, t) \, du = 1. \tag{12.12}$$

As an example, let us discuss the nearest-particle distance in one dimension. Since the radius of the trap is irrelevant in this case, we take $a = 0$. Equation (12.1) reduces to

$$\frac{\partial}{\partial t} \rho(x, t) = D \frac{\partial^2}{\partial x^2} \rho, \qquad x \geq 0,$$
$$\rho(0, t) = 0; \qquad \rho(x, 0) = \rho_0. \tag{12.13}$$

The solution is $\rho(x, t) = \rho_0 \, \mathrm{erf}[x/(2\sqrt{Dt})]$. The long-time behavior is found from the series expansion of the error function,

$$\rho(x, t) \sim \frac{\rho_0 x}{\sqrt{\pi Dt}} e^{-x^2/(4Dt)}, \qquad x \ll \sqrt{Dt}. \tag{12.14}$$

Using (12.12), we find

$$x_{min} \sim \left(\frac{4\pi Dt}{\rho_0^2} \right)^{1/4}. \tag{12.15}$$

Thus, in one dimension the distance to the nearest particle scales only as $t^{1/4}$. Finally, Eqs. (12.11) and (12.14) yield the distribution

$$p(x, t) = \frac{\rho_0 x}{\sqrt{\pi Dt}} \exp\left(-\frac{\rho_0 x^2}{2\sqrt{\pi Dt}} \right). \tag{12.16}$$

An easier way to deal with Eq. (12.1) is by using a quasistatic approximation. The idea is to solve a static (Laplace) equation in the "active" region $a \leq r \leq r_b$, where trapping has a strong effect on the concentration profile, and assume that beyond that region the concentration remains unchanged, $\rho(r \geq r_b) = \rho_0$. The boundary between the two regions scales as $r_b = \sqrt{Dt}$, as expected from the diffusion of the particles. The time dependence is effectively taken into account through this moving boundary. For example, in one dimension the solution to $0 = D \, \partial^2 \rho / \partial x^2$, with the moving boundary condition, is

$$\rho(x, t) = \begin{cases} \rho_0 x / \sqrt{Dt} & x < \sqrt{Dt}, \\ \rho_0 & x > \sqrt{Dt}. \end{cases} \tag{12.17}$$

Using this form of $\rho(x, t)$ for $x < x_b = \sqrt{Dt}$ in (12.11) and (12.12) leads to the correct functional behaviors of $p(x, t)$ and x_{min}.

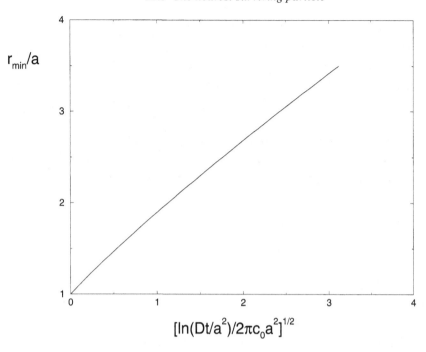

Fig. 12.3. The numerical solution of Eq. (12.20). As $t \to \infty$, $r_{\min} \sim \sqrt{\ln(Dt/a^2)/(2\pi\rho_0)}$.

The quasistatic approximation is particularly useful when the solution to the diffusion equation is complicated. A good example is the Smoluchowski problem in two dimensions. The equation to be solved is

$$\frac{\partial}{\partial t}\rho(r, t) = D\frac{1}{r}\frac{\partial}{\partial r}\left(r\frac{\partial\rho}{\partial r}\right), \qquad r \geq a,$$

$$\rho(a, t) = 0; \qquad \rho(r, 0) = \rho_0.$$

(12.18)

The quasistatic approximation yields

$$\rho(r, t) = \frac{\rho_0}{\ln(\sqrt{Dt}/a)}\ln\left(\frac{r}{a}\right), \qquad a \leq r \leq \sqrt{Dt}. \qquad (12.19)$$

To compute $r_{\min}$, we use Eq. (12.12); $\int_a^{r_{\min}} 2\pi r\rho(r, t)\, dr = 1$. Then, $z_{\min} \equiv r_{\min}/a$ is the solution to the transcendental equation

$$z^2 \ln z - \frac{1}{2}(z^2 - 1) = \frac{1}{2\pi\rho_0 a^2}\ln(Dt/a^2). \qquad (12.20)$$

The numerical solution is plotted in Fig. 12.3. For $t \to \infty$, $r_{\min} \sim \sqrt{\ln(Dt/a^2)/(2\pi\rho_0)}$. In three dimensions, the limit $\rho(r, t \to \infty)$ exists and therefore $r_{\min}$ tends to a constant.

12.4 Mobile traps

The next generalization of the trapping problem is when the traps are no longer static, but instead may diffuse with a diffusion constant D_T. This difficult problem has generated much confusion and contradictory answers had been suggested. Bramson and Lebowitz (1988) have proved rigorously that the asymptotic decay is $\ln \rho \sim -S(t)$.

The problem is indeed far from trivial. To see that, consider the Smoluchowski model with a *single* trap, when the trap is mobile. For simplicity, let us focus first on the one-dimensional case. The concentration of traps satisfies the equation

$$\frac{\partial}{\partial t}\rho(x,t) = D\frac{\partial^2}{\partial x^2}\rho, \qquad x > x'(t),$$

$$\rho[x'(t), t] = 0; \qquad \rho(x, 0) = \rho_0,$$

(12.21)

where $x'(t)$ denotes the position of the trap: it moves as a random walk, with diffusion coefficient D_T. Thus, even in this ostensibly simple case we are led to a diffusion equation with a complicated stochastic boundary condition.

D and D_T are the only physical parameters determining the trapping kinetics. We have seen that, when $D_T = 0$ (the trap is static), the distance from the nearest particle to the trap scales as $(x - x')_{\min} \sim t^{1/4}$. The opposite limit, of a mobile trap $(D_T > 0)$ but static particles $(D = 0)$, can also be solved analytically. The trap sweeps across the line, cleaning it of particles. At time t the region swept by the trap is of the order of $\sqrt{D_T t}$. Thus, even without the exact detailed solution, it is obvious that in this case $(x - x')_{\min} \sim t^{1/2}$. When both the trap and the particles are mobile, we do not know the exact solution. One expects then that $(x - x')_{\min} \sim t^\beta$, where β is a function of $\Delta \equiv D_T/D$. Computer-simulations indeed support this hunch.

In higher dimensions the problem is difficult to solve, even in the limit of static particles. Simulation results for static particles in two dimensions show that $r_{\min} \sim \ln t$. This is in contrast to $r_{\min} \sim (\ln t)^{1/2}$, when the trap is static. In three dimensions $r_{\min} \sim$ constant for all values of Δ, presumably because the trail swept by the trap is not compact.

12.5 Imperfect traps

In the Smoluchowski model with a single static perfect trap, the rate of influx of particles into the trap is initially infinite (in all dimensions). One way to remove this singularity is to consider less than ideal traps. Simply, assume that trapping events are not immediate, but occur at some finite rate proportional to κ. The boundary condition of Eq. (12.1) is then replaced by

$$\frac{\partial}{\partial r}\rho(r,t)|_{r=a} = \kappa\rho(a,t), \tag{12.22}$$

the ideal trap corresponding to the limit $\kappa \to \infty$.

In one dimension, for example, the solution subject to the *radiating boundary condition* of Eq. (12.22) is

$$\rho = \rho_0 \left[\text{erf}\left(\frac{x}{2\sqrt{Dt}}\right) + e^{\kappa^2 Dt + \kappa x}\, \text{erfc}\left(\kappa\sqrt{Dt} + \frac{x}{2\sqrt{Dt}}\right) \right]. \tag{12.23}$$

In the limit of $\kappa \to \infty$ one recovers the solution of the ideal trap, $\rho = \rho_0 \,\text{erf}[x/(2\sqrt{Dt})]$. The initial rate of trapping is now finite: $\kappa\rho_0 e^{\kappa^2 Dt}\, \text{erfc}(\kappa\sqrt{Dt})$. In the long-time asymptotic limit, $t \to \infty$, the trapping rate becomes $\rho_0/\sqrt{\pi Dt}$, the same as for an ideal trap and independent of κ! What happens is that, owing to the low concentration of particles in the vicinity of the trap, the particle nearest to the trap may interact with it several times before other particles get involved, and the trapping efficiency is effectively perfect. The net effect is the same as if reactions were immediate. Similar arguments apply for higher dimensions.

Other kinds of imperfect traps may be considered, with more spectacular results. For example, Weiss and Havlin (1985) have studied non-Markovian traps on the line, characterized by a set of probabilities $\{f_j\}$ of absorbing a given particle upon its jth encounter with the trap. For f_j with an associated finite first moment they reproduce the Donsker–Varadhan asymptotic behavior, but for $f_j \sim 1/j^{1+\alpha}$, where $0 < \alpha < 1$, the survival probability drops off as $1/t^\alpha$.

12.6 Exercises

1. Obtain a classical rate equation for the concentration of particles in the reaction-limited trapping process (with a small trapping rate: $\kappa \to 0$). Use it to show that the concentration then decays in a purely exponential fashion.
2. Compare the early-time behavior of the Rosenstock approximation with the Donsker–Varadhan long-time behavior in one, two, and three dimensions.
3. Verify the saddle-point approximation in Eq. (12.7) and obtain the temporal dependence of the prefactor. (Answer: $\sim t^{1/6}$.)
4. Apply the saddle-point approximation (12.7) to a system of finite length N. Show first that there is a typical maximum length $L_{\max}$ of the trap-free regions that substitutes the upper integration boundary in (12.7) and that increases logarithmically with N.
5. Compute the constant a_d in Eq. (12.9). (Answer:

$$a_d = \frac{d+2}{2d}(2\beta_d^2\gamma_d^{2/d})^{d/(d+2)},$$

where β_d is the first zero of the Bessel function of order $d/2 - 1$, and γ_d is the surface area of the d-dimensional unit sphere.)

6. Show that the Donsker–Varadhan result of Eq. (12.9) generalizes to fractals, with d replaced by the fracton dimension d_s. Analyze also the short-time behavior. (Hint: a particle surrounded by an absorbing spherical boundary of radius R typically gets trapped after time $\tau \sim R^{d_w}$, and its survival probability is $\rho_R \sim \exp(-t/\tau)$.)

7. Write a computer program to simulate the concentration of particles for a single trap, by the exact enumeration method. Use it to investigate the time dependence of r_{min} in one- and two-dimensional lattices, and in the Sierpinski gasket.

8. Use the quasistatic approximation to compute $p(r, t)$, the distribution of distances to the nearest particle, in three dimensions. Show that $r_{min} \to$ constant as $t \to \infty$. (Answer:

$$p(r, t) = 4\pi r^2 \rho_0 \left(1 + \frac{a}{\sqrt{Dt}}\right) \exp\left[-4\pi r^2 \rho_0 \left(1 + \frac{a}{\sqrt{Dt}}\right) \left(\frac{z^3 - 1}{3} - \frac{z^2 - 1}{2}\right)\right],$$

where $z = r/a$.)

12.7 Open challenges

1. What is the crossover time between the early-time regime and the Donsker–Varadhan long-time behavior? Fixman (1984) and Havlin *et al.* (1984d) have designed effective algorithms to investigate this issue. Numerical findings indicate that the Donsker–Varadhan asymptotic behavior sets in very late; when the survival probability has dropped to $\lesssim 10^{-13}$, in two and three dimensions. An analytical derivation of the crossover time is still missing. Note, however that a second crossover time, $t_\times$, from the Donsker–Varadhan behavior to an exponential decay is discussed shortly in Section 12.2. For more details see: Bunde *et al.* (1997) and Phillips *et al.* (1998).

2. For mobile traps and static particles the decay is purely exponential. This case is known as the scavenger problem, and was studied by Blumen *et al.* (1986b). When the particles are also mobile, one expects at early times the Donsker–Varadhan decay, provided that this behavior has already set in. This issue is still rather unclear, in spite of the availability of rigorous results obtained by Bramson and Lebowitz (1988). See also Redner and Kang (1984), and Berezhkovskii *et al.* (1989a).

3. The distance between a single *mobile* trap and the nearest (mobile) particle remains unknown. Even in one dimension there seems to be no way to reduce the many-body problem. An approximate expression based on the analogy with

a three-body problem has been suggested by Schoonover *et al.* (1991). Their conclusions are supported by more recent work of Koza and Taitelbaum (1998). Some exact results have been obtained by Koza *et al.* (1998), and extended by Sánchez (1999). The early-time regime has been studied by Sánchez *et al.* (1998).

4. Little is known about the minimal distance of particles to a single static trap in fractals.

5. A particle surrounded by traps must exhibit anomalous diffusion in order to survive to long times, since the span of the surviving walks is limited by the traps. Simple scaling arguments, similar to the ones used to derive the Donsker–Varadhan behavior, suggest that $d_w = d + 2$ (Weiss and Havlin, 1984). There exist no numerical simulations to confirm this behavior. It would also be of interest to study other statistics of long surviving walks (see, e.g., Kozak (1997)).

6. Trapping in the presence of a bias field offers rich possibilities. Makhnovskii *et al.* (1998a) suggest that bias may help in singling out many-body effects in the trapping model. The problem calls for further theoretical and numerical studies. See also Condat *et al.* (1995). The question of trapping with random local bias (as in the Sinai problem) has been considered by Taitelbaum and Weiss (1994).

12.8 Further reading

- Trapping was studied extensively by many authors, see, e.g., Beeler and Delaney (1963), Balagurov and Vaks (1973), Muthukumar and Cukier (1981), Bixon and Zwanzig (1981), Grassberger and Procaccia (1982), Kirkpatrick (1982), Keyser and Hubbard (1983; 1984), Redner and Kang (1983; 1984), den Hollander (1984), Weiss *et al.* (1985), Havlin *et al.* (1984e), Anlauf (1984), Webman (1984a; 1984b), Klafter *et al.* (1984), Zumofen *et al.* (1984), Shlesinger and Montroll (1984), Blumen *et al.* (1986b), Agmon and Glasser (1986), Glasser and Agmon (1987), ben-Avraham and Weiss (1989), Redner and ben-Avraham (1990), Weiss (1993), Condat *et al.* (1995), and Phillips (1996).

- Trapping by nonideal traps and other kinds of traps: Weiss and Havlin (1985), Ohtsuki (1985), Taitelbaum *et al.* (1990), Taitelbaum (1991), Ben-Naim *et al.* (1993), and Koza and Taitelbaum (1998).

- Several variations of the trapping problem are drawing current interest: trapping in lattices with random barriers (Giacometti and Murthy, 1996); trapping in mixed molecular solids (Parichha and Talapatra, 1996); trapping with correlations in the initial positions of traps and particles (Berezhkovskii and Weiss, 1995; Makhnovskii *et al.*, 1998b); and trapping of ballistic particles (Berezhkovskii *et al.*, 1999).

- Applications of the theory of trapping: trapping of ruthenium complexes on clay with Fe^{3+} traps (van Damme *et al.*, 1986); trapping of excitons in naphthalene (Kopelman, 1987); excitons in aromatic vinyl polymers (Sokolov and Kauffmann, 1998); cosmic rays and synchrotron radiation (Ragot and Kirk, 1997); and optical imaging of biological tissues (Bonner *et al.*, 1987; ben-Avraham *et al.*, 1991; Gandjbakhche *et al.*, 1996).

13

Simple reaction models

The simplest models of diffusion-limited reactions exhibit anomalous kinetics and can be used to illustrate general characteristic principles. Cases of one-species coalescence, $A + A \rightarrow A$, or annihilation, $A + A \rightarrow 0$, demonstrate the effect of diffusion most clearly. We use these models to introduce various analysis techniques, such as scaling and effective rate equations. The two-species reaction $A + B \rightarrow C$ brings to light the important effect of segregation, and emphasizes the role played by fluctuations in the local concentration of reactants. Similarly, the voters model provides us with a striking example of the effect of discrete fluctuations, or fluctuations in number space.

13.1 One-species reactions: scaling and effective rate equations

The *one-species-annihilation* process consists of particles, all belonging to the same species A, diffusing and annihilating upon encounter, according to the reaction scheme

$$A + A \rightarrow 0. \tag{13.1}$$

Reaction is immediate, and all particles diffuse with the same diffusion constant, D. The model is more general than it seems at first sight. For example, it includes the case in which the particles do not necessarily annihilate, but the product of the reaction (13.1) is some inert species of no consequence to the overall kinetics. This makes the model suitable for a variety of chemical reactions, or catalysis on surfaces, when desorption of the product from the surface is rapid. In the closely related *one-species coalescence* process, the reacting particles fuse together to yield a single particle:

$$A + A \rightarrow A. \tag{13.2}$$

179

Again, alternative interpretations of (13.2) are possible (for example, when two particles react one of the particles gets annihilated, or it is converted into an inert species, etc.).

In the reaction-limited regime, these processes may be treated at the level of classical rate equations. The global concentration of A particles, c, obeys the equation

$$\frac{d}{dt}c(t) = -(2 - \mu)kc^2,$$ (13.3)

where $\mu = 0$ for annihilation and $\mu = 1$ for coalescence. The solution is

$$c(t) = \frac{1}{c_0^{-1} + (2 - \mu)kt},$$ (13.4)

where $c_0 \equiv c(0)$ is the initial concentration. In the long-time asymptotic limit the concentration decays as $c(t) = 1/[(2 - \mu)kt]$. Remarkably, this limit is independent of the initial concentration. The similarity between annihilation and coalescence is also emphasized by this result: simply, the concentration in the annihilation process is half that of the concentration in coalescence. Most importantly, our conclusions are completely independent of the dimensionality of space, as is always the case with mean-field classical rate equations.

The situation is quite different in the diffusion-limited regime. The long-time asymptotic behavior may be inferred from the following *scaling argument*. Suppose that the process takes place in a container of dimensionality d. If $d \leq 2$, diffusion is recurrent. That means that a particle with r.m.s. displacement $r(t) \sim (Dt)^{1/2}$ sweeps the volume $S(t) \sim r(t)^d$ compactly. Consequently, after a time t all particles within that volume would have coalesced, or annihilated, leaving behind of the order of one particle. The concentration would therefore be

$$c(t) \sim \frac{\text{number of particles left}}{r(t)^d} \sim \frac{1}{(Dt)^{d/2}}.$$ (13.5)

The rate of decay of concentration now depends on the dimensionality of space. Interestingly, the long-time asymptotic behavior is independent of the initial concentration, as predicted by classical rate equations.

For dimensions $d > 2$, the volume swept by a diffusing particle is no longer compact and the above argument does not hold. Instead, $S(t) \sim t$ and a particle is constantly exploring mostly new territory. The number of particles it may encounter during a small period of time Δt is $c \Delta t$, and therefore the decrease in the total number of particles is $Nc \Delta t$ (N is the total number of particles at time t). Dividing by the volume of the system, V, we obtain the change in the concentration of particles during time Δt: $\Delta c = -\Delta N/V \sim -(N/V)c \Delta t = -c^2 \Delta t$. Therefore, $dc/dt \sim -c^2$, exactly the same as in the classical mean-field case.

In summary, we see that, for the one-species diffusion-limited coalescence or annihilation process, the concentration of particles decays as

$$c(t) \sim \begin{cases} 1/t^{d/2} & d \leq 2, \\ 1/t & d > 2. \end{cases} \tag{13.6}$$

The dependence on dimensionality is due to the physics of diffusion. For $d > d_c = 2$ the mean-field theory of classical rate equations applies, and diffusion ceases to be important. The dimension d_c at which this transition takes place is known as the *upper critical dimension*. The behavior at $d_c = 2$ is not quite like that of the mean-field limit. In our case, from the fact that $c \sim 1/S(t)$ for $d < 2$, we may guess that this is also true at the critical dimension, and therefore for $d = d_c = 2$ one expects $c \sim \ln t/t$. This is confirmed by a more careful analysis. Such logarithmic corrections are typical of systems at the upper critical dimension.

A different approach to dealing with diffusion-limited reactions is the method of *effective rate equations*. The idea is to find a differential equation similar to the classical rate equation for the reaction-limited case, but such that it would predict the anomalous kinetics correctly. One possibility is to allow for a time-dependent rate constant, $k = k(t)$. Thus, for example, using $k(t) \sim 1/t^{1/2}$ in Eq. (13.3) leads to the correct prediction in $d = 1$. In this way, the anomalously slow kinetics is phenomenologically explained by the decrease of the rate constant with time.

The other possibility is to postulate an effective reaction order ν in the equation $dc/dt \sim -c^\nu$. Using $\nu = 3$ reproduces once again the correct behavior in one dimension. This time, the anomalously slow kinetics is explained by the increase in the effective reaction order: the gathering of a larger number of particles to within reaction range of each other is a less likely event.

The effective order can actually be predicted by the following scaling argument. If the dimensionality is $d \leq 2$ so that diffusion is compact, a particle would typically react with one of the particles in its immediate neighborhood. The average distance to neighboring particles is $\ell \sim 1/c^{1/d}$, and the diffusion time required for the reaction is $\tau \sim \ell^2$. Therefore the rate of change in concentration during the time interval $\Delta t = \tau$ is $dc/dt \sim \Delta c/\tau \sim -c/\tau \sim -c^{1+2/d}$. The effective order $\nu = 1 + 2/d$ indeed agrees with the result of Eq. (13.5). Notice that the argument breaks down for $d > 2$, because, when diffusion is no longer compact, a particle would not necessarily react with one of its immediate neighbors.

The methods of scaling and effective rate equations provide us with valuable physical insight. In spite of their phenomenological nature, effective rate equations may sometimes lead to new accurate predictions. For example, consider coalescence or annihilation in the presence of a homogeneous input of particles at rate R (per unit volume per unit time). The change in particle concentration due to input is $(dc/dt)_{\text{input}} = R$. Naively, then, the system could be described by the effective rate equation

$$\frac{d}{dt}c(t) = -kc^{1+2/d} + R. \qquad (13.7)$$

Putting $dc/dt = 0$, this yields the stationary solution $c_s \sim R^{d/(2+d)}$, which happens to be the right answer (see, however, Chapter 16).

On the other hand, the limitations of scaling and effective rate equations are obvious. They may predict the exponent of the power-law decay at long times, but fail to predict the asymptotic value of the decay amplitude. Indeed, notice that, with these approaches, we could not distinguish between coalescence and annihilation. Without doubt, however, the biggest drawback of scaling techniques is that the analysis of new reaction models would often require new ideas, and there is no obvious general recipe that one could follow in every case.

13.2 Two-species annihilation: segregation

At the next level of complexity we may consider two-species annihilation, of the sort

$$A + B \rightarrow 0. \qquad (13.8)$$

Again, instead of annihilation we may imagine that the product is some inert species irrelevant to the kinetics. The appropriate mean-field rate equations, valid for the reaction-limited regime, are

$$\frac{dc_A(t)}{dt} = -kc_Ac_B ; \quad \frac{dc_B(t)}{dt} = -kc_Ac_B. \qquad (13.9)$$

An important observation is that $dc_A/dt - dc_B/dt = 0$, so the difference between the concentrations of the two species is conserved: $c_A(t) - c_B(t) = \text{constant} = c_A(0) - c_B(0)$. With this in mind, it is easy to obtain the concentration of the *minority* species (the smaller of c_A and c_B):

$$c = \frac{c_0\delta_0 e^{-k\delta_0 t}}{(c_0 + \delta_0) - c_0 e^{-k\delta_0 t}}, \qquad (13.10)$$

where c_0 is the initial concentration of the minority species, and $\delta_0 \equiv |c_A(0) - c_B(0)|$. That is, the minority species decays exponentially in time, and the decay rate is proportional to δ_0. In the special case of $c_A(0) = c_B(0)$ the decay is algebraic, $c \sim 1/kt$ (Exercise 6).

The analysis of two-species annihilation in the diffusion-limited regime brings to light the important role of spatial fluctuations in the local concentration of reactants. Imagine that the two species diffuse with equal diffusion constant D, and that initially their concentrations are equal. In a region of space of linear size

r there would be initially of the order of $N_A(0) = c_A(0)r^d \pm \sqrt{c_A(0)r^d}$ A particles (and likewise for B). The second term denotes the typical fluctuation in the number of particles due to their initial random distribution. During time $Dt \sim r^{1/2}$ all the particles within this domain would have reacted with each other. Because the reaction (13.8) conserves the difference between the numbers of particles $N_A - N_B$ (this is the microscopic analogue of the global conservation law for $c_A - c_B$), at the end of this process there would be left in our region of the order of $\sqrt{c_A(0)r^d}$ A particles. (Here we have assumed, without loss of generality, that species A is locally in excess.) Thus, the concentration of A particles at time t is

$$c_A(t) \sim \frac{N_A(t)}{r^d} \sim \frac{\sqrt{c_A(0)r^d}}{r^d} \sim \frac{\sqrt{c_A(0)}}{(Dt)^{d/4}}. \qquad (13.11)$$

The decay is even slower than that for one-species processes, and the amplitude *is* dependent upon the initial concentration of particles.

The above scaling argument suggests also the possibility that, as the reaction proceeds, there evolve alternating domains of A and B particles. This *segregation* effect is indeed observed in simulations (Fig. 13.1). It contributes critically to the slowing down of the process, since reactions can take place only at the boundary between domains.

What is the condition for the stability of a domain? Suppose that we have a domain of A particles, of linear size r. The probability that a B particle annihilates in its attempt to traverse the domain scales like $S_B(t)c_A(t) \sim r^2 r^{-d/2}$ (assuming that $d > 2$ and diffusion is not recurrent). For $d < 4$ this probability vanishes as r grows larger and hence domains are stable, but for $d \geq 4$ there is a finite probability of a particle crossing a domain of a different species without reacting and the system gets effectively mixed. In other words, $d_c = 4$ is the upper critical dimension above which the mean-field approach is valid, and the scaling argument (13.11) applies only for $d \leq 4$. Interestingly, in this case anomalously slow decay takes place even in the physical case of $d = 3$ dimensions.

Segregation is an example of pattern formation, which arises in several reaction–diffusion systems: complex patterns and correlations may evolve at several length scales. Consider the distance between nearest particles in the two-species annihilation model in one dimension (Fig. 13.2). The size of a domain grows like $r \sim t^{1/2}$, and the concentration of particles within it is $c_A \sim t^{-1/4}$. Hence, the number of particles in a domain increases as $N_A \sim rc_A \sim t^{1/4}$, and the typical distance between adjacent particles is

$$\ell_{AA} \sim r/N_A \sim t^{1/4}. \qquad (13.12)$$

For ℓ_{AB}, the distance between the particles at the edge of adjacent domains, we must construct a more careful argument. It can be inferred from the effective rate

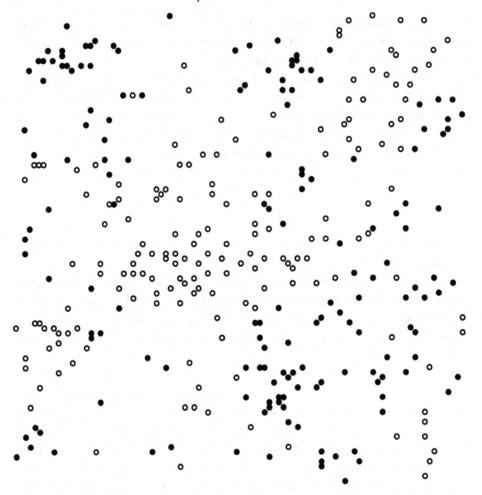

Fig. 13.1. Segregation of species in the A + B → 0 process. Shown is the state of a 100×100 lattice that started off with all the sites filled with either A (•) or B (○) particles, with equal probability and randomly, after 10^6 time steps. The segregation into A-rich and B-rich domains is clearly visible.

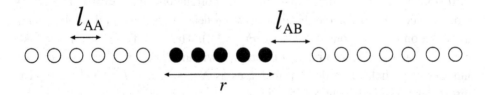

Fig. 13.2. Microscopic length scales in the A + B → 0 process. The domain size r, the gap between domains ℓ_{AB}, and the average distance between particles within a domain ℓ_{AA} are shown in this schematic representation of the two-species annihilation process in one dimension.

equation $dc/dt \sim -(N_{AB}/V)/\tau$. N_{AB} is the number of adjacent A–B pairs, or simply the number of boundaries between domains, where reactions take place; V is the "volume" of the system (its length); and $\tau \sim \ell_{AB}^2/D$ is the typical diffusion time required for the reaction of an A–B pair of particles. Since $N_{AB} \sim V/r \sim t^{-1/2}$, we find $dc/dt \sim -t^{-5/4} \sim -t^{-1/2}/\ell_{AB}^2$, or

$$\ell_{AB} \sim t^{3/8}. \tag{13.13}$$

So, we have identified at least three length scales (r, ℓ_{AA}, and ℓ_{AB}), which grow at different rates.

Finally, we would like to stress the role of fluctuations in the initial distribution of particles in the two-species annihilation process. Imagine a system in which the A and B particles are first generated through the mechanism $C \rightarrow A + B$. Suppose that we begin with only C particles and let them decompose into A–B pairs. Let the products diffuse for some finite period of time τ_0, and only then turn on the annihilation reaction (13.8). A system prepared in this way is said to be *geminate*. Its initial condition is obviously highly correlated: each A particle has a matching B particle within a radius determined by the initial diffusion process, $\ell_0 \sim \sqrt{D\tau_0}$. Therefore, in a region of size $r(t) \gg \ell_0$ the number of A particles is *almost equal* to that of B particles, and following annihilation there would remain in that region only of the order of one particle. Thus $c(t) \sim 1/t^{d/2}$, instead of the result $c \sim 1/t^{d/4}$ found with random initial conditions.

13.3 Discrete fluctuations

We have witnessed the impacts of diffusion and of spatial fluctuations in concentration on the kinetics of diffusion-limited processes. The *voters model* emphasizes yet another source of anomalous kinetics: fluctuations in number space, or, for short, *discrete fluctuations* (van Kampen, 1981; ben-Avraham, 1987). The voters model is similar to the two-species annihilation process of the previous section, but with a different reaction product:

$$A + B \rightarrow \begin{cases} 2A & \text{with probability } \frac{1}{2}, \\ 2B & \text{with probability } \frac{1}{2}. \end{cases} \tag{13.14}$$

One can imagine A and B representing Democrats and Republicans, respectively. When a Democrat and a Republican meet an "interaction" takes place, whereby one may convince the other to shift party allegiance. This would yield a naive model of the political map just before election time.

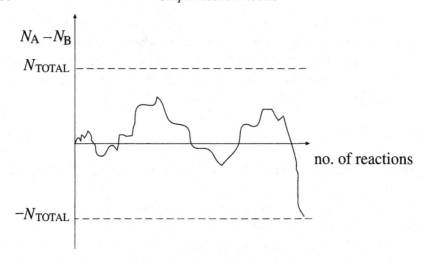

Fig. 13.3. Discrete fluctuations in the voters model. The gap between members of different species, $N_A - N_B$, evolves as a random walk, with the reaction number acting as a time variable. The process stops when all voters have converted to the same species (B, in our example).

Since the two possible processes in (13.14) occur with equal probability, we get the mean-field rate equation

$$\frac{dc_A(t)}{dt} = \frac{1}{2}kc_Ac_B - \frac{1}{2}kc_Ac_B = 0, \tag{13.15}$$

and likewise for dc_B/dt. In other words, $c_A(t)$ and $c_B(t)$ remain constant – nothing happens. (This is more a reflection on the model's simplicity than on the outcome of political activity!)

However, viewed from a different perspective, consider the gap in the *number* of voters, $\Delta(t) \equiv N_A(t) - N_B(t)$. Each time an interaction takes place $\Delta(t)$ changes by ± 2. Thus, $\Delta(t)$ performs an unbiased random walk and the average gap grows as $\langle \Delta(t)^2 \rangle \sim$ number of interactions. If the pool of voters is finite, sooner or later the process will end, when everybody belongs to the same party (Fig. 13.3). The striking difference from the prediction of (13.15) is due to the fluctuations in the number of particles, which we have now taken into account. Notice, however, that there is no contradiction: $\langle \Delta(t) \rangle = \Delta(0)$, just like in (13.15), and, when the pool of voters is infinite, the process never ends.

In some cases, however, discrete fluctuations critically alter the kinetics. As an example, consider the two-species *stochastic* annihilation model:

$$A + B \rightarrow \begin{cases} A & \text{with probability } \frac{1}{2}, \\ B & \text{with probability } \frac{1}{2}. \end{cases} \tag{13.16}$$

The net effect of these reactions is to annihilate equal numbers of A's and B's, so one expects similar kinetics to two-species annihilation. The difference is that, in the latter process, the gap $\Delta = N_A - N_B$ is conserved in each reaction event, whereas in the process (13.16) Δ performs a random walk, as in the case of the voters model.

Consider a geminate system. In a region of size $r \gg \ell_0$ one has initially the same number, $c_0 r^d$, of A and B particles. After a time $t \sim r^2$, when all reactions in the domain have taken place, the stochastic annihilation would have generated a surplus of the order of $\sqrt{c_0 r^d}$ particles of one species (because of the random walk performed by Δ), and the concentration then decays as $c(t) \sim 1/t^{d/4}$. Thus, although we have started with a geminate system, which would normally evolve without segregation, discrete fluctuations give rise to segregation and slow the kinetics drastically.

13.4 Other models

There is obviously an infinity of possible reaction models. For reaction-limited processes this presents no particular problem, because the method of classical rate equations may be successfully employed to tackle any existing process. In the case of diffusion-limited reactions, we lack such a powerful analytical tool and study of each system requires a separate effort. For this reason one focuses on the most basic, abstract reaction schemes, hoping to learn general truths from these simple examples. To be sure, a large number of reaction–diffusion systems have been studied so far, and a comprehensive review is beyond the scope of this book. In the following, we merely identify some general trends of research.

Reaction mechanisms. We have considered only the extreme cases of $\tau_{\mathrm{diff}} \ll \tau_{\mathrm{reac}}$ (reaction-limited) and $\tau_{\mathrm{diff}} \gg \tau_{\mathrm{reac}}$ (diffusion-limited). The intermediate case in which $\tau_{\mathrm{diff}} \approx \tau_{\mathrm{reac}}$ is also of interest. For example, in lattice models a collision of two particles could result in reaction with a probability smaller than unity, or a predetermined number of collisions may be required before reaction takes place.

Transport mechanisms. We have considered only the simplest instances of diffusion. Some generalizations include allowing different species to diffuse with different diffusion constants, diffusion with drift, particles performing CTRWs, or Lévy flights, particles moving ballistically ($r \sim t$), or even *static* particles reacting at a distance. Each of these options generates new behavior.

Reaction schemes. The possibilities here are truly endless, so research is limited to only very basic reaction schemes. We list a few examples in the following paragraphs.

A generalization of the one-species processes of Section 13.1 is the many-body process

$$nA \rightarrow mA. \tag{13.17}$$

Annihilation corresponds to $n = 2$ and $m = 0$, and coalescence corresponds to $n = 2$ and $m = 1$. An interesting effect is that an increase in n lowers the upper critical dimension.

The coalescence process may be generalized to

$$A_k + A_l \rightarrow A_{k+l}. \tag{13.18}$$

Here A_k may represent a particle of mass km_0, a particle of charge ke, or a polymer of k monomers, say, and the reactants coalesce to yield a product of larger mass or charge, or a longer polymer. Allowing for diffusion coefficients that depend on the size of the particles, this becomes a very powerful model with an enormous range of applications. If the diffusion constant is the same for all k, and if we focus on the total number of particles, regardless of k, the process reduces to one-species coalescence. If, on the other hand, we count only particles with *odd* values of the index k, the process degenerates to the one-species annihilation model.

The n-species annihilation model concerns particles belonging to n different species, $A_1, A_2, \ldots, A_n$. The particles diffuse and annihilate upon colliding, provided that they belong to different species:

$$A_i + A_j \rightarrow 0, \qquad i \neq j. \tag{13.19}$$

When $n = 2$, we recover the two-species annihilation model of Section 13.2. As $n \rightarrow \infty$, the distinction between species becomes irrelevant, because most encounters are bound to be between particles of different species. In this case the process is identical to one-species annihilation. Thus, upon varying n, the model interpolates between the different kinetics of two-species and one-species annihilation.

As a last example we mention the model of branching annihilating walks (BAWs). It is similar to one-species annihilation in that, when two walkers meet, they annihilate, but in addition walkers may give birth to n offspring:

$$A + A \rightarrow 0; \qquad A \rightarrow (1+n)A. \tag{13.20}$$

The system exhibits a kinetic phase transition, similar to that of Schlögl's first model, but with critical exponents different than mean-field ones.

13.5 Exercises

1. Generalize the scaling argument for one-species annihilation or coalescence, leading to Eq. (13.6), to reactions in fractal media. Show that d is simply replaced by the fracton dimension d_s. (Hint: see Meakin and Stanley (1984).)

2. Write a computer program for the simulation of $A + A \rightarrow A$ with input in one dimension. Measure the steady-state concentration and confirm that it depends on the input rate as $c_s \sim R^{1/3}$.

3. Obtain the steady-state concentration for coalescence or annihilation with input in dimension $d \leq 2$ from dimensional analysis. (Hint: argue that the only physical parameters determining c_s would be D and R, and recall that the units of R are $[R] = 1/(\text{volume})(\text{time})$.)

4. For the two-species annihilation model, show that $\ell_{AA} \sim t^{1/4}$ for $d \leq 4$, and $\ell_{AA} \sim t^{1/d}$ for $d > 4$. Find the scaling of ℓ_{AB} with time in two dimensions. (Hint: assume that the domains have a smooth boundary, and that near the boundary $\ell_{AA} \sim \ell_{AB}$.) (Answer: $\ell_{AB} \sim t^{1/3}$.)

5. Show that, in the two-species annihilation model, the concentration difference $c_A(t) - c_B(t)$ obeys a diffusion equation. Toussaint and Wilczek used this argument as a basis for a more rigorous derivation of the anomalous decay of Eq. (13.6).

6. Show that the classical-rate-equation prediction for the two-species annihilation model, when the initial concentrations are equal, is $c(t) = c_0/(1 + kc_0t)$, where $c_0 = c_A(0) = c_B(0)$. Notice that, in this case, the mean-field approach fails to reproduce the right asymptotic dependence on the initial concentration of Eq. (13.11).

7. Generalize the scaling argument for two-species annihilation, leading to Eq. (13.11), to reactions on fractals. Show that d is replaced by the fracton dimension d_s. (Hint: see Meakin and Stanley (1984).)

13.6 Open challenges

Given the lack of a general approach for the treatment of diffusion-limited reactions, virtually every new reaction creates an outstanding challenge. The field is wide open to research. Few systems have been analyzed exactly (an example is given in Part IV of this book), but many more, including as basic a model as two-species annihilation in one dimension, are still awaiting exact solution. In other cases we are missing even a heuristic scaling argument to explain the anomalous kinetics observed in numerical simulations.

A recent example is that of two-species annihilation with drift. Janowsky (1995a; 1995b) has found that, in one dimension, when the two species are subject to a background drift, the decay is $c \sim t^{-x}$, with $x \approx \frac{1}{3}$, and slower than $c \sim t^{-1/4}$

without drift (see also Ispolatov *et al.* (1995)). ben-Avraham *et al.* (1995) have generalized this model as follows. On a one-dimensional lattice each site can be in one of the three states: empty (0), occupied by a single A particle, or occupied by a single B particle. Particles may hop only to the nearest site to their right:

$$\cdots A0 \cdots \rightarrow \cdots 0A \cdots, \qquad \cdots B0 \cdots \rightarrow \cdots 0B \cdots,$$

with equal rates for A and B (extreme drift conditions). Hopping is disallowed if the target site is occupied by a particle of the same species. If the target site is occupied by a particle of the opposite species, hopping is allowed and reaction takes place, with an outcome determined by the probabilistic rules

$$\cdots AB \cdots \rightarrow \begin{cases} \cdots 0A \cdots & \text{probability } p, \\ \cdots 0B \cdots & \text{probability } 1 - p, \end{cases}$$

$$\cdots BA \cdots \rightarrow \begin{cases} \cdots 0B \cdots & \text{probability } p, \\ \cdots 0A \cdots & \text{probability } 1 - p. \end{cases}$$

Thus, p may be thought of as a parameter that represents the "persistence" of the hopping particle in determining the output. Notice, however, that, on average, equal amounts of A and B particles get annihilated, regardless of p. Numerical results show that the decay exponent x varies continuously with p, ranging from $x = \frac{1}{4}$ for $p = 0$ to $x = \frac{1}{2}$ for $p = 1$. Janowsky's model corresponds to $p = 0$. So far, there is no theoretical argument to explain $x(p)$.

13.7 Further reading

- For general texts and reviews on diffusion-limited reactions, see van Kampen (1981), Haken (1978; 1983), Nicolis and Prigogine (1980), Liggett (1985), Oppenheim *et al.* (1977), Rice (1985), Kang and Redner (1984), Kuzovkov and Kotomin (1988), Berezhkovskii *et al.* (1989b), Privman (1997), and Mattis and Glasser (1998). See also the conference proceedings of "Models of Non-Classical Reaction Rates" held at the National Institutes of Health, March 25–27, 1991, in honor of the 60th birthday of G. H. Weiss (*J. Stat. Phys.* **65**, nos. 5/6, 1991).

- One-species diffusion-limited reactions: Torney and McConnell (1983), Peliti (1985), Rácz (1985), Rácz and Plischke (1987), Lushnikov (1987), Spouge (1988), Kanter (1990), ben-Avraham *et al.* (1990; 1994), ben-Avraham (1993), Lin (1991), Krapivsky (1993), Argyrakis and Kopelman (1990), Clément *et al.* (1994), Alemany *et al.* (1994), Menyhárd (1994), Privman (1992; 1994a; 1994b), Privman *et al.* (1995; 1996), Schütz (1995), Simon (1995), Hinrichsen (1996), Hinrichsen *et al.* (1995; 1996a; 1996b), Wehefritz *et al.* (1995), Henkel *et al.* (1995; 1997), Balboni *et al.* (1995), Derrida and Zeitak (1996), and Rey and Droz (1997)

- Two-species diffusion-limited reactions: see, for example, Zel'dovich and Ovchinnikov (1977), Bramson and Griffeath (1980), Keizer (1985; 1987), Anacker and Kopelman (1987), Bramson and Lebowitz (1988), ben-Avraham and Doering (1988), Toussaint and Wilczek (1983), Schnörer *et al.* (1989), Kuzovkov and Kotomin (1992; 1993), Lee and Cardy (1994; 1997), Oshanin *et al.* (1996), Oerding (1996), Redner (1997), and Rant (1997).

- Diffusion-limited reactions in fractals and random media: Meakin and Stanley (1984), Clement *et al.* (1990; 1991), Zumofen *et al.* (1991a; 1991b; 1991c), Lindenberg *et al.* (1991), Argyrakis and Kopelman (1992), Lianos and Duportail (1992), Mártin and Braunstein (1993), Hoyuelos and Mártin (1993a; 1994; 1995; 1996), Dickinson *et al.* (1993), Gonzalez *et al.* (1995), Allen and Seebauer (1996), Schütz (1997), and Oshanin and Blumen (1998). Diffusion-reaction kinetics in porous fractal systems modeling real catalysts: Elias-Kohav *et al.* (1991), Gutfraind and Sheintuch (1992), Giona (1992), Coppens and Froment (1995), Andrade *et al.* (1997b), and Rigby and Gladden (1998).

- One-species reaction models find applications in the study of colloids and aerosols (Friedlander, 1977), star formation (Field and Saslow, 1965), vapor-deposited thin films (Family and Meakin, 1989), exciton reactions (Kopelman, 1987; Kopelman *et al.*, 1988), and bacterial growth (Nelson and Shnerb, 1998). Two-species annihilation models have been applied to the question of residual matter in the primordial universe (Toussaint and Wilczek, 1983), and to the interpretation of experiments on electron–hole recombination (Schiff, 1995). There are also numerous experiments with initially separated reactants (see Chapter 14).

- There exists a vast literature on other models of reactions (Section 13.4). Examples include nA $\rightarrow$ mA (Privman and Grynberg, 1992; ben-Avraham, 1993; ben-Avraham and Zhong, 1993; Krapivsky, 1994; Lee, 1994), $A_k + A_l \rightarrow A_{k+l}$ (Spouge, 1988), n-species annihilation (ben-Avraham and Redner, 1986), and BAWs (Sudbury, 1990; Takayasu and Tretyakov, 1992; Jensen, 1993; 1994; ben-Avraham *et al.*, 1994).

14

Reaction–diffusion fronts

The dynamics of diffusion-controlled reactions of the type $A + B \rightarrow C$ has been studied extensively since the pioneering work of von Smoluchowski (1917). Most studies have focused on homogeneous systems, i.e., systems in which both reactants are initially uniformly mixed, and interesting theoretical and numerical results have been obtained, as discussed in Section 13.2. However, experimental confirmation of these beautiful theories has proved elusive, partly because of difficulties in implementing the initially uniformly mixed distributions of reactants.

In recent years it has been realized that reaction–diffusion systems in which the reactants are initially separated are more amenable to experiments, while the kinetics of such systems is as rich as that of initially mixed reactants. The reaction front that forms under initially segregated conditions appears in a host of biological, chemical, and physical processes. It serves as a model for the anomalous kinetics of reactions in reaction–diffusion systems in general, in which domains of reactants frequently form naturally as part of the process.

14.1 The mean-field description

The process $A + B \rightarrow C$, with initially segregated reactants, may be described by a set of reaction–diffusion equations:

$$\frac{\partial}{\partial t} c_A(x, t) = D_A \frac{\partial^2}{\partial x^2} c_A - k c_A c_B, \tag{14.1a}$$

$$\frac{\partial}{\partial t} c_B(x, t) = D_B \frac{\partial^2}{\partial x^2} c_B - k c_A c_B. \tag{14.1b}$$

Here $c_{A,B}$ and $D_{A,B}$ denote the concentrations and the diffusion constants of A and B particles. The particles are initially separated: the A species is uniformly distributed at $x > 0$, and the B species is uniformly distributed at $x < 0$. As noted before, the reaction–diffusion equation is essentially a mean-field approach

192

in which the spatial dimension plays no significant role: the reactants are assumed to maintain spatial homogeneity in all dimensions perpendicular to x (c depends only on x), and the reaction rate assumes a classical form.

The rate of production of C particles at site x and time t, which we call the reaction-front profile, is given by $R(x, t) \equiv kc_A c_B$. The overall reaction rate, integrated over all space, is $R(t) = \int_{-\infty}^{\infty} R(x, t)\, dx$, and the total amount of reactions until time τ is $R_{TOT}(\tau) = \int_0^{\tau} R(t)\, dt$.

Two features characterize the reaction front $R(x, t)$; (a) its width, which is assumed to scale with time as $w \sim t^{\alpha}$, and (b) its "height", i.e., the reaction rate at the center of the front, $h \sim t^{-\beta}$. Using scaling arguments, one can identify α and β as follows (Gálfi and Rácz, 1988). The total amount of reactions until time τ, $R_{TOT}(\tau)$, scales as $\tau^{1/2}$. This arises from the fact that the number of particles that reach the front (through diffusion) is proportional to $t^{1/2}$. From this it follows that $R(t) \sim t^{-1/2}$. On the other hand, the rate can be approximated as $R(t) \sim w(t)h(t)$, from which follows a relation between α and β:

$$\alpha - \beta = -\tfrac{1}{2}. \tag{14.2a}$$

The height can be written as $h(t) = kc_A(x_0, t)c_B(x_0, t)$, where $x_0 \sim w$ is the point where the amount of product is largest. Near the reaction zone it is plausible that $c_A(x_0, t) = c_B(x_0, t) \sim x_0/\sqrt{t}$, as suggested by the concentration profile of particles near a single trap (Chapter 12). Thus, $h(t) \sim x_0^2/t \sim w^2/t$, from which follows another relation between α and β:

$$-\beta = 2\alpha - 1. \tag{14.2b}$$

Hence, one concludes that $\alpha = \tfrac{1}{6}$ and $\beta = \tfrac{2}{3}$. Cornell *et al.* (1991) have argued that the upper critical dimension for this problem is $d = 2$, and these mean-field results are therefore valid for $d \geq 2$.

Experiments and simulations for systems in $d \geq 2$, in which both reactants diffuse, support the above-predicted values for α and β. In one dimension, numerical simulations predict different exponents: $\alpha \approx 0.3$ and $\beta \approx 0.8$. It has recently been argued that the exact results are $\alpha = \tfrac{1}{4}$ and $\beta = \tfrac{3}{4}$, and that the slightly higher numerical values are effective exponents due to logarithmic corrections (Larralde *et al.*, 1992d; Barkema *et al.*, 1996). The difference between one and higher dimensions is due to fluctuations in the densities of reactants, which are neglected in the mean-field approach.

Taitelbaum *et al.* (1991; 1992; 1996) have studied Eqs. (14.1) analytically, and have conducted experiments characterized by small reaction constants, for which the mean-field approach is expected to be valid. Their main result is that several measurable quantities undergo interesting crossovers. For example, the global reaction rate $R(t)$ crosses over from $t^{1/2}$ at early times to $t^{-1/2}$ in the long-time

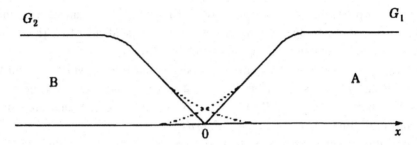

Fig. 14.1. A schematic picture of the concentration profiles of reactants near the origin. The solid lines represent the $G_{1,2}(x, t)$ part of the profile; the dashed lines represent the complete form $G_{1,2}(x, t) \pm \delta c(x, t)$. Note that to the left-hand side of the origin the profile of species A is given solely by $\delta c(x, t)$.

asymptotic regime. Experiments suggest that the center of the front may also change its direction of motion. Ben-Naim and Redner (1992) have studied the solution of (14.1) under steady-state conditions, with input of particles on either side of the reaction front.

14.2 The shape of the reaction front in the mean-field approach

For simplicity, we consider the symmetric case in which both diffusion constants and initial concentrations are equal, i.e., $D_A = D_B \equiv D$ and $c_A(x, 0) = c_B(x, 0) \equiv c_0$ (Larralde *et al.*, 1992c). With the definition $F(x, t) \equiv c_A(x, t) - c_B(x, t)$, Eqs. (14.1) may be rewritten as

$$\frac{\partial}{\partial t} F(x, t) = D \frac{\partial^2}{\partial x^2} F, \qquad (14.3)$$

subject to the same initial conditions as before. The solution is $F(x, t) = c_0 \operatorname{erf}(x/\sqrt{4Dt})$.

Rewrite now the concentrations of A and B particles as (see Fig. 14.1)

$$c_A(x, t) = G_1(x, t) + \delta c_1(x, t), \qquad c_B(x, t) = G_2(x, t) + \delta c_2(x, t), \quad (14.4a)$$

where

$$G_1(x, t) = \begin{cases} F(x, t) & (x > 0) \\ 0 & (x < 0), \end{cases} \qquad (14.4b)$$

and $G_2(x, t) = G_1(-x, t)$. It is easy to see that, under the above conditions, $\delta c_1(x, t) = \delta c_2(x, t) \equiv \delta c(x, t)$. Substituting this form of c into Eq. (14.1) yields

$$\frac{\partial}{\partial t}(\delta c) = D \frac{\partial^2}{\partial x^2}(\delta c) - k \left[c_0 \operatorname{erf}\left(\frac{x}{\sqrt{4Dt}}\right) + \delta c \right] \delta c. \qquad (14.5)$$

The asymptotic solution, to order $(\delta c)/t$, that vanishes as $x \to \infty$, is

$$\delta c(x, t) \sim t^{-1/3} \left(\frac{x}{t^{1/6}} \right)^{-1/4} \exp\left[-\frac{2}{3} \left(\frac{\lambda x}{t^{1/6}} \right)^{3/2} \right], \quad \lambda^{-1} t^{1/6} \ll x \ll (4Dt)^{1/2},$$

(14.6)

where $\lambda = (ka/D)^{1/3}$ and $a \equiv c_0/(\pi D)^{1/2}$.

Using Eq. (14.6), one can write an expression for the reaction-front profile:

$$R(x, t) \simeq \frac{kax}{t^{1/2}} (\delta c) \sim t^{-2/3} \left(\frac{x}{t^{1/6}} \right)^{3/4} \exp\left[-\frac{2}{3} \left(\frac{\lambda x}{t^{1/6}} \right)^{3/2} \right].$$

(14.7)

The width of this reaction front grows as $t^{1/6}$, whereas the height can be identified with the prefactor $t^{-2/3}$, in agreement with the exponents found by scaling arguments. Equation (14.7) provides more detail than scaling does, and it gives the dependences of the form of the reaction front on the parameters c_0, k, and D for the symmetric case.

When one of the reactants is static, the shape of the reaction front is not known analytically. Jiang and Ebner (1990) have given scaling arguments for a front of the form $R(x, t) \sim t^{-\beta} g(x/t^{\alpha}) \exp(-|x|/t^{\alpha})$, with $\alpha = 0$ and $\beta = \frac{1}{2}$. This is well supported by numerical solutions of Eq. (14.1) with $D_B = 0$ (Araujo *et al.*, 1992), shown in Fig. 14.2.

14.3 Studies of the front in one dimension

For dimension $d = 1$, the mean-field approach of Eq. (14.1) is not valid and one must seek alternative descriptions. For the special case of one immobile species ($D_B = 0$), analytical and numerical studies (Larralde *et al.*, 1992d) yield a reaction-front profile of the form

$$R(x, t) = \frac{1}{4t^{3/4}} \left(\frac{2\gamma^2}{\mu\pi} \right)^{1/2} \left(1 + \frac{x - \gamma t^{1/2}}{2\gamma t^{1/2}} \right) \exp\left[-\frac{(x - \gamma t^{1/2})^2}{2\mu t^{1/2}} \right], \quad (14.8)$$

where γ and μ are constants. In this case $\alpha = \frac{1}{4}$ and $\beta = \frac{3}{4}$. The time integral of $R(x, t)$, that is, the total production of C particles at point x up to time t, assumes a simple form:

$$c_C(x, t) = \int_0^t R(x, \tau) \, d\tau = \frac{1}{2} \operatorname{erfc}\left(\frac{x - \gamma t^{1/2}}{\sqrt{2\mu t^{1/2}}} \right).$$

(14.9)

A summary of the front exponents for the cases discussed so far is presented in Table 14.1.

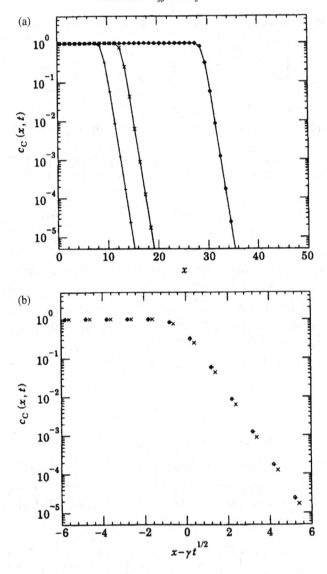

Fig. 14.2. The numerical solution of Eq. (14.1) for the case $D_B = 0$, $D_A > 0$. (a) A plot of $c(x, t)$ as a function of x for $t = 500$, 1000, and 5000. (b) The good scaling of $c(x, t)$ as a function of $(x - \langle x \rangle)$ indicates that $\alpha = 0$ and $\beta = \frac{1}{2}$. After Larralde *et al.* (1992d).

14.4 Reaction rates in percolation

The case of A + B → C with initially separated reactants in fractal substrates has been studied for the $d = 2$ infinite percolation cluster at criticality (Fig. 14.3). It is expected that the total number of products up to time t scales as the mean

Table 14.1. *Width and height exponents for the reaction front in the process*
A + B → C, with initially separated reactants.

	$d = 1$	$d \geq 2$ (mean-field)
Both species moving	$\alpha = \frac{1}{4}$	$\alpha = \frac{1}{6}$
	$\beta = \frac{3}{4}$	$\beta = \frac{2}{3}$
One species static	$\alpha = \frac{1}{4}$	$\alpha = 0$
	$\beta = \frac{3}{4}$	$\beta = \frac{1}{2}$

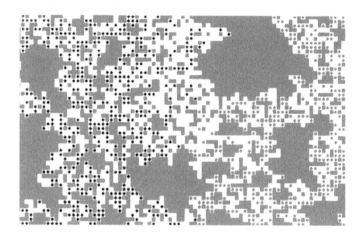

Fig. 14.3. Reaction–diffusion in a two-dimensional infinite percolation cluster at criticality at $t = 2000$. White squares represent sites of the infinite cluster. Black and gray circles represent the A and B particles. Initially all sites of the cluster in the right- and left-hand half-planes were occupied by A and B particles, respectively.

displacement of a random walker in a fractal:

$$R_{\mathrm{TOT}}(\tau) = \int_0^\tau R(t)\, dt \sim \langle r^2 \rangle^{1/2} \sim \tau^{1/d_{\mathrm{w}}}, \qquad (14.10)$$

where d_{w} is the anomalous-diffusion exponent discussed in Chapter 6. From this it follows that the reaction rate is

$$R(t) = \frac{d}{dt} R_{\mathrm{TOT}}(t) \sim t^{-\gamma}, \qquad \gamma = 1 - 1/d_{\mathrm{w}}. \qquad (14.11)$$

One must distinguish between the process limited to the infinite cluster and that in the full percolation system (see Chapter 6 for details). The reaction rate in the infinite cluster is smaller than that in the full percolation system and decreases more slowly. This can be understood by applying scaling arguments similar to the ones made for diffusion.

Until time t the reaction invades a typical length $r^* \sim t^{1/d_w}$, which corresponds to a mass $s^* \sim r^{*d_f} \sim t^{d_f/d_w}$. At any finite time we can divide all clusters into two groups according to their sizes: (a) *active* clusters of mass $s > s^*$, in which particles are not aware of the finiteness of their cluster (this group contains the infinite cluster), and (b) *inactive* clusters of mass $s < s^*$, in which the amount of at least one of the reactants has been depleted and the reaction rate is zero. Thus, in the full percolation system, at any given time there are active clusters of finite size that can contribute to the reaction rate, and the reaction rate is then higher than that for the infinite cluster alone. Moreover, since s^* grows with time, there are always some clusters that become inactive, causing an additional decrease of the rate of reactions in the percolation system.

To quantify the above considerations we look at clusters of mass s and linear size $r \sim s^{1/d_f}$, which are divided by the *front line* into A and B domains. The sites of clusters lying in the front line form what we call the *active front*. The length ℓ_s of the active front in an s-cluster is expected to be

$$\ell_s \sim r^{d_f-1} \sim s^{(d_f-1)/d_f}. \tag{14.12}$$

Next, we assume that the reaction rate per unit length of active front is

$$R_0(t) \sim \begin{cases} t^{-\gamma} & t < t^*, \\ 0 & t > t^*, \end{cases} \tag{14.13}$$

where $t^* = s^{d_w/d_f}$. Therefore, the total contribution of all active clusters of size s to the reaction rate is

$$R_s(t) \sim \varphi_s s^{(d_f-1)/d_f} t^{-\gamma}, \tag{14.14}$$

where φ_s is the number of clusters of size s that intersect the front line. To estimate φ_s, consider a percolation system of size $L \times L$ in which there are $n_s L^2$ clusters of mass s. Only a small fraction of these clusters intersects the front line; namely those in a strip of width $w \sim s^{1/d_f}$ around the front line. Their fraction is w/L. Therefore, $\varphi_s \sim (s^{1/d_f}/L)n_s L^2 \sim s^{1/d_f} n_s$. On substituting this into (14.14) we get

$$R_s(t) \sim s^{1/d_f} n_s s^{(d_f-1)/d_f} t^{-\gamma} = t^{-\gamma} s n_s. \tag{14.15}$$

Thus, the reaction rate in the percolation system is $R(t) = \sum_{s=s^*}^{\infty} R_s(t) = t^{-\gamma}(s^*)^{2-\tau} = t^{-\gamma} t^{-\delta} = t^{-\gamma'}$, where

$$\delta = \gamma' - \gamma = \frac{d_f}{d_w}(\tau - 2). \tag{14.16}$$

Hence, the effect of the inactive clusters changes the reaction-rate exponent, as expected. These results are in agreement with numerical simulations (Fig. 14.4).

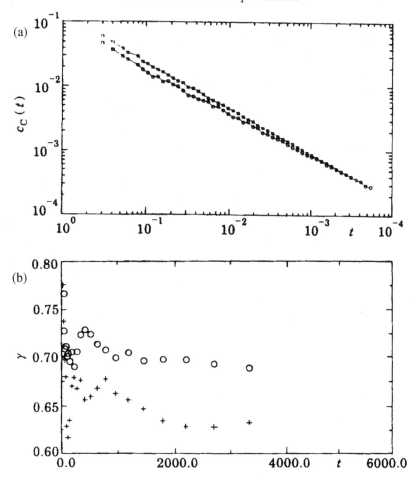

Fig. 14.4. (a) A plot of the rate $c_C(t)$ for the whole percolation system ($\diamond$) and for the infinite percolation cluster ($\circ$). (b) A plot of successive slopes of the data in (a), showing the exponent γ in the infinite cluster ($+$) and in the whole percolation system ($\circ$). After Havlin *et al.* (1995).

Finite-size effects on the reaction–diffusion system can be considered too. For a percolation system of size $L \times L$ we expect that, for the incipient infinite cluster,

$$R(t) = L^{d_f - 1} t^{-\gamma}, \qquad (14.17a)$$

whereas for the full percolation system

$$R(t) = L t^{-\gamma'}. \qquad (14.17b)$$

The prefactor L in (14.17b) assures that, at $t < t^*$, the reaction rate in the full percolation system is larger than that of the infinite cluster. (Indeed, the ratio of the reaction rates at $t = 1$ is L^{2-d_f}.) Finally, we expect that, at time $t^* \sim L^{d_w}$, the two rates (14.17a) and (14.17b) become equal, since no "small" active cluster exists in the system above t^*. On equating the two rates one recovers Eq. (14.16).

14.5 $A + B_{static} \rightarrow C$ with a localized source of A particles

A case that is especially amenable to experiment is the reaction $A + B_{static} \rightarrow C$ with a localized source of the A species. There exist many systems in Nature in which a reactant A is "injected" into a d-dimensional substrate B, with which it reacts to form some inert product C. One such experiment has been performed by injecting iodine into a point of a large silver plate and measuring concentrations in the reaction $I_{2gas} + 2Ag_{solid} \rightarrow 2AgI_{solid}$ (Larralde *et al.*, 1993).

To model the above experiment, we consider N particles of type A that are initially at the origin of a lattice. The B particles are static and distributed uniformly over the lattice sites. Using a quasistatic analytical approximation for trapping with a moving boundary, one can derive expressions for $C(t)$, the time-dependent growth size of the C region, and for $S(t)$, the number of surviving A particles at time t. For extremely short times, $t < t_\times \sim \ln N$, one finds $C(t) \sim t^d$. For $t > t_\times$, the result is

$$C(t) \sim Nf\left(\frac{t}{N^{2/d}}\right), \qquad S(t) = N - C(t). \qquad (14.18)$$

The scaling function $f(u)$ is the solution of the differential equation

$$\frac{df}{d\tau} \sim k_d f^{-2/d}(1 - f), \qquad (14.19)$$

and k_d is a constant that depends only on the dimension d. Figure 14.5 shows simulation data supporting Eq. (14.18).

Suppose now that A particles are injected at rate λ, at the origin of the lattice. For this case we have

$$C(t) \sim \begin{cases} \sqrt{8Dt \ln(\lambda^2 t/2D)} & d = 1, \\ \pi \alpha t & d = 2, \\ \lambda t & d = 3, \end{cases} \qquad (14.20)$$

$$S(t) \sim \begin{cases} \lambda t & d = 1, \\ (\lambda - \pi \alpha)t & d = 2, \\ C_3(\lambda)t^{2/3} & d = 3. \end{cases} \qquad (14.21)$$

Here α is the solution of $\alpha \pi = \lambda \exp[-\alpha/(4D)]$, and $C_3(\lambda) = [\lambda/(4D)][3\lambda/(4\pi)]^{2/3}$. Moreover, it can be seen that both in one and in three dimensions $C(t)$ satisfies the scaling relation

$$C(t) \sim \lambda^{d/(d-2)} g\left(\frac{t}{\lambda^{2/(d-2)}}\right). \qquad (14.22)$$

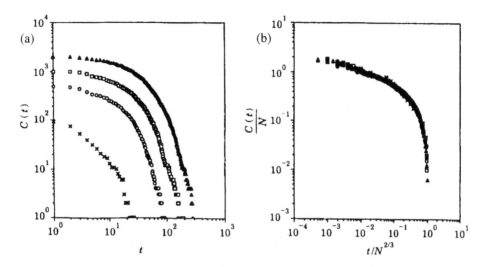

Fig. 14.5. Numerical simulations of A+B → C for the case in which N diffusing particles of type A are initially at the center of a lattice and particles of type B are static and located on each site of the $d = 3$ lattice. (a) A plot of $C(t)$ for $N = 100$ (×), 500 (○), 1000 (□), and 2000 (▲) particles. (b) A plot of $C(t)$ in the scaling form Eq. (14.18). Note that the results are from a *single* Monte Carlo run, showing that fluctuations are negligible in this process.

Equations (14.20) and (14.21) are supported by numerical simulations (see also Fig. 14.6).

In the case of fractals, Eq. (14.18) generalizes to

$$C(t) \sim N f \left(\frac{t}{N^{2/d_s}} \right),$$ (14.23)

where $f(u)$ is the solution of the differential equation

$$\frac{df}{du} \sim k_{d_s} f^{-2/d_s} (1 - f).$$ (14.24)

14.6 Exercises

1. Derive reaction–diffusion equations similar to (14.2) for the case in which one of the reactants is static, and obtain α and β.
2. Show that, for small k and short times, the global reaction rate increases as $t^{1/2}$.
3. Show that (14.6) is a solution of Eq. (14.5).
4. Derive Eq. (14.8). (Hint: follow Larralde *et al.* (1992d).)
5. Simulate the case in which N random walkers start from the origin of a two-dimensional lattice. Every time a particle encounters a new site that has not been visited before by any of the particles the particle is removed. Calculate the

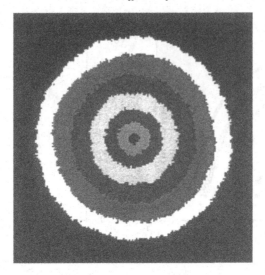

Fig. 14.6. Numerical simulations of $A + B_{static} \rightarrow C_{inert}$ for the case in which A particles are injected at rate $\lambda = 5$ at the origin and particles of type B are static and are located at each site of a $d = 2$ lattice. A plot of the reaction area after $t = 20$, 160, 540, 1280, 2500, 4320, and 6860. After Havlin *et al.* (1995).

number of new sites visited up to time t: $C(t) = N - S(t)$ ($S(t)$ is the number of surviving particles at time t). Take several values of N and confirm the scaling relation of Eq. (14.18). Notice that, if N is large, a single run for each value of N may suffice.

14.7 Open challenges

1. Although, there exists an analytical solution for the front profile in $d \geq 2$ for the case $D_A = D_B$ (the mean-field solution of Eqs. (14.1)), the analogous problem in $d = 1$ remains unsolved.
2. It would be interesting to obtain the mean-field solution for the more general case $D_A \neq D_B$. Even the case $D_B = 0$ and $D_A > 0$ has not yet been solved in the mean-field approximation (Koza, 1998).
3. Consider the reaction $A + B \rightarrow C$ with initially separated reactants, but for which an agent D, which is distributed homogenously, is needed for the reaction to occur. Assume that D loses its power after a reaction takes place. In this case one expects that the reaction front will not remain centered at the origin even for the case $D_A = D_B$ and $\rho_A = \rho_B$, but that it will split into $x_1(t) > 0$ and $x_2(t) < 0$. The dynamics of such a front and its characteristic exponents have not been studied and might be of interest.
4. For the case of constant injection rate in a fractal there exists no analytical

derivation of $C(t)$. The number of distinct sites visited by N random walkers starting from a common origin in a fractal was found to be $S_N(t) \sim (\ln N)^{d_f/\delta} t^{d_s/2}$, with $\delta = d_w/(d_w - 1)$ (Havlin *et al.*, 1992). This result suggests an upper bound for the number of distinct sites visited by random walkers injected at the origin at a constant rate λ, simply by substituting $N = \lambda t$: $C(t) \leq (\ln \lambda t)^{d_f/\delta} t^{d_s/2}$.

5. Reaction fronts in other types of diffusion-limited reaction may prove interesting. See for example Cox *et al.* (1998), and Clifford *et al.* (1998), for fronts in the two-stage reaction $A + B \to R$, $B + R \to S$, when the reactants A and B are initially segregated (R is the desired product, and S is an unwanted product).

14.8 Further reading

- The reaction–diffusion system with initially separated reactants has been studied experimentally by Koo *et al.* (1990), Koo and Kopelman (1991), Taitelbaum *et al.* (1992; 1996), and Yen *et al.* (1996), and theoretically by Gálfi and Rácz (1988), Jiang and Ebner (1990), Cornell *et al.* (1991), Taitelbaum *et al.* (1991; 1996), Taitelbaum and Koza (1998), Araujo *et al.* (1992; 1993), Ben-Naim and Redner (1992), Larralde *et al.* (1992c; 1992d), Chopard *et al.* (1997) and Sinder and Pelleg (1999).

- Reaction fronts under steady-state conditions: Ben-Naim and Redner (1992), Howard and Cardy (1995), and Barkema *et al.* (1996).

- For selected applications of reaction fronts, see Liesegang (1896), Avnir and Kagan (1984), Dee (1986), and Heidel *et al.* (1986).

- Reaction–diffusion processes with initially separated reactants in percolation at criticality were studied by Taitelbaum *et al.* (1991) and Havlin *et al.* (1995).

Part four

Diffusion-limited coalescence:
an exactly solvable model

As discussed in Part III, diffusion-limited reactions are generally studied through a variety of approximation and computer-simulation techniques, since a comprehensive exact method of analysis has not yet been suggested. For this reason, exactly solvable models are of extreme importance: they serve as benchmark tests for the existing approximation and simulation methods; the exact techniques may hint at a more general approach, and serve as a basis for better approximations; and they contribute enormously to our understanding of the field as a whole.

In Part IV, we discuss a particular example of an exactly solvable model – that of diffusion-limited coalescence, $A + A \to A$, in one dimension. The model has been studied extensively by numerous researchers, who have thought up an impressive amount of imaginative, elegant, solutions. The description of these works would require an additional volume, so they are acknowledged only in the bibliography. Instead, we limit ourselves to the method of interparticle distribution functions (IPDF), merely because we took part in its development and we understand it best. It should likewise be noted that neither is our model of choice the only one which can be solved exactly (very few others exist, though). The coalescence model yields an astonishingly wide range of kinetic behavior, well beyond what might be suspected from its stark simplicity. The restriction of the model to one dimension commonly draws criticism. This is dictated by the need to find *exact* solutions. On the other hand, recall that diffusion-limited kinetics is more anomalous the lower the dimension, so there is an advantage in studying one-dimensional models, in which differences from mean-field classical behavior are most pronounced.

In Chapter 15 we detail the coalescence model and the IPDF method of solution. Chapter 16 discusses the case of irreversible coalescence, $A + A \to A$, with and without input of A particles. These serve as examples of ordering in a system far from equilibrium. The reversible reaction $A + A \rightleftharpoons A$ is treated in Chapter 17. It exhibits a kinetic phase transition in the relaxation time for the approach to the equilibrium state. In Chapter 18 we describe how the IPDF method deals with inhomogeneous systems and with multiple-point correlation functions: in some cases the full hierarchy of correlation functions may be obtained exactly. Finally, in Chapter 19 we demonstrate how the exact treatment of the IPDF method can serve as a basis for approximations in closely related models.

15

Coalescence and the IPDF method

The diffusion-limited coalescence model, $A+A \rightarrow A$, can be treated exactly in one dimension. The process is unexpectedly rich, displaying self-critical ordering in a nonequilibrium system, a kinetic phase transition, and a lattice version of Fisher waves. Thus, in spite of its simplicity it sheds light on many important aspects of anomalous kinetics. It also serves as a benchmark test for approximation methods and simulation algorithms. The coalescence model will concern us throughout the remainder of the book. Here we introduce the model and explain the technique which allows its exact analysis.

15.1 The one-species coalescence model

Our basic model is a lattice realization of the one-dimensional coalescence process $A + A \rightarrow A$. The exact analysis can also be extended to the reversible process, $A \rightarrow A + A$, as well as to the input of A particles. The system is defined on a one-dimensional lattice of lattice spacing Δx. Each site may be either occupied by an A particle or empty. The full process consists of the following dynamic rules.

Diffusion. Particles hop randomly to the nearest lattice site with a hopping rate $2D/(\Delta x)^2$. The hopping is symmetric, with rate $D/(\Delta x)^2$ to the right and $D/(\Delta x)^2$ to the left. At long times this yields normal diffusion, with diffusion coefficient D.

Birth. A particle gives birth to another at an adjacent site, at rate $v/\Delta x$. This means a rate of $v/(2\,\Delta x)$ for birth on each side of the original particle. Notice that, while v is a constant (with units of velocity), the rate $v/\Delta x$ diverges in the continuum limit of $\Delta x \rightarrow 0$. This is necessary because of the possibility of recombination of the newly born and the original particle, which also takes place at an infinite rate when $\Delta x \rightarrow 0$. The birth process models the reverse reaction, $A \rightarrow A + A$.

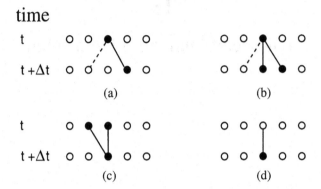

Fig. 15.1. Coalescence-model reaction rules: (a) diffusion; (b) birth; (c) coalescence (following diffusion); and (d) input. The broken lines in (a) and (b) indicate alternative target sites.

Input. Any empty site becomes spontaneously occupied at a rate $R \Delta x$. Thus, R is the average number of particles input to the system per unit length, per unit time.

Coalescence. When a particle lands on another through diffusion, input, or birth, then it disappears (or simply merges with the target particle). This condition implements the coalescence process $A + A \rightarrow A$. Because this coalescence reaction is infinitely fast, the overall process is diffusion-limited. In Chapter 19 we shall consider the consequences of relaxing this condition.

Each of these processes (except coalescence) takes place independently from the others. The various processes are illustrated in Fig. 15.1.

15.2 The IPDF method

The interparticle distribution function (IPDF) method was originally introduced for the solution of the diffusion-limited coalescence process, $A + A \rightarrow A$, in one dimension. The method is, however, more general, so we shall therefore describe it for any dynamic process on the line, with the restriction that lattice sites may be either empty (o) or occupied by a single particle (●). The states of the sites evolve with time according to some dynamic rules that are prescribed by the model in question. We shall further assume that an ensemble average of the state of our system is translationally invariant. (This restriction is not absolutely necessary, as we shall see, but it makes the introduction of the method simpler.)

We define the quantity $E_n(t)$ as the probability that a randomly chosen segment of n consecutive sites is empty, i.e., contains no particles. The E_n can be used to describe important characteristics of the distribution of particles (including the concentration and the distribution of interparticle distances) and to construct kinetic

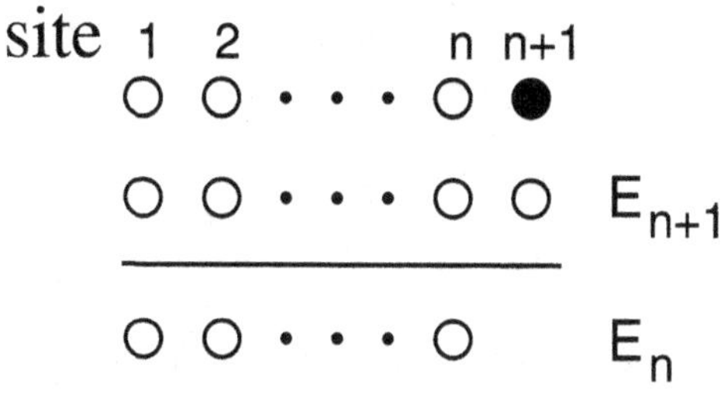

site 1 2   n n+1
○ ○ • • • ○ ●

○ ○ • • • ○ ○ E_{n+1}

_____

○ ○ • • • ○ E_n

Fig. 15.2. Illustration of Eq. (15.2). The event that n consecutive sites are empty (probability E_n) includes two possibilities regarding the $(n + 1)$th site: the site is also empty (probability E_{n+1}), or it is occupied. It follows that the latter occurs with probability $E_n - E_{n+1}$.

equations describing the dynamic rules. For example, the probability that a site is occupied is $1 - E_1$. Thus, the density, or concentration, of particles is expressed as

$$c(t) = (1 - E_1)/\Delta x. \qquad (15.1)$$

E_n gives the probability that, say, sites 1 through n are empty, while E_{n+1} gives the probability that sites 1 through $n + 1$ are empty. The event that 1 through $n + 1$ are empty contains the event that 1 through n are empty. Thus, the probability that a segment of n sites is empty but that there is a particle at the adjacent site $n + 1$ is (Fig. 15.2)

$$\text{Prob}(\overbrace{\circ \cdots \circ}^{n} \bullet) = E_n - E_{n+1}. \qquad (15.2)$$

From the $E_n(t)$ one may also derive $p_n(t)$ – the probability that the nearest neighbor to (say, the right of) a given particle is n lattice spacings away, at time t. That is, p_1 is the probability that the nearest neighbor lies in the site next to the particle, p_2 is the probability that the nearest neighbor is two sites away, etc. The p_n are normalized, $\sum p_n = 1$, and the average distance between particles is the reciprocal of the concentration:

$$\langle x \rangle = \sum_{n=1}^{\infty} n p_n \, \Delta x = \frac{1}{c}. \qquad (15.3)$$

Choose a lattice site at random. The probability that the next n sites are empty, E_n, may be written in terms of the p_n. The probability that the chosen point lies within a gap of length m is proportional to $m p_m$, which can be normalized with the help of Eq. (15.3), yielding the probability distribution $c \, \Delta x \, m p_m$. The probability that there are k lattice spacings until the next particle, given that the point is in the gap of length m, is $1/m$ if $1 \le k \le m$, and zero otherwise. Thus,

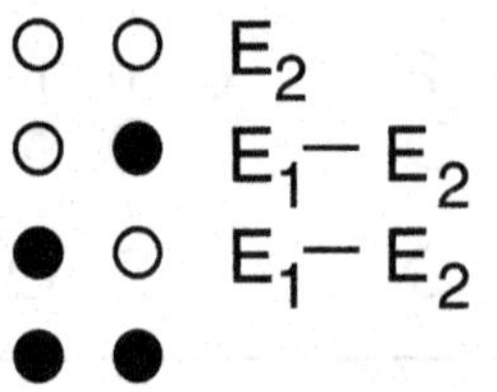

$$
\begin{array}{ll}
\circ\ \circ & E_2 \\
\circ\ \bullet & E_1 - E_2 \\
\bullet\ \circ & E_1 - E_2 \\
\bullet\ \bullet &
\end{array}
$$

Fig. 15.3. The probability of finding two adjacent particles is computed by listing all four possibilities for the occupancy of two adjacent sites. The respective probabilities add up to one.

the (unconditional) probability that there are exactly k lattice spacings to the next particle is

$$
\sum_{m=k}^{\infty} \frac{1}{m} c\, \Delta x\, m p_m = c\, \Delta x \sum_{m=k}^{\infty} p_m. \tag{15.4}
$$

Finally, the probability that the next n sites are empty, E_n, is the probability that $k > n$:

$$
E_n = c\, \Delta x \sum_{k=n+1}^{\infty} \sum_{m=k}^{\infty} p_m. \tag{15.5}
$$

This can be inverted to yield

$$
c\, \Delta x\, p_n = E_{n-1} - 2E_n + E_{n+1}. \tag{15.6}
$$

This equation does not apply for p_1, because E_n is not defined for $n = 0$. To obtain the probability that two adjacent sites are occupied, we simply enumerate all their possible states (the sum of the probabilities is 1), and use Eq. (15.2) to get (Fig. 15.3)

$$
\mathrm{Prob}(\bullet\bullet) = 1 - \{\mathrm{Prob}(\circ\circ) + \mathrm{Prob}(\circ\bullet) + \mathrm{Prob}(\bullet\circ)\} = 1 - 2E_1 + E_2. \tag{15.7}
$$

Then, p_1 is equal to the conditional probability $\mathrm{Prob}(\bullet\bullet)$, given that the first site is occupied, or

$$
c\, \Delta x\, p_1 = 1 - 2E_1 + E_2. \tag{15.8}
$$

In fact, a similar argument can be made for an alternative derivation of Eq. (15.6) (Fig. 15.4).

For some two-state models with simple dynamics the various probabilities discussed above are sufficient to write down closed kinetic equations – equations that involve only the E_n's.

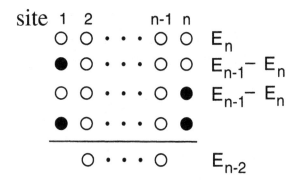

Fig. 15.4. An alternative derivation of Eq. (15.6). The probabilities of all four possibilities of occupation of the sites at the edges of an nth segment (while the intervening sites are empty) add up to E_{n-2}. From this, one can deduce the probability of having two particles separated by a specific number of empty sites (the fourth possibility), and Eq. (15.6) follows.

15.3 The continuum limit

To achieve a spatial continuum limit, we define the spatial coordinate $x = n\,\Delta x$. The probabilities $E_n(t)$ are replaced by the two-variable function $E(x, t)$, and, in the limit of $\Delta x \to 0$, Eq. (15.1) becomes

$$c(t) = -\left.\frac{\partial E(x, t)}{\partial x}\right|_{x=0}. \qquad (15.9)$$

Likewise, the probabilities $p_n(t)$ are replaced by probability densities $p(x, t)$, which are related to the density of an empty interval, $E(x, t)$, through Eq. (15.6):

$$c(t)p(x, t) = \frac{\partial^2 E(x, t)}{\partial x^2}. \qquad (15.10)$$

We refer to p_n, or to $p(x, t)$, as the "interparticle distribution function", which is the origin of the name of our technique. The IPDF method allows one to compute $c(t)$ – the time dependence of the concentration. The IPDFs themselves are studied because they convey more detailed information than the average density, and because of their relation to the Smoluchowski problem. IPDFs were first introduced by Kopelman (Kopelman, 1986; 1988; Kopelman *et al.*, 1988; Argyrakis and Kopelman, 1990), for the analysis of simulations and experimental results of diffusion-limited reactions, and have since continued to attract considerable interest. Our method is often referred to as "the method of empty intervals", after the E_n probabilities, but we prefer to emphasize the role played by the IPDFs, and the fact that they can be computed through this formalism.

15.4 Exact evolution equations

To implement the IPDF method, we construct a closed kinetic equation for the evolution of the E_n. Consider the changes in E_n due to the various processes occurring during a small time interval Δt.

Diffusion. We may have an empty segment of n sites, and site $n+1$ occupied (with probability $E_n - E_{n+1}$, Eq. (15.2)). If the particle in the $(n+1)$th site hops into the empty segment the probability E_n decreases by $D/(\Delta x)^2(E_n - E_{n+1})$. Likewise, we may have an empty segment of $n-1$ sites and site n occupied (with probability $E_{n-1} - E_n$). If the particle at the edge hops to site $n+1$ then E_n increases by $D/(\Delta x)^2(E_{n-1} - E_n)$. Therefore, the total change in E_n due to diffusion is

$$(\partial_t E_n)_{\text{diffusion}} = 2\frac{D}{(\Delta x)^2}(E_{n-1} - 2E_n + E_{n+1}), \qquad (15.11)$$

where the additional factor of 2 accounts for the possibility that both processes may take place on either side of the segment, independently.

Birth. The birth process brings about a decrease in E_n in a way similar to the diffusion process. In the case of an empty segment of n sites followed by an occupied $(n+1)$th site, the particle at the edge may give birth to a particle just inside the empty segment. Thus, the change in E_n due to birth (accounting for both sides of the segment) is

$$(\partial_t E_n)_{\text{birth}} = -\frac{v}{\Delta x}(E_n - E_{n+1}). \qquad (15.12)$$

Input. Whenever a particle is input into an empty n-sites segment, E_n decreases. Since the rate of input into each of the n sites is $R\,\Delta x$, we have

$$(\partial_t E_n)_{\text{input}} = -R\,n\,\Delta x\,E_n. \qquad (15.13)$$

Coalescence. In a sense, the process of immediate coalescence is implicit in each of the three processes already discussed above. For example, in the case of diffusion, when the particle at site n clears the nth interval by hopping to $n+1$, we have not cared about the state of $n+1$. However, when site $n+1$ is occupied, such a move is possible only if coalescence takes place immediately (otherwise there is a finite probability that the move cannot be effected). Coalescence enters our analysis more explicitly by imposing a boundary condition, as follows. Because E_0 is not defined, Eq. (15.11) is valid only for $n > 1$. We need to consider $(\partial_t E_1)_{\text{diffusion}}$ separately. However, according to Eq. (15.1), $\partial_t E_1 = -\partial_t c\,\Delta x$ is the rate of change in the *number* of particles, due to diffusion. This happens only when either of two adjacent particles hops into its neighbor, decreasing the number of particles by one.

Then, using the hopping rate and Eq. (15.7),

$$(\partial_t E_1)_{\text{diffusion}} = 2\frac{D}{(\Delta x)^2}(1 - 2E_1 + E_2). \tag{15.14}$$

To make this consistent with Eq. (15.11) for the case of $n = 1$, we require the boundary condition

$$E_0 = 1. \tag{15.15}$$

Notice that Eqs. (15.12) and (15.13) for the birth and input processes pose no additional constraints.

Combining all of the various contributions to changes in E_n, we get

$$\partial_t E_n = 2\frac{D}{(\Delta x)^2}(E_{n-1} - 2E_n + E_{n+1}) - \frac{v}{\Delta x}(E_n - E_{n+1}) - R n \Delta x \, E_n, \tag{15.16}$$

with the boundary condition Eq. (15.15). Another boundary condition,

$$E_\infty(t) = 0, \tag{15.17}$$

holds as long as the system is not completely empty. Finally, an initial condition may be derived from the configuration of the lattice at time $t = 0$. For example, an initially empty lattice corresponds to $E_n(0) = 1$, while a homogeneously random distribution of particles with density c_0 leads to $E_n(0) = (1 - c_0 \Delta x)^n$, etc.

Although Eq. (15.16) can be tackled through standard approaches for difference equations, it is convenient to pass to the continuum limit, as discussed in Section 15.2. We then have

$$\frac{\partial E(x, t)}{\partial t} = 2D \frac{\partial^2 E}{\partial x^2} + v \frac{\partial E}{\partial x} - Rx E, \tag{15.18}$$

with boundary conditions

$$E(0, t) = 1, \qquad E(\infty, t) = 0. \tag{15.19}$$

This lends itself to an exact solution for $E(x, t)$, which is then used to find $c(t)$ and $p(x, t)$.

15.5 The general solution

To solve Eq. (15.18) we expand $E(x, t)$ in the form

$$E(x, t) = \sum_\lambda E_\lambda(x)e^{-\lambda t}, \tag{15.20}$$

where the $E_\lambda(x)$ are eigenfunctions of

$$-\lambda E_\lambda(x) = 2D \frac{\partial^2 E_\lambda(x)}{\partial x^2} + v \frac{\partial E_\lambda(x)}{\partial x} - Rx E_\lambda(x). \tag{15.21}$$

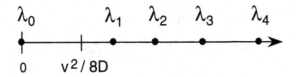

Fig. 15.5. The spectrum of eigenvalues for Eq. (15.21). The limit $R \to 0$ is *not* trivial, because spectrum appears then in the gap $(0, v^2/(8D))$ (see Chapter 17).

On making the substitution $E_\lambda(x) = F_\lambda(x)e^{-vx/(4D)}$, one obtains Airy's equation

$$2D \frac{\partial^2 F_\lambda(x)}{\partial x^2} = \left[Rx + \left(\frac{v^2}{8D} - \lambda \right) \right] F_\lambda(x), \tag{15.22}$$

with the solution

$$F_\lambda(x) = \text{Ai}\left[\left(\frac{R}{2D} \right)^{1/3} x + \frac{v^2/(8D) - \lambda}{(2DR^2)^{1/3}} \right]. \tag{15.23}$$

$\text{Ai}(z)$ is Airy's function, satisfying $\text{Ai}''(z) = z\text{Ai}(z)$. The linearly independent solution $\text{Bi}(z)$ is excluded because it grows faster than $e^{vx/(4D)}$, violating the boundary condition $E(x, t) \to 0$ as $x \to \infty$.

The steady-state solution is obtained from Eq. (15.23) by setting $\lambda = 0$ and using the boundary condition $E(0, t) = 1$ (or $F_0(0) = 1$):

$$E_0(x) = e^{-vx/(4D)} \frac{\text{Ai}\{[r/(2D)]^{1/3}x + [v^2/(8D)](2DR^2)^{-1/3}\}}{\text{Ai}\{[v^2/(8D)](2DR^2)^{-1/3}\}}. \tag{15.24}$$

From this we obtain the steady-state concentration, using Eq. (15.9):

$$c_s = \frac{v}{4D} - \left(\frac{R}{2D} \right)^{1/3} \frac{\text{Ai}'\{[v^2/(8D)](2DR^2)^{-1/3}\}}{\text{Ai}\{[v^2/(8D)](2DR^2)^{-1/3}\}}. \tag{15.25}$$

The steady-state interparticle distribution function $p_s(x) = p(x, \infty)$ may also be computed, using Eq. (15.10).

The transient solutions correspond to $\lambda > 0$. In this case the boundary condition $E(0, t) = 1$, combined with $F_0(0) = 1$, implies that $F_\lambda(0) = 0$ for $\lambda > 0$. Applying this to Eq. (15.23), we find a discrete spectrum for nonvanishing R and D:

$$\lambda_n = v^2/(8D) + (2DR^2)^{1/3}|a_n|, \tag{15.26}$$

where a_n is the nth zero of the Airy function $\text{Ai}(z)$. These are all negative $(a_1 = -2.3381\ldots, a_2 = -4.0879\ldots,$ etc.), and are tabulated in the literature. The spectrum of eigenvalues is illustrated in Fig. 15.5. In following chapters, we shall examine particular cases of the general solution in further detail.

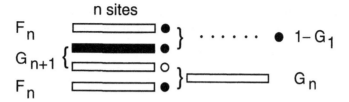

Fig. 15.6. A hint for Exercise 5. Empty/solid rectangles symbolize nth segments with even/odd numbers of particles. Empty/solid circles represent empty/occupied sites, as before. F_n represents the desired quantity: a sequence of n sites containing an even number of particles, followed by an occupied site.

15.6 Exercises

1. Write a computer algorithm for the simulation of diffusion-limited coalescence in a small lattice, of $L = 40$ sites, say. Perform simulations for the cases (i) $v = 0, R = 0$; (ii) $v = 0, R > 0$; and (iii) $v > 0, R = 0$. In each case, construct space–time plots of the system's evolution (print out the state of the lattice at successive time steps, in sequence). From these diagrams it should be clear that the steady state of case (ii) is an equilibrium state (that is, it is symmetric under time-reversal), whereas the steady state of case (iii) is a nonequilibrium state (not symmetric). (Examples of the diagrams requested can be seen in subsequent chapters.)

2. Find a practical interpretation of the probability difference $E_1 - E_3$. (Answer: $E_1 - E_3 = \text{Prob}(\circ \bullet \bullet) + \text{Prob}(\circ \bullet \circ) + \text{Prob}(\circ \circ \bullet)$.)

3. *Coalescence with bias.* Imagine that diffusion of the particles is asymmetric; they step to the right at rate $2pD/(\Delta x)^2$, and to the left at rate $2qD/(\Delta x)^2$ $(p + q = 1)$. Assume for simplicity that there is no input and no reverse reactions. Show that one recovers Eq. (15.4), as well as the boundary condition (15.5). (Hint: argue that there is reflection symmetry, in spite of the bias! See Chapter 17, for a more detailed treatment of bias.)

4. *One-species annihilation.* Consider coalescence, but with the difference that, when two particles meet, they annihilate immediately (the particles are removed from the system). This models the reaction $A + A \to 0$. Why is it that one cannot write a closed equation for the evolution of the $\{E_n\}$ in this case?

5. Consider the probability that n consecutive sites contain an *even* number of particles (including zero) at time t, $G_n(t)$. Express the concentration of particles as a function of $G_1(t)$. Show also that the probability of having an even (odd) number of particles in n consecutive sites, followed by a particle at the $(n+1)$th site, is $[\pm(G_n - G_{n+1}) + 1 - G_1]/2$. (Hint: see Fig. 15.6.)

6. Derive an evolution equation for G_n for the annihilation process with input, but

without back reactions. Derive also boundary conditions and write the equations in the continuum limit. (Answer: In the discrete version,

$$\partial_t G_n(t) = \frac{2D}{(\Delta x)^2}(G_{n-1} - 2G_n + G_{n+1}) - Rn\,\Delta x(2G_n - 1),$$

with boundary conditions $G_0(t) = 1$ and $G_\infty(t) \le 1$.)

15.7 Open challenges

1. In view of the lack of an exact general approach to diffusion-limited reaction kinetics, special cases of exactly solvable models attract tremendous interest. It would be useful to find other models that can be solved exactly. Are there any other models that can be treated exactly with the IPDF technique?
2. The restriction of the IPDF method to one dimension is very limiting. It seems impossible to generalize the method even to two dimensions, without resorting to approximations. On the other hand, it may be possible to adapt the technique to reactions in the Cayley tree, and possibly to reactions in some finitely ramified fractals. This has not been attempted yet.

15.8 Further reading

- The coalescence process, A + A → A, and the closely related annihilation process, A + A → 0, can be treated exactly in many other ways aside from the IPDF method. See, for example, Bramson and Griffeath (1980), Torney and McConnell (1983), Peliti (1985), Rácz (1985), Lushnikov (1987), Spouge (1988), Amar and Family (1990), Kanter (1990), Krapivsky (1993), Privman (1992; 1994a; 1994b), Privman *et al.* (1995; 1996), Schütz (1995), Simon (1995), Hinrichsen (1996), Hinrichsen *et al.* (1995; 1996a; 1996b), Henkel *et al.* (1995; 1997), Balboni *et al.* (1995), Derrida and Zeitak (1996), Santos *et al.* (1996), and Rey and Droz (1997). A review of quantum-field-theory methods, with extensive annotated references to diffusion-limited reactions in general, is given by Mattis and Glasser (1998).
- On the IPDF method and the coalescence process, see Doering and ben-Avraham (1988), ben-Avraham *et al.* (1990), and ben-Avraham (1995; 1997).

16

Irreversible coalescence

The simple coalescence process $A + A \rightarrow A$ serves as an example of exactly solved anomalous kinetics. The long-time asymptotic behavior is characterized by ordering, resembling the phenomenon of self-organized criticality. When there is a steady input of A particles, the system attains a nontrivial stationary behavior. Comparison of the processes with and without input shows that it is impossible to describe the kinetics by writing an autonomous ordinary differential equation. Hydrodynamic rate equations also fail to reproduce the exact asymptotic kinetics.

16.1 Simple coalescence, $A + A \rightarrow A$

The simplest case for the process is when there is no input ($R = 0$) and there are no back reactions ($v = 0$). In this case one has the irreversible coalescence process $A + A \rightarrow A$ alone, as illustrated in Fig. 16.1. This process has a trivial steady state with zero concentration of particles. With $R = v = 0$, Eq. (15.18) reduces to

$$\frac{\partial}{\partial t} E(x, t) = 2D \frac{\partial^2}{\partial x^2} E, \tag{16.1}$$

with the boundary conditions of Eq. (15.19). The time-dependent solution to this diffusion equation is complicated by the unusual boundary condition of $E(0, t) = 1$, and it is simpler to consider the second derivative of Eq. (16.1) with respect to x, yielding

$$\frac{\partial \rho(x, t)}{\partial t} = 2D \frac{\partial^2 \rho}{\partial x^2}, \tag{16.2}$$

where $\rho(x, t) = c(t)p(x, t)$, from Eq. (15.10). This satisfies the more convenient boundary conditions $\rho(0, t) = \rho(\infty, t) = 0$. The Green function for Eq. (16.2) is

$$G(x, x', t) = \frac{1}{(8\pi Dt)^{1/2}} \left\{ \exp\left[\frac{(x - x')^2}{8Dt}\right] - \exp\left[\frac{(x + x')^2}{8Dt}\right] \right\}, \tag{16.3}$$

217

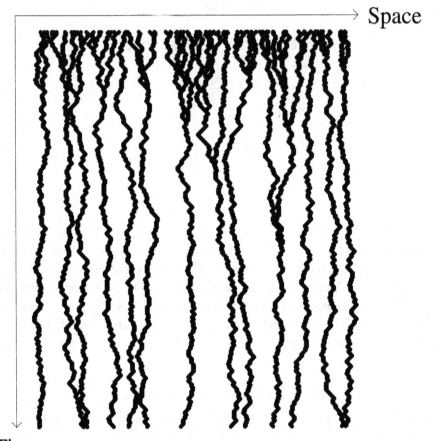

Space

Time

Fig. 16.1. The space–time diagram for irreversible coalescence, $A + A \rightarrow A$. "World-lines" represent the paths of diffusing particles (generated in a computer simulation of the process). Coalescence events correspond to the merging of two world-lines.

so that

$$\rho(x, t) = \int_0^\infty dx'\, G(x, x', t)\rho(x', 0). \qquad (16.4)$$

Given an initial IPDF $p(x, 0)$, and using Eqs. (16.3) and (16.4) and the relation $c(t) = \int_0^\infty dx\, \rho(x, t)$, one obtains a full solution for the time-dependent concentration $c(t)$ and the IPDF $p(x, t)$.

The long-time asymptotic limit $(t \rightarrow \infty)$ is easily computed from

$$G(x, x', t) \rightarrow \frac{2x'}{(8\pi Dt)^{1/2}} \frac{2x}{8Dt} \exp\left(-\frac{x^2}{8Dt}\right). \qquad (16.5)$$

on substituting this into Eq. (16.4) and using the expression $c(t)^{-1} = \int_0^\infty dx\, xp(x, t)$, we obtain

$$c(t) \to \frac{1}{(2\pi Dt)^{1/2}} \quad \text{as} \quad t \to \infty, \tag{16.6}$$

and

$$p(x, t) \to \frac{x}{4Dt} \exp\left(-\frac{x^2}{8Dt}\right) \quad \text{as} \quad t \to \infty, \tag{16.7}$$

independent of the initial conditions. The dimensionless, or *scaling* interparticle distance, $z = c(t)x$, approaches the *stationary* distribution

$$p(z, t) \to \frac{\pi}{2} z \exp\left(-\frac{\pi}{2}\frac{z^2}{2}\right) \quad \text{as} \quad t \to \infty. \tag{16.8}$$

The transient behavior depends strongly on the initial distribution $p(x, 0)$. For example, starting with a completely *random* distribution, $p^{\text{ran}}(x, 0) = c_0 \exp(-c_0 x)$, the transient behavior is

$$\frac{c^{\text{ran}}(t)}{c_0} \approx 1 - \left(\frac{8c_0^2 Dt}{\pi}\right)^{1/2} + \mathcal{O}(c_0^2 Dt), \tag{16.9}$$

with an infinite initial reaction rate, $dc^{\text{ran}}/dt|_{t=0} = \infty$. This happens because the random initial configuration places many particles right next to each other. In contrast, for an ordered, *periodic* initial distribution, for which $p^{\text{per}}(x, 0) = \delta(x - c_0^{-1})$, the reaction proceeds at a transcendentally small rate until $t = \mathcal{O}(1/c_0^2 D)$;

$$\frac{c^{\text{per}}(t)}{c_0} \approx 1 - \left(\frac{8c_0^2 Dt}{\pi}\right)^{1/2} \exp\left(-\frac{1}{8c_0^2 Dt}\right)[1 + \mathcal{O}(c_0^2 Dt)], \tag{16.10}$$

and $dc^{\text{per}}/dt|_{t=0} = 0$.

A very interesting initial configuration is the *scaling* distribution of Eq. (16.8), $p^{\text{sc}}(x, 0) = (\pi/2)c_0^2 x \exp[-(\pi/2)c_0^2 x^2/2]$. This distribution falls between the two extremes of initial order (the periodic distribution) and disorder (the random distribution). In this case, the integral in Eq. (16.4) is easily evaluated in closed form, yielding

$$\frac{c^{\text{sc}}(t)}{c_0} = \frac{1}{(1 + 2\pi c_0^2 Dt)^{1/2}}. \tag{16.11}$$

For this initial condition the interparticle distribution *remains* in its scaling form at all times:

$$p^{\text{sc}}(x, t) = \frac{\pi}{2} c(t)^2 x \exp\left(-\frac{\pi}{2}\frac{c(t)^2 x^2}{2}\right). \tag{16.12}$$

It is clear why the concentration falls between the two extremes in this case. There

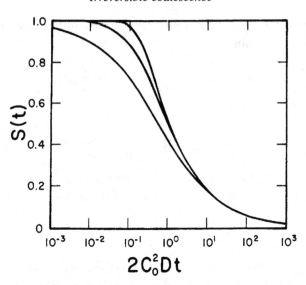

Fig. 16.2. Concentration decay in irreversible coalescence. A plot of the survival probability $S(t) = c(t)/c(0)$ versus the dimensionless time variable $2c(0)^2 Dt$ for the various initial particle distributions discussed in the text: periodic (top), an invariant scaling distribution (middle), and completely random (bottom). After ben-Avraham *et al.* (1990).

is little probability of particles starting off close to each other (the density vanishes as $x \to 0$), so the reaction rate is less than that under totally random initial conditions. On the other hand, there *is* a nonvanishing probability of particles initially being found arbitrarily close, so the rate is greater than that for the periodically ordered distribution, for which this probability is zero.

In Fig. 16.2 we show the temporal evolution of the concentration for the random, periodic, and scaling distributions discussed above. In Fig. 16.3 we plot the evolution of $p(z, t)$ as a function of time, starting from a totally random initial configuration and continuing up to the asymptotic long-time stationary distribution of Eq. (16.8).

The IPDF of Eq. (16.12) displays an interesting microscopic structure for this nonequilibrium state. In thermal equilibrium one expects the maximum entropy distribution of particles, characterized by an exponential IPDF, $p(x) = ce^{-cx}$. (In fact, this is the case with reversible coalescence, as we shall see in the next chapter.) For irreversible coalescence the scaling form of the IPDF vanishes near $x = 0$, indicating that there is an effective mutual repulsion of the particles. The probability of large gaps decays faster than exponentially (proportional to a power of $\exp(-x^2)$). This simple interacting model thus serves as an example of *dynamic self-ordering* in a system far from equilibrium.

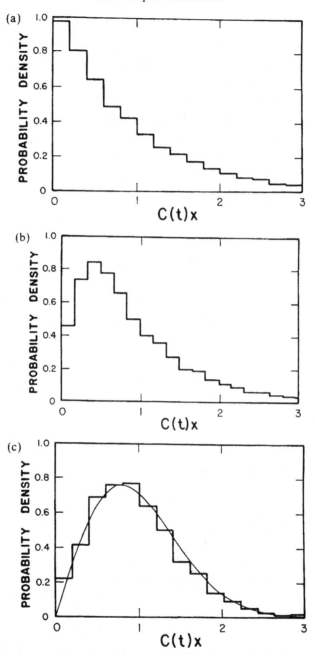

Fig. 16.3. The time-dependent IPDF for the process of irreversible coalescence. $p(z, t)$ is shown as a function of $z = c(t)x$ for a system starting off with a completely random distribution at time $t = 0$ (a), at an intermediate time $2c(0)^2 Dt = 0.066$ (b), and in the long-time asymptotic regime $2c(0)^2 Dt = 4.2$ (c). The histograms represent data from numerical simulations, and the smooth curve in (c) is the analytical result of Eq. (16.8). After ben-Avraham *et al.* (1990).

16.2 Coalescence with input

We now turn to the case of coalescence $A + A \rightarrow A$ with input ($R > 0$), but with no back reactions ($v = 0$), as illustrated in Fig. 16.4. The solution is obtained by straightforward substitution of $v = 0$ into the results for the general case of Chapter 15. In contrast to the trivial long-time asymptotic limit of the pure coalescence process of the previous section, when there is a constant input the system reaches a nontrivial stationary steady state, with a nonvanishing concentration c_s. Note, however, that this is a *nonequilibrium* steady state, since the dynamic rules of coalescence and input are strictly irreversible, so there is no detailed balance. Indeed, the temporal asymmetry of the process is evident from Fig. 16.4.

On substituting $v = 0$ into Eq. (15.25), we find the steady-state concentration

$$c_s(R, D) = \frac{|Ai'(0)|}{Ai(0)} \left(\frac{R}{2D}\right)^{1/3} = (0.729\,01\ldots)\left(\frac{R}{2D}\right)^{1/3}, \tag{16.13}$$

(where $Ai(0) = 0.355\,02\ldots$ and $Ai'(0) = -0.258\,81\ldots$). The fact that c_s is proportional to $(R/D)^{1/3}$ can be deduced from a scaling argument (really just dimensional analysis). Assume that the initial concentration plays no role in the steady state. Then, the only physical parameters influencing the process are the diffusion constant D, with dimensions of (length)^2/(time), and the input rate R, with dimensions of (length $\times$ time)$^{-1}$. To obtain the correct dimension of c_s, (length)$^{-1}$, one must combine them as $(R/D)^{1/3}$. In contrast, notice that, for the reaction-limited process, a classical rate equation would predict $c_s \sim R^{1/2}$. We will discuss classical rate equations and their application in due course.

The stationary IPDF computed from Eq. (15.24) with $v = 0$ is

$$p_s(x) = \frac{1}{c_s(R, D)} \frac{\partial^2 E_0(x)}{\partial x^2} = \left(\frac{R}{2D}\right)^{1/3} \frac{Ai''[(R/2D)^{1/3}x]}{|Ai'(0)|}. \tag{16.14}$$

This stationary distribution is plotted in Fig. 16.5. It is interesting that the probability of there being large gaps between particles falls off as $\exp(-x^{3/2})$, which is *slower* than $\exp(-x^2)$ of the IPDF for pure coalescence. This can be understood in view of the random input of A particles, which is effectively a disordering agent. Notice, however, that the random input does not manage to induce complete disorder, and the stationary IPDF displays a greater ordering than a totally random distribution of particles (a purely exponential IPDF). Moreover, the probability of there being particles near $x = 0$ remains depleted, regardless of the input rate, i.e., the effective repulsion due to coalescence dominates the short length scale behavior. The steady state of the coalescence process with input serves as an example of *static self-ordering* in a system far from equilibrium.

The transient behavior of the system is represented by the eigenfunctions of

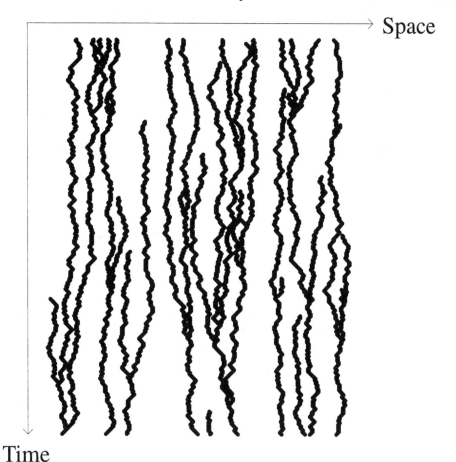

Fig. 16.4. The space–time diagram of irreversible coalescence with input. Results from simulations after the steady state has set in are shown. Input events appear as fresh starts of new world-lines. While the stationary nature of the system's state is evident, it is also evident that this is *not* an equilibrium state: there is a distinct directionality along the time axis.

Eq. (15.21). From Eq. (15.26), we see that, for $v = 0$, the eigenvalues still form a discrete spectrum, $\lambda_n = (2DR^2)^{1/3}|a_n|$. Hence, at long times the transient behavior is dominated by the eigenfunction E_1, and there is an exponential approach to steady state with decay rate $\lambda_1 = (2DR^2)^{1/3}|a_1| = (2.3381\ldots)(2DR^2)^{1/3}$.

16.3 Rate equations

Let us now investigate the usefulness of rate equations for describing the coalescence process. Consider first the simplest case of irreversible coalescence with

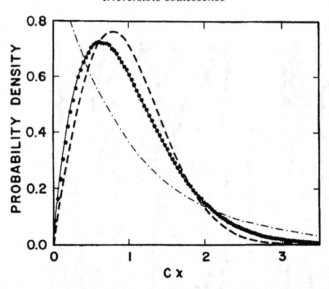

Fig. 16.5. The stationary IPDF for irreversible coalescence with input. Monte Carlo simulation results (points) and the theoretical curve of Eq. (16.14) (solid line) are plotted in comparison with the invariant scaling IPDF of irreversible coalescence without input (broken line) and a completely random distribution (dash–dotted line). After ben-Avraham *et al.* (1990).

no input, as in Section 16.1. For a spatially homogeneous system at macroscopic length scales the system may be described by the classical rate equation

$$\frac{dc(t)}{dt} = -kc(t)^2,   \tag{16.15}$$

where k is an effective reaction rate. This, however, has the solution $c(t) = 1/(c_0^{-1} + kt)$, which contradicts the long-time behavior of the diffusion-controlled process, $c(t) \sim t^{-1/2}$.

Since the neglect of microscopic spatial variations proves to be fatal, one should attempt a hydrodynamic description:

$$\frac{\partial c(x, t)}{\partial t} = D \frac{\partial^2 c(x, t)}{\partial x^2} - kc^2(x, t),   \tag{16.16}$$

where the spatially homogeneous $c(t)$ is replaced by $c(x, t)$. Although the reaction term is fashioned after the classical approach, microscopic fluctuations are now taken into account: they are assumed to dissipate diffusively. Nevertheless, Eq. (16.16) still fails to reproduce the true kinetics. To see this, suppose that the system is enclosed within a volume L. Define a spatial-average global concentration by

$$c(t) = L^{-1} \int_0^L dx\, c(x, t). \tag{16.17}$$

By performing this spatial averaging on Eq. (16.16), we obtain

$$\frac{dc(t)}{dt} = -kL^{-1} \int_0^L dx\, c^2(x, t) \leq -k \left(L^{-1} \int_0^L dx\, c(x, t) \right)^2 = -kc^2(t), \tag{16.18}$$

where the diffusion term vanishes (assuming that there is zero flux at the boundaries), and we have invoked the Cauchy–Schwartz inequality. It follows that the mean-field decay, $c(t) \sim 1/kt$, is an *upper bound* on the decay of the spatially averaged concentration given by the nonlinear partial differential equation (16.16). This is, however, in clear contradiction to the exact solution, which is *slower* than $1/t$. The attempt to separate the effect of diffusion from that of reaction, as is done in reaction–diffusion equations, leads in this case to wrong results.

It should be noted that this kind of calculation does give the correct behavior of the diffusion-limited two-species annihilation process $A + B \rightarrow$ inert. There *macroscopic* segregation of the two species takes place, and the hydrodynamic equation is apparently able to account for the effects of *long-range* spatial inhomogeneities on the reaction rate. In the case of one-species coalescence, though, the nonequilibrium spatial structure at *microscopic* length scales persists throughout the process, rendering the hydrodynamic approach ineffective.

Let us now test the idea of effective rate equations. The effective reaction terms are supposed to account for the effects of the diffusion mechanism, the spatial inhomogeneities and correlations, and the reactions combined. In the case of coalescence, the asymptotic kinetics can be properly described by a classical rate equation with a reaction of effective order three:

$$\frac{dc(t)}{dt} = -\pi D c(t)^3 \qquad \text{as } t \rightarrow \infty. \tag{16.19}$$

where the prefactor is chosen so as to reproduce the exact long-time behavior, $c(t) \rightarrow (2\pi Dt)^{-1/2}$. In fact, if the gaps between nearest particles are initially distributed according to the scaling distribution $p^{sc}(x, 0) = (\pi/2)c_0^2 x \exp[-(\pi/2)c_0^2 x^2/2]$, Eq. (16.19) is exact at *all* times, since it yields $c(t) = c_0/(1 + \pi c_0^2 Dt)^{1/2}$, i.e., it is in agreement with Eq. (16.11). If the initial IPDF is other than p^{sc}, Eq. (16.19) is valid only in the long-time asymptotic limit. Thus, there *can* be an autonomous polynomial rate equation for the irreversible coalescence process in one dimension, valid for all times, but its existence depends on the *microscopic* initial conditions.

Consider now the irreversible coalescence process with input. As discussed above, in the absence of input ($R = 0$) the concentration (eventually) obeys $dc/dt = -\pi Dc^3$. On the other hand, if diffusion is turned off ($D = 0$) the concentration obeys $dc/dt = R$. It is sensible to hypothesize an effective autonomous polynomial rate equation for the combined processes, of the form $dc/dt = -\pi Dc^3 + R$. However, this is incompatible with the correct stationary concentration of Eq. (16.13). Notice, though, that it predicts the correct – and nonclassical – scaling of c_s with D and R.

Near the nonempty stationary states an approximate rate equation can be derived on the basis of the exact concentration and the relaxation spectrum. The asymptotic approach to a nonempty stationary state is given by $c(t) = c_s + \delta c\, e^{-\lambda_1 t}$, with $c_s = [|\text{Ai}'(0)|/\text{Ai}(0)](R/2D)^{1/3}$ and $\lambda_1 = |a_1|(2DR^2)^{1/3}$. Hence, the simplest autonomous, first-order equation which correctly captures both the nontrivial concentration and its relaxation is

$$\frac{dc(t)}{dt} = -\alpha Dc^3 + \beta R, \qquad (16.20a)$$

where

$$\alpha = \frac{2|a_1|\text{Ai}(0)^2}{3\text{Ai}'(0)^2}, \qquad \beta = \frac{|a_1||\text{Ai}'(0)|}{3\text{Ai}(0)}. \qquad (16.20b)$$

The difference in the reaction kinetics in the presence of *both* diffusion–reaction and input can be traced to the spatial structure of the nonequilibrium states. The IPDFs, which are different for the reactions with and without input, determine the rate at which particles interact; hence, it is not surprising that the reactions proceed at different rates with different microscopic configurations even if the macroscopic concentrations coincide.

We may combine our knowledge of the system's behavior in the limits $D/R \to 0$ and ∞, as well as the stationary case, to construct a rate equation that reproduces all the correct dynamics. Such an equation must be of the form

$$\frac{dc(t)}{dt} = (-\alpha Dc^3 + \beta R)F\left(\frac{c}{c_s}\right), \qquad (16.21)$$

where c_s is the exact stationary concentration of Eq. (16.13), and the "scaling" function $F(z)$ satisfies

$$F(0) = \frac{1}{\beta}, \quad F(1) = 1, \quad F(\infty) = \frac{\pi}{\alpha}. \qquad (16.22)$$

Since no polynomial can satisfy the conditions of Eq. (16.22), our equation must be *nonpolynomial*. The claim is that it would describe the time-dependent concentration after initial transients have died away. Perhaps it would even be valid for some restricted class of time-dependent problems; for example, if R were modulated periodically.

16.4 Exercises

1. An alternative approach to the method of empty intervals is the following. If we label the particles at time $t = 0$, and agree that, following a coalescence event, the new particle carries the label of the particle that originally came from the left, say, then we can simply study the survival probability of labeled particles. This problem is equivalent to that of a single random walker with an absorbing point at the origin. Show that this leads to Eq. (16.2), with the same interpretation of $\rho(x, t)$, and with the boundary condition $\rho(0, t) = 0$. (Hint: see Doering and ben-Avraham (1988).)

2. Obtain the long-time asymptotic behavior of Eqs. (16.6) and (16.7) with the *Ansatz* that $E(x, t) = E(z)$ as $t \to \infty$, where $z = x/\sqrt{Dt}$. (Hint: derive an ordinary differential equation for $E(z)$ and solve it.)

3. *Cooperative input.* Suppose that the input rate is affected by some cooperative effect and depends on the global concentration; $R = Ac(t)^\gamma$ ($A, \gamma > 0$ are constants). Compute the steady state as a function of γ. Discuss the singular case of $\gamma = 3$. Notice that the spectral decomposition of Section 16.2 is not valid, since $R = R(t)$.

4. Analyze the cooperative input of the previous exercise (a) in the reaction-limited case, and (b) in the diffusion-limited case, by means of an effective classical rate equation.

5. Solve the equations for the annihilation process (see Exercises 15.4–15.6) with and without input. Show that, in the long-time asymptotic limit, $c_{anni} = 2^{-1} c_{coal}$ if $R = 0$ and $c_{anni} = 2^{-2/3} c_{coal}$ if $R > 0$.

6. Prove that, for pure coalescence, if the initial distribution is the scaling distribution, Eq. (16.12), then the effective classical rate equation (16.19) is satisfied at all times.

7. The following example proves that no autonomous *first-order* rate equation can possibly describe the dynamics of coalescence with arbitrarily *fast* changes in the input rate. Consider an experiment in which $R = 0$ from some large negative time until $t = 0$, so that the interparticle distances are distributed according to the scaling distribution of Eq. (16.12), with $c_0 = c(0) \neq 0$. At time $t = 0$, R is suddenly switched to the value $R_0 > 0$ such that the stationary concentration for input rate R_0 is exactly $c(0)$, i.e., $c_s(R_0, D) = c(0)$. If the concentration obeys a first-order equation, then $c(t) = c(0)$ for all $t \geq 0$. Show, however, that $dc/dt > 0$ at $t = 0^+$. Thus, at least a second-order differential equation is necessary in order to describe this situation.

16.5 Open challenges

1. In spite of the exact results, the IPDF for irreversible coalescence is not well understood. The depletion of probability near $x = 0$ is explained by the effective repulsive interaction between particles, arising from coalescence, but what is the origin of the sharpening of the IPDF's tail at long distances (from e^{-x}, for random distributions, to e^{-x^2}, in the long-time asymptotic limit)?

2. Consider cooperative input, like in Exercise 3, but instead of being a response to the *global* concentration, the input is proportional to some power of a *local* concentration, defined for example as $c_\ell(x, t) \equiv (2\ell)^{-1} \int_{-\ell}^{\ell} c(x + x', t)\, dx'$. The study of such a system may be especially interesting in the marginal case of $R(x, t) \sim c_\ell(x, t)^3$. We expect that, owing to fluctuations, there will be regions where the concentration explodes, while in other regions it is quelched to zero, and there might emerge a nontrivial dynamical phase transition. Its study would require techniques for dealing with inhomogeneous systems (see Chapter 18).

3. We have seen that coalescence with input organizes itself into a nonequilibrium steady state. The situation is similar to the famous sand-pile model of self-organized criticality. The question is whether one could identify analogous avalanche events, and study the corresponding critical exponents of the coalescence model analytically.

16.6 Further reading

- Rácz (1985), Doering and ben-Avraham (1988; 1989), ben-Avraham *et al.* (1990), Clément *et al.* (1994), and Rey and Droz (1997).

17

Reversible coalescence

The case of reversible coalescence, when the back reaction $A \to A + A$ is allowed, is special in that the steady state is a true equilibrium state. The process can then be analyzed by standard thermodynamics techniques and one expects simple, classical behavior. It is therefore surprising to find that the approach to equilibrium is characterized by a dynamical phase transition in the typical time of relaxation. This phase transition can be exactly analyzed in finite lattices, providing us with a unique opportunity to study finite-size effects in a dynamical phase transition.

17.1 The equilibrium steady state

We now turn to the case of reversible coalescence, when the back reaction $A \to A + A$ occurs at rate $v > 0$ (and with no input, $R = 0$). The process is illustrated in Fig. 17.1. After a short transient the system arrives at a steady state with a finite concentration of particles. Because coalescence is now reversible, this steady state is in fact an *equilibrium* state, which satisfies *detailed balance*. The statistical time-reversible invariance of this equilibrium state can be seen in Fig. 17.1, in which the direction of time is ambiguous (compare it with Figs. 16.1 and 16.4).

The equilibrium state is a state of maximum entropy, and therefore the particles follow a completely random (Poisson) distribution. This is characterized by an exponential IPDF, $p_{eq}(x) = c_{eq} \exp(-c_{eq}x)$. Alternatively, the state may be described by the lack of correlation among the occupation probabilities of different sites: each site is occupied with probability $c_{eq} \Delta x$, independent of other sites. With this in mind we may compute c_{eq} by equating the rate of coalescence to the rate of birth of new particles (provided that they are being born into *empty* sites):

$$2\frac{D}{(\Delta x)^2}(c_{eq}\,\Delta x)^2 = \frac{v}{\Delta x}(c_{eq}\,\Delta x)(1 - c_{eq}\,\Delta x), \qquad (17.1)$$

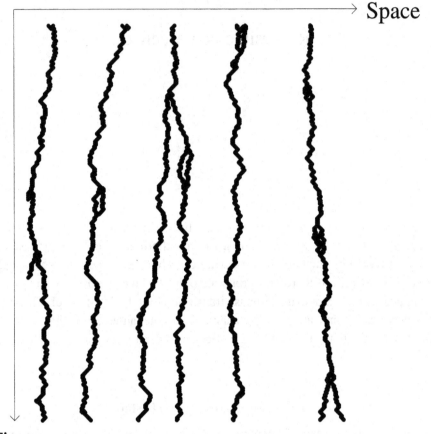

Space

Time

Fig. 17.1. The space–time diagram for reversible coalescence, $A + A \rightleftharpoons A$. The computer-generated state of the system after equilibrium has been achieved is shown. Merging of world-lines represents coalescence, whereas birth events manifest themselves as bifurcations of the world-lines. The equilibrium nature of this state is evident from the time-reversal symmetry of the diagram.

which has the solution

$$c_{eq} = \frac{v}{2D + v\,\Delta x}. \tag{17.2}$$

A different approach is to start from the eigenvalue equation (15.21) and set $R = 0$. We have, recalling that $E_\lambda(x) = F_\lambda(x)e^{-vx/(4D)}$,

$$0 = 2D\frac{\partial^2 F_\lambda(x)}{\partial x^2} - \left(\frac{v^2}{8D} - \lambda\right)F_\lambda(x). \tag{17.3}$$

The equilibrium state is obtained by setting $\lambda = 0$. The boundary condition at

$x = 0$ implies $E_0(0) = F_0(0) = 1$, and we find

$$E_{eq}(x) = E_0(x) = e^{-vx/(2D)}.\tag{17.4}$$

Using Eqs. (15.9) and (15.10), we derive

$$c_{eq} = \frac{v}{2D},\tag{17.5}$$

and

$$p_{eq}(x) = \frac{v}{2D}e^{-vx/2D} = c_{eq}e^{-c_{eq}x}.\tag{17.6}$$

Thus, the stationary IPDF is indeed exponential, corresponding to the maximum-entropy case of thermal equilibrium. The equilibrium concentration of Eq. (17.5) is of course the continuum limit of Eq. (17.2).

17.2 The approach to equilibrium: a dynamical phase transition

For the transient behavior, we want to solve for the eigenvalues $\lambda > 0$. If one were to take a naive $R \to 0$ limit of the spectrum in Eq. (15.26), one would have predicted a gap between the stationary eigenvalue $\lambda = 0$ and the first decaying solution $\lambda = v^2/(8D)$. It turns out, however, that there is a *continuous* spectrum in the interval $(0, v^2/(8D))$.

For $\lambda > 0$, $F_\lambda(x)$ need not necessarily vanish as $x \to \infty$, because $E_\lambda = e^{-vx/(4D)}F_\lambda$, and the boundary conditions require only that $E_\lambda \to 0$ as $x \to \infty$. Suppose first that $\lambda > v^2/(8D)$. Then $F_\lambda(0) = 0$, and the solution of Eq. (17.3) is

$$F_\lambda(x) = \sin\left[\left(\frac{\lambda}{2D} - \frac{v^2}{16D^2}\right)^{1/2}x\right], \qquad \lambda > \frac{v^2}{8D}.\tag{17.7a}$$

For $\lambda = v^2/(8D)$, we have

$$F_\lambda(x) = x, \qquad \lambda = \frac{v^2}{8D},\tag{17.7b}$$

whereas for $\lambda < v^2/(8D)$ (but still $\lambda > 0$)

$$F_\lambda(x) = \sinh\left[\left(\frac{v^2}{16D^2} - \frac{\lambda}{2D}\right)^{1/2}x\right], \qquad \lambda < \frac{v^2}{8D}.\tag{17.7c}$$

The eigenfunctions for $\lambda > 0$ are thus

$$E_\lambda(x) = \begin{cases} e^{-vx/(4D)} \sin\left[\left(\dfrac{\lambda}{2D} - \dfrac{v^2}{16D^2}\right)^{1/2} x\right] & \lambda > \dfrac{v^2}{8D}, \\[3mm] xe^{-vx/(4D)} & \lambda = \dfrac{v^2}{8D}, \\[3mm] e^{-vx/(4D)} \sinh\left[\left(\dfrac{v^2}{16D^2} - \dfrac{\lambda}{2D}\right)^{1/2} x\right] & 0 < \lambda < \dfrac{v^2}{8D}. \end{cases} \tag{17.8}$$

From this one can infer the approach to equilibrium for various initial conditions. If the initial IPDF falls off as $\exp(-c_0 x)$ as $x \to \infty$, with $c_0 > v/(4D) = c_{eq}/2$, then the time-dependent solution cannot contain any modes with $\lambda < v^2/(8D)$, because these modes decay more slowly than does $\exp(-c_{eq}x/2)$. In this case the most slowly decaying component corresponds to $\lambda = v^2/(8D) = Dc_{eq}/2$ and $c(t) - c_{eq} \sim \exp(-Dc_{eq}^2 t/2)$ as $t \to \infty$. If the initial IPDF falls off as $\exp(-c_0 x)$ as $x \to \infty$, with $c_0 < v/(4D) = c_{eq}/2$, then it must contain a mode with $\lambda = c_0 v - 2c_0^2 D = 2Dc_0(c_{eq} - c_0)$. This would be the most slowly decaying mode, so that $c(t) - c_{eq} \sim \exp[-2Dc_0(c_{eq}-c_0)t]$ as $t \to \infty$. Thus, there is a sharp transition in the dynamics of the approach to equilibrium, i.e., in the exponential relaxation time:

$$\tau = -\lim_{t\to\infty} t^{-1} \ln(|c(t) - c_{eq}|), \tag{17.9}$$

governed by the spatial decay of the initial IPDF. This transition is shown in Fig. 17.2.

In fact, given a specific initial IPDF, one can obtain an explicit solution for the transient dynamics. The special case of an initially random distribution, $p(x, 0) = c_0 \exp(-c_0 x)$, has been worked out exactly (Burschka *et al.*, 1989). This particular case is natural in the sense that it corresponds to the system's equilibrium steady state at some fixed values of the parameters D and v. The long-time behavior, computed from the exact expression valid for all times, is

$$c(t) - c_{eq} \sim \begin{cases} -(c_{eq} - 2c_0) \exp[-2Dc_0(c_{eq} - c_0)t] & c_0 < c_{eq}/2, \\[3mm] -\dfrac{1}{(2\pi Dt)^{1/2}} \exp\left(-\dfrac{Dc_{eq}^2 t}{2}\right) & c_0 = c_{eq}/2, \\[3mm] \dfrac{2}{\sqrt{\pi}}[c_{eq}^{-2} - (c_{eq} - 2c_0)^{-2}] \dfrac{1}{(2Dt)^{3/2}} \exp\left(-\dfrac{Dc_{eq}^2 t}{2}\right) & c_0 > c_{eq}/2, \end{cases} \tag{17.10}$$

in agreement with our discussion of general initial conditions. In Fig. 17.3 we show the excellent agreement between the exact solution and computer simulations obtained with initially exponential IPDFs.

Spatial correlations in the microscopic distribution of particles provide the physical mechanism for the slow-relaxation "phase". When there are large gaps

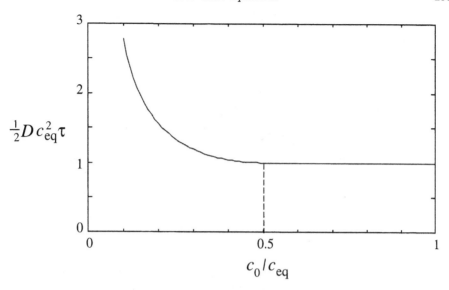

Fig. 17.2. The exponential relaxation time τ (in reduced units) as a function of the initial concentration c_0. A sharp transition in the behavior of τ occurs when c_0 is half of the equilibrium concentration, c_{eq}.

between neighboring particles, then the relaxation is dominated by the time taken to fill these gaps. These large empty regions can be filled only from the sides, and a simple argument provides the correct decay time: concentration fronts drift into empty regions at speed v, and a typical initial interparticle distance is c_0^{-1}; thus, the time taken for the typical gap to be filled is $\tau = (c_0 v)^{-1} = (2Dc_{eq}c_0)^{-1}$, the correct value of Eq. (6.10), as $c_0 \to 0$. This argument is clearly not valid for concentration perturbations above the equilibrium concentration, and a relaxation time uniform in initial conditions is not surprising. What is surprising is that the mechanism described above exerts its influence at a specific critical initial concentration, $c_0 = c_{eq}/2$, well below c_{eq}.

17.3 Rate equations

The dynamical phase transition in the reversible coalescence process (when R is strictly zero) has strong implications for the possible description of the system in terms of classical rate equations. The typical decay to equilibrium, as given by Eq. (17.10) for the case $c_0 > c_{eq}/2$, is not purely exponential or purely algebraic – which would be predicted by any finite-dimensional description of the process, *even arbitrarily close to equilibrium*. Moreover, the purely exponential approach to equilibrium (when $c_0 < c_{eq}/2$) depends on the initial condition. This behavior can never be accounted for by a finite system of ordinary differential equations with

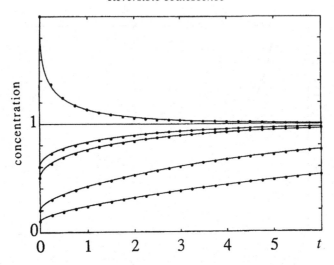

Fig. 17.3. The approach to equilibrium in reversible coalescence, shown for various values of the initial concentration: $c_0/c_{eq} = 0.1, 0.2, 0.5, 0.6$, and 2.0 (bottom to top). The Monte Carlo results (dots) are plotted together with the exact results (solid curves). The units are c_{eq} for concentration, and $1/(2Dc_{eq}^2)$ for time. After Burschka *et al.* (1989).

coefficients independent of the initial state. Below the transition point, $c_0 < c_{eq}/2$, where the *initial* state is far enough from equilibrium, spatial correlations among the particle positions persist forever.

17.4 Finite-size effects

The transition in the relaxation time discussed above is well defined, and takes place at a precise critical point, because the system is infinite. In finite systems the abrupt transition is replaced by a smooth transition about the region of the original critical point. This effect is of extreme practical importance, since real-life systems (and computer simulations) are always finite.

We shall now examine what becomes of our dynamical phase transition when the system is finite (Doering and Burschka, 1990). Suppose that the reversible coalescence process takes place in a lattice of length L, with periodic boundary conditions: particles that hop out of one edge re-enter the lattice at the opposite edge. This is equivalent to a ring of perimeter L, and ensures that the system preserves translation invariance. The analysis is almost identical to that of the infinite system and the process is still described by Eq. (17.3). The only difference is that the boundary condition $E(\infty, t) = 0$ is replaced by $E(L, t) = 0$, stating as before that the system is never empty. With this boundary condition, the stationary solution of Eq. (17.3) becomes

$$E_{eq}(x) = (e^{-vx/(2D)} - e^{vL/(2D)})/(1 - e^{-vL/(2d)}), \qquad (17.11)$$

and the equilibrium concentration in finite volume L is

$$c_{eq} = \frac{v}{2D}(1 - e^{-vL/(2D)}). \qquad (17.12)$$

The transient behavior is obtained from the eigenfunctions with $\lambda > 0$. The only case which satisfies the boundary condition $E(L, 0) = 0$ is that of Eq. (17.7a), with

$$\lambda_n = \frac{v^2}{8D}\left[1 + \left(\frac{4\pi n D}{vL}\right)^2\right], \qquad n = 0, 1, 2, \ldots. \qquad (17.13)$$

Suppose that the system is initially prepared at a random distribution of concentration c_0. The approach to equilibrium can then be written as

$$c(t) - c_{eq} = \frac{2}{L}\sum_{n=0}^{\infty} a_n e^{-\lambda_n t}, \qquad (17.14)$$

where

$$a_n = \frac{1 - (-1)^n e^{(c_c - c_0)L}}{1 - e^{c_0 L}}\left(\frac{1}{1 + [c_c L/(n\pi)]^2} - \frac{1}{1 + (1 - c_0/c_c)^2[c_c L/(n\pi)]^2}\right), \qquad (17.15)$$

and $c_c \equiv v/(4D)$ is the critical concentration of the phase transition in the bulk ($L = \infty$).

In Fig. 17.4 we plot the exact relaxation dynamics for various initial concentrations, with the leading bulk dependence (of $\exp[-v^2 t/(8D)]$) removed, for clarity. The decay is eventually dominated by the first term in the sum (17.15), but remnants of the infinite-volume transition are visible, for $c_0 < c_c$, for times short relative to the "crossover time" $t_\times$. A reasonable measure of this crossover time is the inflection point of the curves in Fig. 17.4:

$$0 = \left.\frac{d^2 \ln(|c(t) - c_{eq}|)}{dt^2}\right|_{t=t_\times}. \qquad (17.16)$$

As one can see from Fig. 17.4, $t_\times$ grows as the initial concentration approaches the critical bulk concentration. Define $\epsilon = (c_c - c_0)/c_c \geq 0$, measuring the distance from the critical concentration. When the equilibrium number of particles is large, $vL/(2D) \gg 1$, and, for a system close to the transition, one finds from Eqs. (17.15) and (17.16)

$$t_\times(L, \epsilon) \approx (5\pi^2)^{-1}\frac{L^2}{2D}\ln\left(\frac{1}{\epsilon}\right). \qquad (17.17)$$

Amazingly, $t_\times$ is *unbounded* for finite systems near the transition at $\epsilon = 0$!

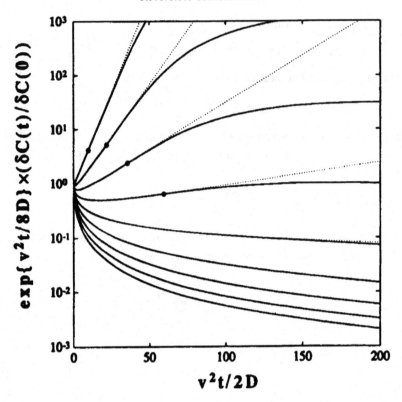

Fig. 17.4. Crossover between infinite and finite-size behavior in the relaxation of reversible coalescence. The leading-time behavior of $c(t)$ is factored out to highlight the sharp transition around $c_0 = c_{eq}/2$. The dotted lines correspond to the behavior in an infinite system (cf. Fig. 17.3). Below the transition point the anomalous decay persists until the crossover time defined by the inflection point, Eq. (17.16), and indicated by a dot on the curves. The initial concentrations are given by $c_0 = a[v/(2D)]/\{1 - \exp[-vL/(2D)]\}$, with $a = 0.1, 0.2, \ldots, 0.9$ (top to bottom). After Doering and Burschka (1990).

17.5 Exercises

1. In the equilibrium state, the state of different sites is uncorrelated and $E_n = E_1^n$. Use this information to work out the IPDF of the equilibrium state for reversible coalescence directly from Eq. (15.16), with $R = 0$ and $\partial_t E_n = 0$. (Answer: $E_n = (2D)^n/(2D + v\,\Delta x)^n$.)

2. *The approach to equilibrium in reversible coalescence.* Suppose that the initial distribution is random, $E(x, 0) = \exp(-c_0 x)$. Find the decomposition of this initial condition into the eigenfunctions of Eq. (17.8). Work out the long-time asymptotic limit and confirm (17.10).

3. For a fractal initial distribution of particles, for which $E(x, 0) \sim x^{-\epsilon}$ as $x \to \infty$, show that the approach to equilibrium is $c(t) - c_{eq} \sim t^{-\epsilon}$. What is ϵ if the

particles are initially distributed as in the Cantor set, with $d_f = \ln 2/\ln 3$?

4. Solve the problem of irreversible coalescence (without input) in a ring of perimeter L. Show that the long-time asymptotic solution yields a *finite* concentration, $c(\infty) = 1/L$ (the last particle cannot be eliminated!). Prove also that the initial decay is $\sim 1/\sqrt{t}$ and that at long times it crosses over to an exponential decay. Estimate the crossover time.

17.6 Open challenges

1. Finite-size effects have been studied only for reversible coalescence. It may prove useful to derive analytical solutions for other cases of coalescence in finite segments (cf. Exercise 4).

2. The study of finite-size effects on the dynamical phase transition in reversible coalescence has considered only periodic boundary conditions. It is a well-known fact that finite-size effects are strongly dependent on boundary conditions. It will therefore be interesting to investigate other possibilities, such as reflecting, absorbing, and radiating boundary conditions.

17.7 Further reading

- Burschka *et al.* (1989), Doering and Burschka (1990), and ben-Avraham *et al.* (1990).
- On the effect of exotic (fractal) initial distributions of particles, see Alemany (1997).

18

Complete representations of coalescence

A stochastic particles system is truly fully characterized only when the infinite hierarchy of multiple-point density correlation functions – the probability of finding any given number of particles at some specified locations, simultaneously – is known. The IPDF method is capable of handling this complicated question, and, in fact, in several cases the complete exact solution may be thus obtained. Such studies reveal a peculiar property of "shielding", particular to reversible coalescence, whereby a particle at the edge of the system seems to shield the rest of the particles from the imposed boundary conditions.

We begin with an analysis of inhomogeneous systems, when translational symmetry is broken, at the simple level of point densities (i.e., the particle concentration). An interesting application is to the study of Fisher waves, and the effect of internal fluctuations on this well-known mean-field model for invasion of an unstable phase by a stable phase.

18.1 Inhomogeneous initial conditions

Until now we have discussed only translationally symmetric systems. The method of interparticle distribution functions can be generalized to inhomogeneous situations (Doering *et al.*, 1991). To this end, $E_n(t)$ need simply be replaced by $E_{n,m}(t)$ – the probability that the sites $n, n + 1, \ldots, m$ are empty at time t. Thus, for example, the density of particles at site n at time t is (cf. Eq. (15.1))

$$\rho_n(t) = (1 - E_{n,n}(t))/\Delta x, \tag{18.1}$$

since $E_{n,n}(t)$ is the probability that site n is empty. Likewise, the probability that sites n through m are empty, but that site $m + 1$ is occupied, is $E_{n,m} - E_{n,m+1}$ (cf. Eq. (15.2)), and $p_{n,m}(t)$ – the probability that sites n and m are occupied, but that there are no particles in between – is (Fig. 18.1),

$$p_{n,m}(t) = E_{n,m}(t) + E_{n+1,m-1}(t) - E_{n+1,m}(t) - E_{n,m-1}(t). \tag{18.2}$$

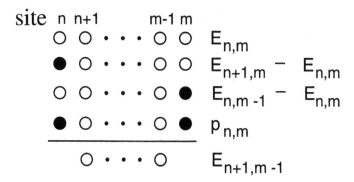

Fig. 18.1. The derivation of Eq. (18.2). The probabilities of all four possibilities of occupation of the sites at the edges of an empty $[n, m]$th segment add up to $E_{n+1,m-1}$. From this, one can deduce the probability of having two particles separated by a specific number of empty sites (the fourth possibility), and Eq. (18.2) follows.

We can now construct an evolution equation for the coalescence process on the line. Because the system may now be inhomogeneous, we let the rates of the various processes (D, v, and R) depend on the location, n. Following the procedure of Section 15.4, we find

$$\frac{\partial E_{n,m}(t)}{\partial t} = \frac{D_n}{(\Delta x)^2}[E_{n+1,m}(t) - E_{n,m}(t)]$$

$$+ \frac{D_m}{(\Delta x)^2}[E_{n,m-1}(t) - E_{n,m}(t)]$$

$$- \frac{D_{n-1}}{(\Delta x)^2}[E_{n,m}(t) - E_{n-1,m}(t)]$$

$$- \frac{D_{m+1}}{(\Delta x)^2}[E_{n,m}(t) - E_{n,m+1}(t)] \tag{18.3}$$

$$- \frac{v_{n-1}}{2\,\Delta x}[E_{n,m}(t) - E_{n-1,m}(t)]$$

$$- \frac{v_{m+1}}{2\,\Delta x}[E_{n,m}(t) - E_{n,m+1}(t)]$$

$$- \sum_{k=n}^{m} R_k\,\Delta x\, E_{n,m}(t).$$

This equation is good for $m > n$. For $m = n$ one derives a special equation, which describes the change in the number of particles in the system. This can be absorbed in (18.3) if one defines

$$E_{n+1,n}(t) = E_{n,n-1}(t) = 1, \tag{18.4}$$

which serves as a boundary condition.

As before, it is more convenient to pass to the continuum limit, $\Delta x \to 0$, in the same fashion as in Chapter 15. The density of particles then becomes (taking the limit of Eq. (18.1))

$$\rho(x) = -\frac{\partial E(x, y, t)}{\partial y}\bigg|_{y=x}, \tag{18.5}$$

and the IPDF is obtained from the continuum limit of (18.2),

$$p(x, y, t) = -\frac{\partial^2 E(x, y, t)}{\partial x \, \partial y}. \tag{18.6}$$

The evolution equation for empty intervals, Eq. (18.3), becomes

$$\frac{\partial E(x, y, t)}{\partial t} = \frac{\partial}{\partial x} D(x) \frac{\partial E}{\partial x} + \frac{\partial}{\partial y} D(y) \frac{\partial E}{\partial y} - \frac{1}{2} v(x) \frac{\partial E}{\partial x} + \frac{1}{2} v(y) \frac{\partial E}{\partial y}$$
$$- \left(\int_x^y R(z) \, dz \right) E, \tag{18.7}$$

subject to the boundary condition (18.4),

$$\lim_{y \downarrow x \text{ or } x \uparrow y} E(x, y, t) = 1. \tag{18.8}$$

This formalism allows one to deal with general inhomogeneous problems concerning coalescence. We shall now apply it to a specific case of interest.

18.2 Fisher waves

Fisher studied the dynamics of the invasion of a stable phase into an unstable phase. For this purpose, he focused on the reaction–diffusion equation

$$\frac{\partial \rho}{\partial t} = D \nabla^2 \rho + k_1 \rho - k_2 \rho^2, \tag{18.9}$$

where $\rho(x, t)$ in this context is a local concentration characterizing the state of the system in space ($x \in R^d$). D describes diffusion in the system, and k_1 and k_2 are rates of growth and death processes described by the local dynamics. There are two homogeneous steady states: (a) an unstable state, $\rho = 0$, and (b) a stable state, $\rho = k_1/k_2$.

Consider a system that consists initially of a plane, say $x_1 = 0$, separating a stable half-space (with $\rho = k_1/k_2$, for $x_1 < 0$) from an unstable half-space (with $\rho = 0$, for $x_1 > 0$), as illustrated in Fig. 18.2a. As the system evolves, the stable phase will invade the unstable phase, and the front will broaden (Fig. 18.2b). It can be shown that eventually the wave front approaches a stationary profile $\rho(x, t) = f(x_1 - ct)$, traveling at constant speed $c = 2\sqrt{k_1 D}$ (Fig. 18.2c).

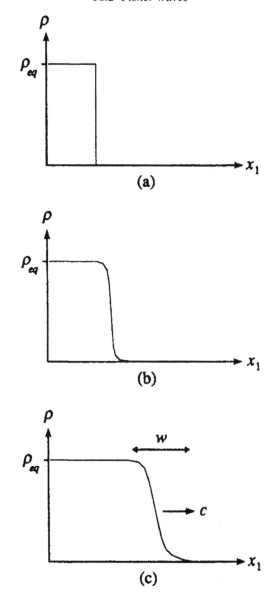

Fig. 18.2. The Fisher wave. (a) The initial profile with a sharp interface between the stable phase (left) and the unstable phase (right). (b) The wave front begins to broaden as the front moves to the right. (c) The asymptotic wave-front speed is realized, and the wave front's width is w. After Riordan *et al.* (1995).

The coalescence process provides us with a unique opportunity to test how internal noise, and dimensionality, may affect the mean-field picture of Fisher waves. Indeed, reversible coalescence without input, and with homogeneous diffusion and back-reaction rates $(D(x) = D, v(x) = v)$ is described at the mean-field level by the Fisher equation, Eq. (18.9). If the system is initially prepared with a half-space at the equilibrium concentration of Chapter 17, and a half-space with zero concentration;

$$\rho_0(x) \equiv \rho(x, t = 0) = \begin{cases} v/(2D) & x < 0, \\ 0 & x > 0, \end{cases} \tag{18.10}$$

then the analogy is complete.

In one dimension, we can write an exact evolution equation for the empty intervals (a special case of Eq. (18.7));

$$\frac{\partial E(x, y, t)}{\partial t} = D \frac{\partial^2 E}{\partial x^2} + D \frac{\partial^2 E}{\partial y^2} - \frac{v}{2} \frac{\partial E}{\partial x} + \frac{v}{2} \frac{\partial E}{\partial y}, \tag{18.11}$$

in the half-plane $x < y$, with the boundary condition of Eq. (18.8). A general, closed solution (given $E(t = 0)$) is possible (see the exercises). For ρ_0 of Eq. (18.10), one gets

$$\rho(x, t) = \frac{v}{4D} \operatorname{erfc}\left(\frac{x - vt/2}{(4Dt)^{1/2}}\right). \tag{18.12}$$

Thus, the front propagates at speed $v/2$, but the wave-front profile does not achieve a stationary shape. Instead, it broadens proportionally to $\sqrt{t}$ (see, however, Section 18.4).

The spreading of the front is clearly the result of internal fluctuations. It is natural to ask what happens in higher dimensions. As the dimensionality increases the internal noise becomes less important and eventually the mean-field Fisher equation should adequately describe the system. Unfortunately, the exact solution of the IPDF method is limited to $d = 1$, and the question must be studied numerically.

Computer simulations reveal an interesting picture (Riordan *et al.*, 1995). In $d = 2$ the wave front propagates at constant speed and it is still found to broaden as $t^{0.272}$, slower than it does in one dimension. In $d = 3$ the spreading of the front is very slow, and its is difficult to decide numerically whether the width grows with a very low exponent (as $t^{0.10}$), or logarithmically ($\sim\sqrt{\ln t}$). Finally, in $d \geq 4$, the wave front seems to approach a stationary form that is well described by the mean-field Fisher equation. Thus, simulations suggest that $d = 3$ is the upper critical dimension beyond which internal fluctuations may be safely neglected.

There has recently been a surge of activity regarding Fisher waves and the effects induced by discreteness (Mai *et al.*, 1996; Brunet and Derrida, 1997; Kessler *et al.*, 1998; Ebert and van Saarloos, 1998). The point of view of these studies is slightly different than the one expounded above: essentially, they examine the question of passage to the mean-field limit as a function of the number of particles per

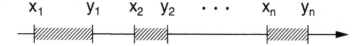

Fig. 18.3. The joint probability of empty intervals $E_n(x_1, y_1, x_2, y_2, \ldots, x_n, y_n, t)$ is the probability that the nonoverlapping (shaded) intervals (x_1, y_1), (x_2, y_2), $\ldots$, (x_n, y_n) are empty at time t.

site. In contrast, in our coalescence model we allow at most *one* particle per site, and the passage to the mean-field limit is investigated as a function of the spatial dimension. Indeed, an interesting open challenge is that of how to design a coalescence model with arbitrary occupancies, and to solve it exactly.

18.3 Multiple-point correlation functions

The IPDF method can also handle multiple-point correlation functions – the probability of finding particles at $x_1, x_2, \ldots, x_n$ at time t (Doering and Burschka, 1990). Let $E_n(x_1, y_1, x_2, y_2, \ldots, x_n, y_n, t)$ be the joint probability that the intervals $[x_i, y_i]$ $(i = 1, 2, \ldots, n)$ are empty at time t. The intervals are nonoverlapping, and ordered: $x_1 < y_1 < \cdots < x_n < y_n$ (Fig. 18.3). Then, the n-point correlation function is given by

$$\rho_n(x_1, \ldots, x_n, t) = (-1)^n \frac{\partial^n}{\partial y_1 \cdots \partial y_n} E_n(x_1, y_1, \ldots, x_n, y_n, t)|_{y_1=x_1,\ldots,y_n=x_n}.$$
(18.13)

The E_n satisfy the partial differential equation:

$$\frac{\partial}{\partial t} E_n(x_1, y_1, \ldots, x_n, y_n, t) = D\left(\frac{\partial^2}{\partial x_1^2} + \frac{\partial^2}{\partial y_1^2} + \cdots + \frac{\partial^2}{\partial x_n^2} + \frac{\partial^2}{\partial y_n^2} \right) E_n$$
$$- \frac{v}{2}\left[\left(\frac{\partial}{\partial x_1} - \frac{\partial}{\partial y_1} \right) + \cdots + \left(\frac{\partial}{\partial x_n} - \frac{\partial}{\partial y_n} \right) \right] E_n$$
$$- R[(y_1 - x_1) + \cdots + (y_n - x_n)] E_n,$$
(18.14)

with the boundary conditions

$$\lim_{x_i \uparrow y_i \text{ or } y_i \downarrow x_i} E_n(x_1, y_1, \ldots, x_n, y_n, t) = E_{n-1}(x_1, y_1, \ldots, \cancel{x_i}, \cancel{y_i}, \ldots, x_n, y_n, t),$$
(18.15a)

$$\lim_{y_i \uparrow x_{i+1} \text{ or } x_{i+1} \downarrow y_i} E_n(x_1, y_1, \ldots, x_n, y_n, t) = E_{n-1}(x_1, y_1, \ldots, \cancel{y_i}, \cancel{x_{i+1}}, \ldots, x_n, y_n, t).$$
(18.15b)

For convenience, we use the notation that crossed-out arguments (e.g. $\cancel{x_i}$) have been removed. The E_n are tied together in a hierarchical fashion, through the boundary conditions: one must know E_{n-1} in order to compute E_n. Notice that, at the bottom of the hierarchy, $E_1(x_1, y_1, t) \equiv E(x, y, t)$ is simply the empty-interval probability discussed in Sections 18.1 and 18.2, and it satisfies its own boundary conditions (Eq. (18.8), and possibly others).

As a trivial example, consider the homogeneous steady state of reversible coalescence ($v > 0$, $R = 0$). Recall that this is in fact an *equilibrium* state, which satisfies detailed balance. The particles are simply distributed completely randomly in a state that maximizes their entropy. One obtains

$$E_{n,\text{eq}} = \exp\{-\gamma[(y_1 - x_1) + \cdots + (y_n - x_n)]\}, \tag{18.16}$$

and

$$\rho_{n,\text{eq}}(x_1, x_2, \ldots, x_n) = \gamma^n, \tag{18.17}$$

where $\gamma \equiv v/(2D)$ is the particle concentration at equilibrium.

A more interesting case is that of irreversible coalescence ($v, R = 0$). It can be shown (ben-Avraham, 1998a) that the E_n then satisfy the recursion relation

$$E_n(x_1, y_1, \ldots, x_n, y_n, t) = + \sum_{j=1}^{n} E(x_1, y_j, t) E_{n-1}(\cancel{x_1}, y_1, \ldots, x_j, \cancel{y_j}, \ldots, x_n, y_n, t)$$

$$- \sum_{j=2}^{n} E(x_1, x_j, t) E_{n-1}(\cancel{x_1}, y_1, \ldots, \cancel{x_j}, y_j, \ldots, x_n, y_n, t). \tag{18.18}$$

If $E_1 = E$ is known, one can generate the E_n, and then the ρ_n, using Eq. (18.13). For example, the long-time distribution is obtained by using the asymptotic solution $E(x, y; t) = \text{erfc}[(y - x)/\sqrt{8Dt}]$. For $n = 2$ we get

$$\frac{\rho_2(x_1, x_2; t)}{\rho_{\text{asymp.}}^2} = 1 - e^{-2\xi^2} + \sqrt{\pi}\xi e^{-\xi^2}\,\text{erfc}(\xi), \tag{18.19}$$

where we used the notation $\xi = (x_2 - x_1)/\sqrt{8Dt}$, and, for $n = 3$,

$$\frac{\rho_3(x_1, x_2, x_3; t)}{\rho_{\text{asymp}}^3} = 1 - e^{-2\xi_{21}^2} - e^{-2\xi_{32}^2} - e^{-2\xi_{31}^2} + 2e^{-\xi_{21}^2 - \xi_{32}^2 - \xi_{31}^2}$$

$$+ \sqrt{\pi}\xi_{21}(e^{-\xi_{21}^2} - e^{-\xi_{32}^2 - \xi_{31}^2})\,\text{erfc}(\xi_{21}) \tag{18.20}$$

$$+ \sqrt{\pi}\xi_{32}(e^{-\xi_{32}^2} - e^{-\xi_{21}^2 - \xi_{31}^2})\,\text{erfc}(\xi_{32})$$

$$+ \sqrt{\pi}\xi_{31}(e^{-\xi_{31}^2} - e^{-\xi_{21}^2 - \xi_{32}^2})\,\text{erfc}(\xi_{31}),$$

where now $\xi_{ij} = (x_i - x_j)/\sqrt{8Dt}$ (notice that $\xi_{31} = \xi_{32} + \xi_{21}$ is *not* an independent variable).

In Fig. 18.4 we show the two-point correlation function in the long-time asymptotic limit (Eq. (18.19)). We see that the two points become uncorrelated

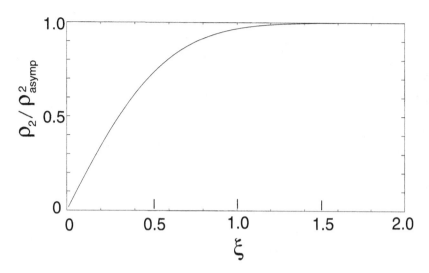

Fig. 18.4. The two-point correlation function for the process of coalescence in the long-time asymptotic limit. After ben-Avraham (1998a).

as the distance between them increases, but that there is an effective strong repulsive interaction (due to the coalescence reaction) between nearby particles. Interestingly, the two-point correlation is a monotonic function of the distance, whereas a simple convolution of the distances between nearest particles predicts an oscillating tail (Alemany and ben-Avraham, 1995).

The three-point correlation function, Eq. (18.20), is a bit harder to illustrate. Instead of a full description, in Fig. 18.5 we compare $\rho_3(x_1, x_2, x_3)$ with $\rho_2(x_1, x_2)\rho_2(x_2, x_3)/\rho(x_2)$ (in the spirit of the truncation *Ansatz* that might be used in a Kirkwood approximation), along the line $x_3 - x_2 = x_2 - x_1$. Again, we see that, as the distance between the three particles increases, they become rapidly uncorrelated, but that the approximation *Ansatz* fails for short distances, due to reactions.

18.4 Shielding

We have recently studied reversible coalescence in the half-line ($x \geq 0$) with a trap at the origin, and we have obtained the full hierarchy of multiple-point correlation functions. We found a most curious property of "shielding". In the (*nonequilibrium*) steady state of the system the particles are distributed randomly, exactly like in the *equilibrium* state of the homogeneous, infinite system (Eqs. (18.16) and (18.17)). The system is then fully characterized by $p(z)$ – the density distribution

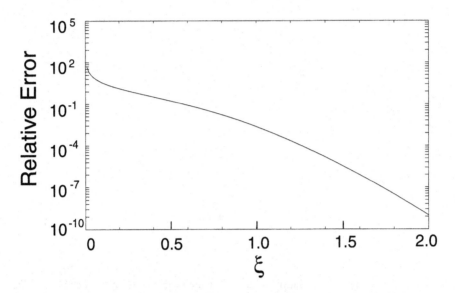

Fig. 18.5. The three-point correlation function for the process of coalescence in the long-time asymptotic limit. The relative error made by adopting a Kirkwood approximation, $(\rho_{\mathrm{Kirkwood}} - \rho_3)/\rho_3$, for the line $\xi_{21} = \xi_{32} \equiv \xi$ is plotted. The Kirkwood approximation used is $(\rho_{\mathrm{Kirkwood}} = \rho_2(\xi)^2/\rho_{\mathrm{asymp}}$ (see the text). After ben-Avraham (1998a).

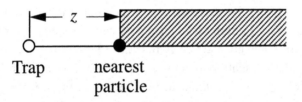

Fig. 18.6. A schematic illustration of the shielding effect. The particles in the shaded area are distributed randomly and independently from each other, just like in equilibrium. The gap z between the trap and the leading (shielding) particle follows the probability density $p(z)$.

function of the distance between the trap and the nearest particle to the trap, z. The nearest particle effectively shields the remaining particles from the trap (Fig. 18.6).

For the case of Fisher waves, we have found that again the leading particle shields all other particles. The particles remain distributed as in equilibrium, at all times, but the leading particle, at the edge of the wave, performs a *biased random walk* with drift velocity $v/2$. This gives rise to a propagation speed of the wave, $v/2$, and to an apparent broadening proportional to $\sqrt{Dt}$, as described by $\rho_1 \equiv \rho$ of Eq. (18.12). The broadening is the result of an ensemble average, in the laboratory reference frame, of all possible realizations of the process: each particular realization is distributed as in Fig. 18.6, exhibiting a sharp edge (and a sharp cutoff in the reactions), but the leading particles of different realizations are scattered about $x = (v/2)t$. Indeed, Kessler *et al.* (1998) suggest that the reference frame attached to the leading particle is more appropriate for the study of front propagation than is the laboratory reference frame.

18.5 Exercises

1. Show that Eq. (18.7) is indeed the continuous limit of Eq. (18.3).
2. For translationally invariant systems, reduce Eq. (18.7) to Eq. (15.18). (Hint: argue first that $\partial E(x, y, t)/\partial x = -\partial E(x, y, t)/\partial y$.)
3. Show, from the mean-field Fisher equation, that the stationary wave-front profile satisfies
$$0 = D \frac{d^2 f(z)}{dz^2} + c \frac{df(z)}{dz} + k_1 f(z) - k_2 f(z)^2.$$
4. *The general solution of Eq. (18.11)* (Doering *et al.*, 1991). Begin by changing variables to $\xi = [v/(2D)](y + x)$, $\zeta = [v/(2D)](y - x)$, and $\tau = [v^2/(2D)]t$, and obtain
$$\frac{\partial E(\xi, \zeta, \tau)}{\partial \tau} = \frac{\partial^2 E}{\partial \xi^2} + \frac{\partial^2 E}{\partial \zeta^2} + \frac{\partial E}{\partial \zeta},$$
in the half-plane $-\infty < \xi < \infty$ and $0 < \zeta < \infty$, with the boundary condition $\lim_{\zeta \downarrow 0} E(\xi, \zeta, \tau) = 1$. Reduce this further to an elementary heat equation, with the transformation $F = e^{\tau/4 + \zeta/2}(E - e^{-\zeta})$. (Answer: $\partial_\tau F = \partial_\xi^2 F + \partial_\zeta^2 F$, with boundary condition $\lim_{\zeta \downarrow 0} F(\xi, \zeta, \tau) = 0$. The general solution is now straightforward. ...)

18.6 Open challenges

1. Find an explicit form for ρ_n in the long-time asymptotic limit of irreversible co-alescence. A good starting point would be the recursion relation of Eq. (18.18).
2. Find the complete hierarchy of n-point correlation functions for other cases of interest. For example, for the stationary state in the presence of input, or for finite systems (under various conditions).
3. What are the exact conditions needed to sustain shielding in the coalescence system? The Fisher-wave case illustrates the fact that the system need not necessarily be stationary. In fact, it can easily be demonstrated that shielding holds for the trapping problem *at all times*, provided that the initial state is the equilibrium distribution (to the right of the trap). On the other hand, we have recently shown (Donev *et al.*, 1999) that, if the trap acts also as a source, giving

birth to particles at the site next to itself, then the shielding property breaks down.

4. The steady state of the trapping problem can be solved exactly also in the reaction–diffusion-equation approximation (ben-Avraham, 1998b). It would be interesting to use the complete exact solution (of the full $\{\rho_n\}$ hierarchy) as a basis for systematic improvements of the reaction–diffusion approach.

5. The trapping problem is better suited for numerical studies than is the Fisher-wave problem, since it has a stationary limit. It would be of interest to perform numerical simulations of trapping (by a hyperplane) in dimensions higher than one, and compare the results with the prediction from the reaction–diffusion equation. This would provide a more sensitive way of determining the upper critical dimension beyond which the mean-field reaction–diffusion approach is applicable.

18.7 Further reading

- The IPDF method for inhomogeneous systems: Doering *et al.* (1991). The influence of homogeneous external noise on the one-species coalescence reaction may also be treated exactly with the IPDF method. The interplay of the external and intrinsic (internal) noises gives rise to a new microscopic length scale (Doering, 1992). Inhomogeneous situations of coalescence and coagulation, including trapping, have been considered by Schütz (1995; 1997), and Schütz and Mussawisade (1998).

- Fisher waves in the coalescence process: Doering *et al.* (1991), Riordan *et al.* (1995), and ben-Avraham (1998c). The role of discreteness in front propagation: Mai *et al.* (1996), Brunet and Derrida (1997), Kessler *et al.* (1998), and Ebert and van Saarloos (1998).

- Coalescence with a trap: ben-Avraham (1998b; 1998d). Coalescence with a mobile trap: Donev *et al.* (1999).

19

Finite reaction rates

When the rate of coalescence is finite the typical time for reaction competes with the typical diffusion time, and a crossover between the reaction-limited regime and the diffusion-limited regime is observed. The model cannot be solved exactly, but it can be approached through an approximation based on the IPDF method. The kinetic crossover is well captured by this approximation.

19.1 A model for finite coalescence rates

Until now we have dealt with infinite coalescence rates: when a particle hops into an occupied site the coalescence reaction is immediate. Thus, the typical reaction time is zero and the process is diffusion-limited. We want now to discuss the case in which the coalescence rate (and the typical time for the coalescence reaction) is finite. In this case a competition arises between the typical transport time and the typical reaction time.

The model we have in mind is the following: when a particle attempts to hop onto a site that is already occupied, the move is allowed and coalescence takes place with probability k ($0 \leq k \leq 1$). The attempt is rejected, and the state of the system remains unchanged, with probability $(1-k)$. The case of $k = 1$ corresponds to an infinite coalescence rate, which we have studied so far. The opposite limit, of $k = 0$, describes diffusion of the particles with hard-core repulsion, but no reactions take place.

Suppose that $0 < k \ll 1$. Reactions then require a large number of collisions, and the kinetics is dominated by the long reaction times (the reaction-limited regime). On the other hand, as the concentration of particles drops, the effective rate of reaction increases and the process eventually becomes diffusion-limited. Thus, for small k, we expect an interesting crossover between classical kinetics, $c \sim 1/t$, and diffusion-limited kinetics, $c \sim 1/\sqrt{t}$.

We may now attempt to write down an evolution equation for the probability of empty intervals, as in Section 15.4. Let us begin with simple coalescence, and postpone momentarily the discussion of back reactions and input. Thus,

$$\partial_t E_n = \frac{2D}{(\Delta x)^2}(E_{n-1} - 2E_n + E_{n+1}) - (1-k)\frac{2D}{(\Delta x)^2} \text{Prob}(\overbrace{\circ \circ \cdots \circ}^{n-1} \bullet\bullet). \quad (19.1)$$

This looks like Eq. (15.11), but a correction has been added due to the finite probability of reaction. When $n-1$ consecutive sites are empty and a particle at the nth site attempts to hop out of the interval, the move is disallowed – with probability $(1-k)$ – if the $(n+1)$th site is occupied. Thus, the growth rate of E_n is smaller than that of Eq. (15.11). Unfortunately, $\text{Prob}(\circ \circ \cdots \circ \bullet\bullet)$ cannot be expressed in terms of the $\{E_n\}$, and an approximation is necessary if one is to obtain a closed equation.

19.2 The approximation method

In the Kirkwood spirit, we use the approximation

$$\text{Prob}(\overbrace{\circ \circ \cdots \circ}^{n-1} \bullet\bullet) \approx \frac{\text{Prob}(\overbrace{\circ \circ \cdots \circ}^{n-1} \bullet)\,\text{Prob}(\bullet\bullet)}{\text{Prob}(\bullet)} = \frac{(1 - 2E_1 + E_2)(E_{n-1} - E_n)}{1 - E_1}.$$
$$(19.2)$$

The evolution equation for arbitrary k then becomes

$$\partial_t E_n = \frac{2D}{(\Delta x)^2}(E_{n-1} - 2E_n + E_{n+1}) - \frac{2D}{(\Delta x)^2}(1-k)\frac{1 - 2E_1 + E_2}{1 - E_1}(E_{n-1} - E_n).$$
$$(19.3)$$

Notice that, for $k = 1$, this properly reduces to Eq. (15.11).

Equation (19.3) is valid for $n > 1$. For $n = 1$, we have the exact equation

$$\partial_t E_1 = \frac{2D}{(\Delta x)^2}k(1 - 2E_1 + E_2), \quad (19.4)$$

which simply states that sites become empty at the same rate as particles coalesce. Comparing Eqs. (19.3) and (19.4), we see that they may be combined by requiring the usual boundary condition, $E_0(t) = 1$. Equation (19.3) is then valid also for $n = 1$. Again, if the system is not empty, $E_\infty(t) = 0$.

In the continuum limit, Eq. (19.3) becomes

$$\partial_t E = 2D\,\partial_x^2 E - 2D(1-k)\frac{\partial_x^2 E|_{x=0}}{\partial_x E|_{x=0}}\,\partial_x E, \quad (19.5)$$

with boundary conditions $E(0, t) = 1$ and $E(x \to \infty, t) = 0$. We see that the finite reaction rate gives rise to a nonlocal, nonlinear term that complicates the solution.

The only approximation made here is in Eq. (19.2), where some correlations between the states of consecutive intervals are neglected. We argue that Eq. (19.3) (and (19.5)) is asymptotically correct both in the early-time and in the long-time regimes, and hence it may provide a reasonable interpolation for intermediate times. If the starting configuration of the system is random, as we shall indeed assume, then $E_n(0) = E_1(0)^n$ and the state of consecutive intervals is uncorrelated. This situation will persist, and Eq. (19.3) will remain valid, until the drop in concentration is noticeable. After very long times, on the other hand, the concentration of particles becomes very small. As a result, adjacent occupied sites are extremely rare and the correction term in Eq. (19.3) becomes negligible.

19.3 Kinetics crossover

Equation (19.3) (or (19.5)) is difficult to analyze. We proceed by an alternative approach, based on an approximation of Eq. (19.3). This approximation is a mere mathematical convenience, necessary to obtain a closed analytic expression. Indeed, Eq. (19.3) itself may be integrated numerically to any desired degree of accuracy.

We first sum Eq. (19.3) over the index n, from 1 to ∞, to yield

$$\partial_t \sum_{n=1}^{\infty} E_n = \frac{2D}{(\Delta x)^2}(1 - E_1) - \frac{2D}{(\Delta x)^2}(1 - k)\frac{1 - 2E_1 + E_2}{1 - E_1}, \qquad (19.6)$$

where we have used the boundary conditions $E_0 = 1$ and $E_\infty = 0$. The RHS can be made a function of E_1 alone, with help of Eq. (19.4). For the LHS, we make the approximation $\sum E_n \approx A/(1 - E_1)$, where A is a constant. This is justified since $E_n \approx 1$ for all n up to a characteristic $\langle n \rangle = 1/(1 - E_1)$, and falls sharply to zero for $n > \langle n \rangle$. In fact, at the beginning of the process, when the initial distribution is random, $A = 1$. At very late times the system behaves asymptotically as if $k = 1$, in which case we know that $A = 2/\pi$, exactly. Our approximation consists of the assumption that $A = 2/\pi$ *at all times*, to match the exact long-time asymptotic solution. Equation (19.6) then becomes

$$\frac{d}{d\tau}\frac{2}{\pi C} = C + \frac{1 - k}{kC}\frac{dC}{d\tau}, \qquad (19.7)$$

where $\tau \equiv (2D/(\Delta x)^2)t$ and $C \equiv (1 - E_1) = c\,\Delta x$ are dimensionless variables.

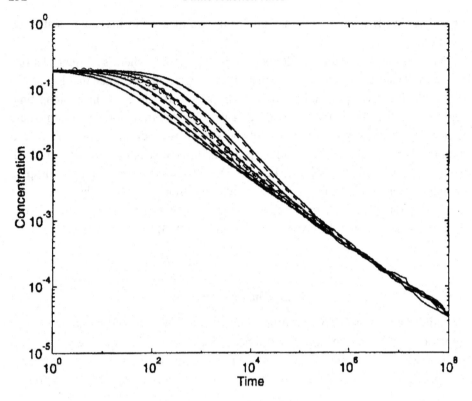

Fig. 19.1. Concentration decay for coalescence with a finite reaction rate. The results from computer simulations (solid curves) and from Eq. (19.8) (broken curves) are plotted for various values of the reaction probability, $k = 0.005, 0.01, 0.02, 0.04, 0.08$, and 0.16 (from top to bottom). For $k = 0.02$, the results from numerical integration of Eq. (19.3) (circles) are also shown. After Zhong and ben-Avraham (1995).

This has the solution

$$C = \frac{1 - k + \sqrt{\left(\dfrac{2k}{\pi C_0} + 1 - k\right)^2 + \dfrac{4k^2}{\pi}\tau}}{2\left(\dfrac{k}{\pi C_0^2} + \dfrac{1 - k}{C_0} + k\tau\right)}, \tag{19.8}$$

where $C_0 \equiv C(t = 0)$.

In Fig. 19.1 we compare the concentration of particles computed from a numerical integration of Eq. (19.3) with the analytical expression of Eq. (19.8) and with computer simulations. The agreement between Eq. (19.8) and the numerical integration is excellent – little is lost in the "mathematical" approximation. More importantly, the agreement between theory and simulations is to within better than 9%.

We can use our result to estimate the two crossover times evident in Fig. 19.1. First expand $C(\tau)$ in powers of τ:

$$C(\tau) = C_0 - \frac{\pi k C_0^3}{2k + \pi(1-k)C_0}\tau + \mathcal{O}(\tau^2). \qquad (19.9)$$

The crossover time between the early-time regime and the intermediate regime, τ_1, is obtained by comparing the linear term in τ with C_0:

$$\tau_1 \sim \frac{2}{\pi C_0^2} + \frac{1-k}{kC_0}. \qquad (19.10)$$

Even in the strictly diffusion-limited case, when $k = 1$, there is an early-time regime in which reactions go unnoticed. Since coalescence is immediate, the crossover time equals the typical time that two nearest particles will take to reach each other. The average distance between particles is $1/c_0$, and, since the particles diffuse, $t_1 \approx 1/(Dc_0^2)$, in agreement with the estimate above. The second term on the RHS predicts that the crossover time τ_1 will increase proportionally to $1/(kc_0)$, which is characteristic of the reaction-limited process (classically, $dc/dt = -kc^2$ and $c = c_0/(1 + kc_0t)$).

Next, we expand Eq. (19.8) in powers of $1/\sqrt{\tau}$,

$$C(\tau) = \frac{1}{\sqrt{\pi\tau}} + \frac{1-k}{2k\tau} + \mathcal{O}\left(\frac{1}{\tau^{3/2}}\right). \qquad (19.11)$$

The leading term corresponds to the long-time asymptotic limit, at which the concentration decays as $c = 1/\sqrt{2\pi Dt}$, just like in the exact solution for $k = 1$. Interestingly, the second term, too, does not retain any memory of the initial density, C_0, but has some k-dependence. By comparing the leading term with the first correction we get an estimate for τ_2, the time for crossover between the intermediate regime and the long-time regime:

$$\tau_2 \sim \frac{\pi(1-k)^2}{4k^2}. \qquad (19.12)$$

This can be explained heuristically, as follows. The long-time asymptotic regime occurs because the reaction probability k effectively renormalizes to 1. If the density of particles is C, the number of sites between neighboring particles is on the average $1/C$. Because of diffusion, it takes nearest particles of the order of $1/C^2$ steps to meet each other. During this time, each of the $1/C$ intermediate sites is visited of the order of $(1/C^2)/(1/C) = 1/C$ times. In particular, two neighboring particles will collide about $1/C$ times before wandering away from each other to interact with other particles. Thus, the particles will almost surely react, before meeting other partners, if $(1/C)k \approx 1$. That is, the crossover occurs when $C \approx k$, or, since in the long-time asymptotic regime $C = 1/\sqrt{\pi\tau}$, $\tau_2 \sim 1/k^2$. Thus, the small-k dependence of τ_2 is reproduced.

19.4 Finite-rate coalescence with input

Consider now the case of finite-rate coalescence with input. Because of its nontrivial steady state, this system may impose a stricter test on the approximation method. With the approximation (19.2), the evolution equation is

$$\partial_t E_n = \frac{2D}{(\Delta x)^2}(E_{n-1} - 2E_n - E_{n+1}) - \frac{2D}{(\Delta x)^2}(1 - k)\frac{1 - 2E_1 + E_2}{1 - E_1}(E_{n-1} - E_n)$$
$$- Rn\,\Delta x\,E_n,$$
(19.13)

where the last term represents input in the usual way.

Passing to the continuum limit, the steady state satisfies the (approximate) equation

$$0 = 2D\,\partial_x^2 E - 2D(1 - k)\omega\,\partial_x E - xRE,$$
(19.14)

where $\omega = \partial_x^2 E|_{x=0}/\partial_x E|_{x=0} = -\partial_x^2 E|_{x=0}/c_s$ is a constant. To determine ω, let us look at the discrete steady-state equation for $n = 1$,

$$0 = \frac{2D}{(\Delta x)^2}k(1 - 2E_1 + E_2) - \Delta x\,RE_1.$$
(19.15)

The continuum limit is problematic. We obtain $\omega = -R\,\Delta x/(2Dkc_s)$ in a somewhat inelegant but effective way, by retaining $\Delta x\,R$ finite.

The solution to Eq. (19.14) is then

$$E_s(x) = \exp\left(-\frac{x}{\kappa c_s}\right)\frac{\mathrm{Ai}[r^{1/3}x + r^{-2/3}/(\kappa c_s)^2]}{\mathrm{Ai}[r^{-2/3}/(\kappa c_s)^2]},$$
(19.16)

where $\kappa = 4Dk/[(1 - k)R\,\Delta x]$ and $r = R/(2D)$. From the relation $c_s = -\partial_x E_s|_{x=0}$, we then obtain a transcendental equation for the steady-state concentration, c_s:

$$c_s = \frac{1}{\kappa c_s} - r^{1/3}\frac{\mathrm{Ai}'[r^{-2/3}/(\kappa c_s)^2]}{\mathrm{Ai}[r^{-2/3}/(\kappa c_s)^2]}.$$
(19.17)

In Fig. 19.2 we plot c_s as a function of the reaction probability k, for fixed r ($r^{1/3} = 0.04$), as obtained from computer simulations and from numerical integration of the discrete steady-state equation. The agreement between simulations and theory is quite good. Also, for the range shown, the agreement between Eq. (19.17) and the numerical integration is to within better than 4%.

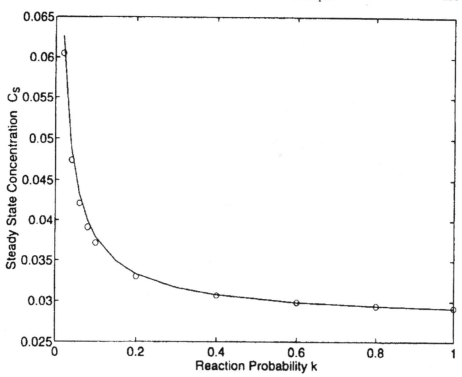

Fig. 19.2. The stationary concentration c_s as a function of the reaction probability k for coalescence with particle input, at input rate $r = (0.04)^3$. Numerical integration of the discrete steady-state equation (19.15) (circles) is compared with computer-simulation results (solid curve). After Zhong and ben-Avraham (1995).

When $c \ll r^{1/3}/\kappa$, one derives from Eq. (19.17)

$$c_s = \frac{1}{2}\left(\tilde{c}_s + \sqrt{\tilde{c}_s^2 + \frac{2(1-k)\,\Delta x}{k}r}\,\right), \qquad \tilde{c}_s = -\frac{\mathrm{Ai}'(0)}{\mathrm{Ai}(0)}r^{1/3}, \qquad (19.18)$$

where $\tilde{c}_s$ is the steady-state concentration when $k = 1$ (and is exact). Thus, we observe a crossover from the diffusion-limited behavior, $c_s \sim \tilde{c}_s \sim (R/2D)^{1/3}$, to the classical result, $c_s \sim \frac{1}{2}\sqrt{(1-k)R\,\Delta x/(kD)}$, as $k \to 0$.

The IPDF is obtained from $p(x) = (1/c_s)\,\partial^2 E/\partial x^2$, using Eq. (19.16) with the c_s found from numerical integration (or any of the approximation formulae). In Fig. 19.3 we compare the IPDFs obtained in this way with those from computer simulations for $r^{1/3} = 0.04$ fixed, for various values of k. The agreement between theory and simulations is best for large k. From Figs. 19.2 and 19.3 one can see that the IPDF is a much more sensitive test for approximations than is the concentration alone.

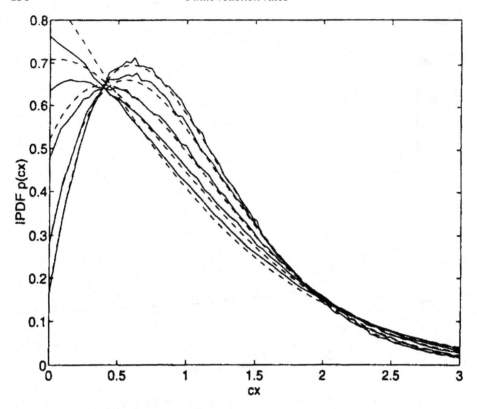

Fig. 19.3. The scaled IPDF $p(c_s x)$ for the same process as that in Fig. 19.2 for $k = 0.02$, 0.04, 0.08, 0.20, and 0.40 ($p(0)$ is smaller for larger k), computed from the second derivative of (19.16) (broken curves) and compared with Monte Carlo simulations (full curve). After Zhong and ben-Avraham (1995).

An interesting result is that $p(x = 0)$ is zero only when $k = 1$. This is in contrast to the case of pure coalescence, in which $p(0) = 0$ in the long-time asymptotic limit, regardless of the value of k. Recall that, in the latter case, k effectively renormalizes to 1 when the concentration decreases. This cannot happen in the case of input, when c approaches a stationary *finite* value.

19.5 Exercises

1. Prove that, when the occupation probabilities of different sites are statistically independent, the approximation (19.2) becomes exact.

2. Compute the constant of proportionality A in $\sum E_n = A/(1 - E_1)$ for (a) a completely random distribution, $E_n = E_1^n$, and (b) the scaling distribution of Eq. (16.7). In the latter case, simplify the calculation by working in the continuum limit. (Answer: (a) $A = 1$, (b) $A = 2/\pi$.)

3. Perform numerical simulations of coalescence with a finite reaction rate and monitor the IPDF at various stages of the process. Integrate Eq. (19.3) numerically and compare the analytical IPDF with simulations. Why is the agreement good for very early and very late times, but not as good for the intermediate-time regime?

4. Compute the *exact* equilibrium state for reversible coalescence ($v > 0$) with finite reaction rate k. Show that the "Kirkwood" approximation yields the exact long-time asymptotic result. (Hint: for the first part use the approach of Section 17.1.)

5. Analyze the annihilation process, $A + A \to 0$, by means of the approximation

$$
\text{Prob}(\overbrace{\circ\cdots\circ}^{n}\bullet\bullet\overbrace{\circ\cdots\circ}^{m}) \approx \frac{\text{Prob}(\overbrace{\circ\cdots\circ\,\bullet}^{n})\,\text{Prob}(\bullet\bullet)\,\text{Prob}(\bullet\,\overbrace{\circ\cdots\circ}^{m})}{\text{Prob}(\bullet)\,\text{Prob}(\bullet)}
$$

$$
= \frac{1 - 2E_1 + E_2}{(1 - E_1)^2}(E_n - E_{n+1})(E_m - E_{m+1}),
$$

following the IPDF method.

19.6 Open challenges

1. In spite of considerable theoretical and numerical work, the crossover between the diffusion-limited and reaction-limited regimes in the coalescence model continues to elude exact analysis.

2. The approximation technique proposed here could be improved systematically, by introducing a closure scheme further down the line. This approach has not yet been investigated fully.

3. In the limit of $k = 0$ the coalescence model turns into diffusion with excluded-volume interactions. Can one modify the approximation method in this chapter and use it to study the question of tracer diffusion?

19.7 Further reading

- Braunstein *et al.* (1992), Hoyuelos and Mártin (1993b), Mártin *et al.* (1995), Privman *et al.* (1993), and Zhong and ben-Avraham (1995).
- Similar approximations to the one presented here, based on the IPDF technique, have been applied to the problem of multiple-particle coalescence: $mA \to nA$ (ben-Avraham, 1993; ben-Avraham and Zhong, 1993), and to the contact process (Ben-Naim and Krapivsky, 1994).

Appendix A

The fractal dimension

Formally, let $r(S)$ be a similarity transformation that maps all points $x \in S$ onto new points $x' = rx$. The set S is called *self-similar* with respect to the scaling ratio $r < 1$ if S is equal to the union of $n(r)$ replicas of $r(S)$. If this is the case, one may further define the self-similarity dimension, D_S:

$$D_S = \frac{\ln n(r)}{\ln(1/r)}. \tag{A.1}$$

For example, the Koch curve (see Chapter 1) is self-similar, with $r = \frac{1}{3}$, $n(r) = 4$, and $D_S = \ln 4 / \ln 3 = d_f$.

This is only one of several ways of defining dimensionality. The box dimension, D_B, of a set of points $S \in R^d$ is defined as

$$D_B = \lim_{\epsilon \to 0} \frac{\ln N(\epsilon)}{\ln(1/\epsilon)}, \tag{A.2}$$

where $N(\epsilon)$ is the *minimal* number of d-dimensional boxes of size ϵ needed to cover the set of points S. It is a special case of the more general *Hausdorff dimension*. Consider a covering set of boxes of linear size ϵ_i. The *Hausdorff measure* is

$$H_p(\epsilon) = \inf \sum_i \epsilon_i^p, \tag{A.3}$$

where the infimum is the minimal value obtained when taking into account all possible covering sets, subject to the constraint $\epsilon_i \le \epsilon$. The Hausdorff dimension, D_H, is then given by the limit

$$\lim_{\epsilon \to 0} H_p(\epsilon) = \begin{cases} 0 & p > D_H, \\ \text{finite} & p = D_H, \\ \infty & p < D_H. \end{cases} \tag{A.4}$$

In the special case that $\epsilon_i = \epsilon$, this reduces to the box dimension.

For a self-similar object, $N(r\epsilon) = n(r)N(\epsilon)$, and it follows from Eqs. (A.1) and (A.2) that $D_B = D_S$. While generally $D_B \geq D_H$, for the more interesting cases the two dimensions are equal and one has $D_H = D_B = D_S \equiv d_f$.

These concepts are easily generalized to self-affine fractals. For example, the *affinity transformation* $r(S)$ maps all points x of the set $S \in R^d$ onto new points x':

$$x' = \begin{pmatrix} x'_1 \\ \vdots \\ x'_d \end{pmatrix} = \begin{pmatrix} r_1 & 0 & \cdots & 0 \\ 0 & r_2 & & 0 \\ \vdots & & \ddots & \vdots \\ 0 & \cdots & & r_d \end{pmatrix} \begin{pmatrix} x_1 \\ \vdots \\ x_d \end{pmatrix}. \qquad (A.5)$$

The set S is (diagonal) self-affine with respect to r if S is equal to the union of $n(r)$ replicas of $r(S)$. In this fashion, one obtains different self-similarity dimensions in each spatial direction:

$$(D_S)_i = \frac{\ln n(r)}{\ln(1/r_i)}, \qquad (A.6)$$

etc.

Appendix B

The number of distinct sites visited by random walks

One of the basic properties of random walks, arising in many applications, is that of the number of distinct lattice sites visited during an n-step walk, $S(n)$. The concept is related to that of first-passage events in relaxation processes and diffusion-limited reactions, such as annealing of defects, trapping of excitons, and ecological models for the spread of populations and diseases. Perhaps the most important physical example is that of a walker moving among randomly distributed traps, at concentration c (Chapter 12). The walker's survival probability after n-steps is $\langle (1 - c)^{S(n)} \rangle$.

The number of distinct sites visited by N random walkers starting from a common origin, $S_N(n)$, is also of interest. Consider first a single random walker. Asymptotic results for $S_1(n)$ in hypercubic lattices are as follows (Weiss, 1994):

$$
\langle S_1(n) \rangle \sim
\begin{cases}
\left(\dfrac{8n}{\pi} \right)^{1/2} \left[1 + \dfrac{1}{4n} - \dfrac{1}{32n^2} + \mathcal{O}\left(\dfrac{1}{n^3} \right) \right], & d = 1, \\[2ex]
\dfrac{\pi n}{\ln n}, & d = 2, \\[2ex]
n, & d \geq 3.
\end{cases}
\tag{B.1}
$$

Biased random walks have a finite probability of visiting new sites at any given time step, and in their case $S_1(n) \sim n$ (in all dimensions). The number of distinct sites visited by a regular random walk has been studied also in fractals: $\langle S_1(n) \rangle \sim n^{d_s/2}$, when the fracton dimension is $d_s = 2d_f/d_w < 2$.

The distribution of $S_1(n)$ in $d > 1$ is not well known. An important result is that the distribution is asymptotically Gaussian. Its variance is $\sigma^2[S_1(n)] \sim n^2/(\ln^3 n)$, in $d = 2$, and $\sigma^2[S_1(n)] \sim n \ln n$, in $d = 3$ (Jain and Pruitt, 1971).

The results of Eq. (B.1) can be generalized for single Lévy flights, in which the single-jump probability has the form $p(r) \sim r^{-\beta-d}$. For $\beta < d + 1$, the variance of a single jump diverges and one expects more distinct sites to be visited than in the case of regular walks. Indeed, it is found that (Gillis and Weiss, 1970)

$$
\langle S_1(n) \rangle \sim
\begin{cases}
n, & \beta < d, \\
\dfrac{n}{\ln n}, & \beta = d, \\
n^{1/\beta}, & d < \beta < d+1.
\end{cases}
\tag{B.2}
$$

For $\beta > d + 1$, $\langle S_1(n) \rangle$ behaves just like in the case of short-range jumps and Eq. (B.1) is valid. For $d = 1$ and $\beta = 2$, which is the critical β between long and short jumps, it is found that $\langle S_1(n) \rangle \sim \sqrt{n \ln n}$.

The generalization of the single-walker problem to the number of distinct sites visited by N random walkers, $S_N(n)$, has been studied for d-dimensional Euclidean lattices (Larralde *et al.*, 1992a; 1992b) and fractals (Havlin *et al.*, 1992), as well as for Lévy flights (Berkolaiko *et al.*, 1996). In the N-walkers problem, each site is counted just once, when the first walker reaches it. Subsequent visits by other walkers do not contribute to $S_N(n)$. Thus, $S_N(n)$ cannot simply be computed as the sum of N uncorrelated walks.

For d-dimensional Euclidean lattices there exist distinct time regimes separated by two crossover times, $n_\times$ and $n'_\times$, where $n_\times \sim \ln N$ (all d), while $n'_\times$ is dimension-dependent: $n'_\times = \infty$ ($d = 1$), e^N ($d = 2$), and N^2 ($d \geq 3$).

- Regime I: $n \ll n_\times$. During this brief regime, $S_N(n) \sim n^d$.
- Regime II: $n_\times \ll n \ll n'_\times$. In this intermediate-time regime, $S_N(n) \sim n^{d/2}(\ln U)^{d/2}$, where $U = N S_1(n)/n^{d/2}$.
- Regime III: $n \gg n'_\times$. The asymptotic long-time regime, in which the walkers become practically independent and $S_N(n) \sim N S_1(n)$, $d \geq 2$.

The behavior of $S_N(n)$ in fractals with fracton dimension $d_s < 2$ falls into just two time regimes (just like in one-dimensional space) demarcated by $n_\times \sim \ln N$.

- Regime I: $n \ll n_\times$. $S_N(n) \sim n^{d_\ell}$.
- Regime II: $n \gg n_\times$. $S_N(n) \sim (\ln N)^{d_\ell/\delta_\ell} n^{d_s/2}$.

These results for fractals encompass the case of regular Euclidean lattices, as can be verified following the substitutions $d_s = d_f = d_\ell = d$ and $\delta_\ell = d_w^\ell/(d_w^\ell - 1) = 2$. When $d_s < 2$ the walkers can never become independent, so there is no counterpart to regime III of regular ($d > 2$) space.

The above behavior can be explained through simple scaling arguments. Let $P_{\min}(n)$ be the smallest nonzero occupation probability, for occupation of the fractal sites by a single walker at time n. Furthermore, let z be the mean coordination number of the fractal lattice, then $P_{\min}(n) \sim z^{-n}$. Consider now N independent walkers starting from the same site. As long as $N P_{\min} \gg 1$ we expect that $S_N(n)$ would be proportional to the volume in chemical ℓ-space, i.e.,

$S_N(n) \sim n_\ell^d$. The condition $N P_{\min}(n_x) \sim 1$ implies that the crossover time is $n_x \sim \ln N$.

The second regime can be understood as follows. The probability of finding a random walk outside a chemical distance ℓ is given by

$$\int_\ell^\infty P(n, \ell') \ell'^{d_\ell - 1} \, d\ell' \sim n^{-d_s/2} \int_\ell^\infty \exp\left[-a\left(\frac{\ell'^{1/d_w^\ell}}{n}\right)^{\delta_\ell}\right] \ell'^{d_\ell - 1} \, d\ell'$$

$$\sim \exp\left[-a\left(\frac{\ell^{1/d_w^\ell}}{n}\right)^{\delta_\ell}\right].$$

Let us assume that this probability is of the order of $1/N$, i.e., only a few particles are farther than ℓ from the origin. It follows that

$$\ell \sim (\ln N)^{1/\delta_\ell} n^{1/d_w^\ell}.$$

Assuming that, up to that chemical distance, ℓ, all sites are visited with a finite probability, one recovers the behavior in regime II: $S_N(n) \sim (\ln N)^{d_\ell/\delta_\ell} n^{d_s/2}$. Note that, for fractals with $d_s < 2$, it is necessary to work in ℓ-space, rather than with the Euclidean metric. Indeed, the result was originally proposed by Havlin *et al.* (1992), but their considerations took place in Euclidean space. It was rederived, working in chemical ℓ-space, and brought to its final form by Drager and Klafter (1999).

Finally, consider the asymptotic forms for the expected number of distinct sites, $\langle S_N(n) \rangle$, visited by N noninteracting n-step symmetric Lévy flights in one dimension (Berkolaiko *et al.*, 1996). The probability for a step of length r is $1/r^{1+\beta}$ (for large r), and we consider all $\beta > 0$. Different asymptotic results are obtained for different ranges of β. When n is fixed and $N \to \infty$, one finds that $\langle S_N(n) \rangle \sim (Nn^2)^{1/(1+\beta)}$ for $\beta < 1$, and $\langle S_N(n) \rangle \sim N^{1/(1+\beta)} n^{1/\beta}$ for $\beta > 1$. The last case includes $\beta > 2$, for which the variance of the step distribution is finite and one would expect the same result as for regular random walks. This is, however, not the case, for the few walkers which make the biggest jumps dominate the number of distinct sites.

$\langle S_N(n) \rangle$ for Lévy flights in dimensions $d \geq 2$ with a step distribution $p(r) \sim 1/r^{d+\beta}$ was studied by Berkolaiko and Havlin (1997). The asymptotic results are as follows. For $d = 2$ and $N \to \infty$, $\langle S_N(n) \rangle \sim \beta N^{2/(2+\beta)} n^{4/(2+\beta)}$, when $\beta < 2$, and $\langle S_N(n) \rangle \sim N^{2/(2+\beta)} n^{2/\beta}$, when $\beta > 2$. For $d = 2$ and $n \to \infty$, $\langle S_N(n) \rangle \sim Nn$ for $\beta < 2$ and $\langle S_N(n) \rangle \sim Nn/\ln n$ for $\beta > 2$. The last limit corresponds to the result obtained earlier for regular walks (bounded jumps).

Appendix C

Exact enumeration

The problem of diffusion on a given lattice may be approached numerically by using the *exact-enumeration* technique (Majid *et al.*, 1984). This eliminates the need to perform Monte Carlo simulations of random walks, and increases accuracy.

The method is based on the master equation for the probability distribution of a walker at the discrete time t:

$$P(j, t+1) = \sum_{\text{nn}\{j\}} W(\text{nn}\{j\}, t) P(\text{nn}\{j\}, t).$$ \hfill (C.1)

Here, $\text{nn}\{j\}$ denotes the set of sites that are accessible from site j in one step (usually, the nearest neighbors of site j). $W(\text{nn}\{j\})$ represents the transition probabilities for steps to those sites.

In the exact-enumeration algorithm the initial probabilities $P(j, 0)$ are stored in a matrix, $\mathbf{M}$, which mirrors the substrate. The master equation (C.1) is then used to compute $P(j, 1)$ and the result is stored in a second, similar matrix, $\mathbf{M}'$. In the next time step the role of the matrices is reversed: the input (at time $t = 1$) is taken from $\mathbf{M}'$, and the computed probabilities (at time $t = 2$) are stored back in $\mathbf{M}$. The probability distribution at later times is obtained by iterating these steps.

From the probability density one may compute various averages of interest. Suppose, for example, that the mean-square displacement of walks originating at site "0" is required. We then run the exact-enumeration algorithm, with $P(j, 0) = \delta_{0, j}$, and obtain the desired result, using

$$\langle r^2(t) \rangle = \sum_j r_j^2 P(j, t),$$ \hfill (C.2)

where r_j is the distance between sites 0 and j.

The transition probabilities, $W(\text{nn}\{j\})$, depend on the kind of walker one wishes to study. In the early stages of research on diffusion in percolation clusters de Gennes coined the suggestive name of "the ant in the labyrinth". Since then a

263

distinction among different kinds of ants has appeared, for walkers with different characteristic transition probabilities.

- *The blind ant.* This walker blindly attempts to step into one of the nearest-neighbor sites, not realizing that some sites might be blocked. An attempt to step into a blocked site fails, and the walker remains in its original position. Notice that, in this case, nn$\{j\}$ includes site j itself, in addition to its nearest neighbors. The transition probabilities are

$$W(\text{nn}\{j\}) = \begin{cases} 1/z, & \text{for transition to each of the open} \\ & \text{nearest-neighbor sites,} \\ (z - z_j)/z, & \text{for remaining at site } j. \end{cases} \quad (C.3)$$

Here z is the coordination number of the underlying lattice, and z_j is the actual number of open nearest-neighbor sites of j. (For example, the Sierpinski gasket in two dimensions is embedded in the triangular lattice, with $z = 6$, but the actual number of nearest neighbors accessible from a site in the gasket is $z_j = 4$; whereas in bond percolation on the square lattice $z = 4$, but z_j varies from site to site.)

- *The myopic ant.* This walker notices that only z_j sites are reachable, and refrains from attempting to step into blocked sites. Thus, the myopic ant never rests, but moves into one of the open sites with equal probabilities; $W(\text{nn}\{j\}) = 1/z_j$.

Convergence to the asymptotic regime occurs more rapidly with the blind ant than it does with the myopic ant, though the reason for this effect is not entirely clear. On the other hand, for random media that are not embedded in a lattice (for example, in continuum percolation, or trees with unrestricted random branching) the blind-ant algorithm cannot be implemented, and the myopic ant proves more useful in such cases. The transition probabilities can easily be adapted to other situations, such as cases with drift, sources, and traps.

When the exact-enumeration algorithm is implemented for long times, rounding errors may become a problem as $P(j, t)$ becomes increasingly small for sites at the diffusion front. This problem is overcome by rescaling the probabilities at every time step: a factor of z, for the blind ant, or $\langle z_j \rangle$, for the myopic ant, is appropriate. The matrix $\mathbf{M}$ is normalized at the last moment, when $P(j, t)$ is required.

Normally, the length of an exact-enumeration computation is $t V$, where V is the volume (number of sites) of the substrate. However, notice that, at time t, only t^{d_f} sites have been visited by the walker. There is no need to compute the "changes" for the remaining sites of the lattice! If this procedure is followed, the length of the computation scales like $\int_0^t t^{d_f} dt \sim t^{d_f+1}$. For times $t < V^{1/d_f}$ this represents a considerable reduction in the computation time. In some cases the substrate itself may be constructed as the walker advances. For example, in the case of percolation clusters new sites may be added by the Leath algorithm.

Appendix D

Long-range correlations

In this appendix we describe several methods for analyzing data sequences in order to detect what type of correlations, if any, are present in the data. We also discuss methods for generating artificial sequences with specified long-range correlations.

D.1 The fast Fourier transform

Given time-series data of length L, we divide each sequence of length L into $K = L/N$ nonoverlapping subsequences of size N, starting from the beginning of the time series, and K nonoverlapping subsequences, starting from the end of the sequence. For each subsequence we compute the Fourier transform

$$q_f \equiv \sum_{k=0}^{N-1} u_k \exp\left(\frac{2\pi i k f}{N}\right), \tag{D.1}$$

and the power spectrum

$$S(f) = |q_f|^2 + |q_{N-f}|^2. \tag{D.2}$$

We then average $S(f)$ of each sequence over the K subsequences obtained when starting from the beginning of the series, and over the K subsequences obtained when starting from end. If the series has long-range correlations, then

$$S(f) \sim f^{-\beta}, \tag{D.3}$$

and consequently a log–log plot of $S(f)$ versus f yields a straight line of slope $-\beta$.

D.2 Detrended fluctuation analysis (DFA)

This method was developed to study nonstationary time-series data, having large linear trends (Peng *et al.*, 1994). The DFA method is based on the following steps.

266

(i) For each numerical sequence $\{u_i\}$, compute the running sum

$$y(n) \equiv \sum_{i=0}^{n} u_i, \qquad y(0) \equiv 0. \tag{D.4}$$

This can be represented graphically as a one-dimensional landscape or walk.

(ii) Divide the entire sequence of length L into L/ℓ nonoverlapping boxes, each containing ℓ numbers and define a "local trend" in each box as the ordinate of the linear least-squares fit for the displacement in that box.

(iii) Calculate the variance about the local trend for each box, and calculate the average of these variances over all boxes of size ℓ. To improve statistics, a modified DFA consisting of a sliding box of size ℓ that starts at i and ends at $i + \ell - 1$ has been suggested (Buldyrev *et al.*, 1995b). One then computes the least-squares fit $y_{i,\ell}(n) = na + b$, such that

$$E_{i,\min}(\ell) \equiv \sum_{n=i}^{i+\ell-1} \left[y(n) - y_{i,\ell}(n) \right]^2 \tag{D.5}$$

is minimal.

(iv) Finally, we average $E_{i,\min}(\ell)$ over all positions of the observation box from $i = 1$ to $i = L - \ell$ and define the "detrended fluctuation function" as

$$F_D^2 \equiv \frac{1}{\ell(L - \ell)} \sum_{i=1}^{L-\ell} E_{i,\min}(\ell). \tag{D.6}$$

For sequences with power-law long-range correlations, F_D itself can be approximated by a power law:

$$F_D \sim \ell^{\alpha}. \tag{D.7}$$

The DFA function F_D has the same scaling as $w(\ell)$ of Eq. (4.15), and may be regarded as arising from a random walk with long-range correlations between steps i and j (Eq. (4.11)): $\langle e_i \cdot e_j \rangle \approx A/|i - j|^{\gamma}$, where $\alpha = (2 - \gamma)/2$. For very long sequences, it can be shown that α and β of Eq. (D.3) are related:

$$\alpha = (\beta + 1)/2. \tag{D.8}$$

More advanced DFA for nonlinear trends assumes $y_{i,\ell}$ to be nonlinear in n. More details can be found on the net, at http://reylab.bidmc.harvard.edu/.

D.3 The wavelet method

The wavelet method is useful in the study of nonstationary time series. It is based on the application of a wavelet transform to the time series in question, $S(t)$:

$$T_\Psi(t_0, a) \equiv \frac{1}{a} \int_{-\infty}^{\infty} S(t) \Psi\left(\frac{t - t_0}{a}\right) dt, \qquad (D.9)$$

where the wavelet function Ψ has a width a and is centered at t_0. The wavelet transform T is sometimes called a "mathematical microscope", because it allows the study of properties of the signal at any arbitrary length scale a. For small a the wavelet function is localized and therefore high-frequency properties are detected. At large a the high frequencies are filtered out and only large-scale fluctuations are detected.

One class of commonly used wavelet functions are the derivatives of a Gaussian: $\Psi^{(n)} \equiv (d^n/dt^n)e^{-t^2/2}$. With this choice, the wavelet transform can eliminate local polynomial trends in a nonstationary signal. $\Psi^{(1)}$ is piecewise orthogonal to segments of the time series with constant local average (Fig. D1). This results in small fluctuations of the wavelet transform around zero, with large spikes where segments of different local averages join together. Likewise, $\Psi^{(2)}$ and higher-order derivatives can eliminate the influences of linear as well as higher-order polynomial trends, respectively.

The sum of the squares of the local maxima of the modulus of T is expected to scale as $a^{2\alpha-1}$. The -1 comes from the $1/a$ prefactor in Eq. (D.9).

The wavelet approach is also useful for studying multifractality or multiaffinity (Vicsek and Barabási, 1991). In this case, one evaluates the scaling of the partition function $Z_q(a)$, which is defined as the sum of the qth power of the local maxima of the moduli of the wavelet-transform coefficients at scale a (Muzy *et al.*, 1991). For small scales we expect

$$Z_q(a) \sim a^{\tau(q)}. \qquad (D.10)$$

For certain values of q, the exponents $\tau(q)$ can be identified. In particular, $\tau(2)$ is related to the scaling exponent of the Fourier power spectrum, $S(f) \sim 1/f^\beta$, since $\beta = 2 + \tau(2)$. For positive q, $Z_q(a)$ reflects the scaling of large fluctuations and strong singularities, whereas for negative q, $Z_q(a)$ reflects the scaling of small fluctuations and weak singularities (Vicsek, 1991; Takayasu, 1990). Thus, the scaling exponents $\tau(q)$ can reveal various aspects of the dynamics of the system.

Monofractal signals display a linear $\tau(q)$ spectrum, $\tau(q) = qH - 1$, where H is the global Hurst exponent. For multifractal signals, $\tau(q)$ is a nonlinear function, $\tau(q) = qh(q) - D(h)$, where $h(q) \equiv d\tau(q)/dq$ is not constant. In other words, the fractal dimension $D(h)$ and $\tau(q)$ are related to each other through a Legendre transform.

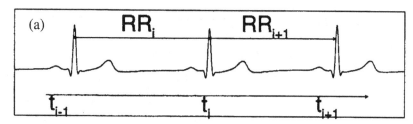

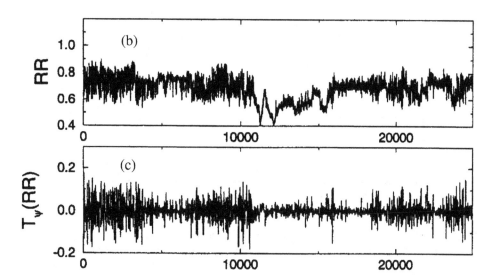

Fig. D.1. Filtering out trends with the wavelet-transform method. (a) A segment of an electrocardiogram showing heart beat-to-beat (R–R) intervals. (b) A plot of R–R time series against the consecutive beat number for 6 h ($\simeq 2.5 \times 10^4$ beats). Nonstationarity (patchiness) is evident for long and short time scales. (c) The wavelet transform $T_\Psi(\text{R–R})$ of the R–R signal in (b). Nonstationarities related to constants and linear trends have been filtered out by using $\Psi^{(2)}$. After Ivanov *et al.* (1996).

D.4 The generation of long-range correlations

Consider a sequence of N independent (uncorrelated) random numbers $\{u_j\}_{j=1,\ldots,N}$:

$$\langle u_j u_{j+n} \rangle = \delta_{n,0}. \tag{D.11}$$

Our goal is to use this sequence for the generation of a new one, $\{g_j\}_{j=1,\ldots,N}$, with long-range power-law correlations (Makse *et al.*, 1996):

$$C_n \equiv \langle g_j g_{j+n} \rangle \sim n^{-\gamma}. \tag{D.12}$$

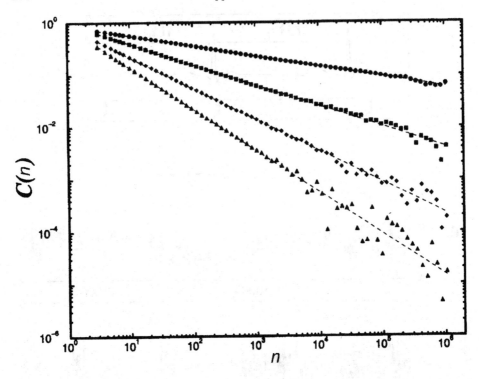

Fig. D.2. Long-range correlations in computer-generated series. A log–log plot of $\langle g_j g_{j+n} \rangle$ versus n for series of $N = 4096$ points, averaged over 500 realizations, for $\gamma = 0.2, 0.4, 0.6,$ and 0.8 (from top to bottom). The slopes of the linear fits are $0.19, 0.37, 0.57,$ and 0.75, respectively.

We introduce a "response" function, ϕ_j, that relates these two sequences:

$$g_j = \sum_{k=1}^{N} \phi_{j-k} u_k. \tag{D.13}$$

Comparison of the Fourier transforms of Eqs. (D.12) and (D.13) yields

$$\hat{\Phi}_q = \sqrt{\hat{C}_q}/|\hat{u}_q|, \tag{D.14}$$

where $\hat{\Phi}_q$, etc. denote the transforms of the corresponding functions. The numerical algorithm used to construct the correlated sequence, $\{g_j\}$, consists of the following steps.

(i) Generate the sequence, $\{u_j\}$, of uncorrelated random numbers and calculate its Fourier transform, $\hat{u}_q$.

(ii) Calculate the Fourier transform, $\hat{C}_q$, of the desired correlation function C_n.

(iii) Using Eq. (D.14), calculate the Fourier transform, $\hat{\Phi}_q$, of the response function, and compute the inverse Fourier transform to obtain ϕ_j.

(iv) Use Eq. (D.13) to get the desired correlated sequence, $\{g_j\}$.

Note that the discrete Fourier transform

$$\hat{C}_q = \frac{1}{N} \sum_{j=0}^{N-1} C_j e^{i2\pi q j/N},$$ (D.15)

must be supplemented by periodic boundary conditions, for Eq. (D.14) to be valid (and likewise for the other functions). Examples of long-range correlations in series generated by this method are shown in Fig. D2.

References

The chapter(s) where a reference is cited in the book is indicated within square brackets at the end of each reference.

Acedo, L., and Yuste, S. B. (1998). Short-time regime propagator in fractals, *Phys. Rev. E* **57**, 5160. [*Chap. 5*]

Acharyya, M., and Stauffer, D. (1998). Effects of boundary conditions on the critical spanning probability, *Int. J. Mod. Phys. C* **9**, 643. [*Chap. 2*]

Adler, J. (1984). Series expansions, *Computers Phys.* **8**, 287. [*Chap. 2*]

Adler, J., Meir, Y., Aharony, A., Harris, A. B., and Klein, L. (1990). Low-concentration series in general dimension, *J. Stat. Phys.* **58**, 511. [*Chap. 2, 6*]

Agmon, N., and Glasser, M. L. (1986). Complete asymptotic expansion for integrals arising from one-dimensional diffusion with random traps, *Phys. Rev. A* **34**, 656. [*Chap. 12*]

Aharony, A., and Stauffer, D. (1984). Possible breakdown of the Alexander–Orbach rule at low dimensionalities, *Phys. Rev. Lett.* **52**, 2368. [*Chap. 6, 7*]

Aharony, A., Alexander, S., Entin-Wohlman, O., and Orbach, R. (1987). Scattering of fractons, the Ioffe–Regel criterion, and the 4/3 conjecture, *Phys. Rev. Lett.* **58**, 132. [*Chap. 6*]

Aharony, A., Blumenfeld, R., and Harris, A. B. (1993). Distribution of the logarithms of current in percolating resistor networks. 1. Theory, *Phys. Rev. B* **47**, 5756. [*Chap. 6*]

Aizenman, M. (1997). On the number of incipient spanning clusters, *Nucl. Phys. B* [FS] **485**, 551. [*Chap. 2*]

Alemany, P. A. (1997). Novel decay laws for the one-dimensional reaction–diffusion model $A + A \rightarrow (2 - \epsilon)A$ as consequence of initial distributions, *J. Phys. A* **30**, 3299. [*Chap. 17*]

Alemany, P. A., and ben-Avraham, D. (1995). Inter-particle distribution functions for one-species diffusion-limited annihilation, $A + A \rightarrow 0$, *Phys. Lett. A* **206**, 18. [*Chap. 18*]

Alemany, P. A., Zanette, D. H., and Wio, H. S. (1994). Time-dependent reactivity for diffusion-controlled annihilation and coagulation in two dimensions, *Phys. Rev. E* **50**, 3646. [*Chap. 13*]

Alexander, S. (1981). Anomalous transport properties for random-hopping and random-trapping models, *Phys. Rev. B* **23**, 2951. [*Chap. 8*]

Alexander, S. (1983). Linear problems on percolation clusters, in *Percolation Structures and Processes*, G. Deutscher, R. Zallan, and J. Adler, eds., *Ann. Israel Phys. Soc.* **5**, Adam Hilger, Bristol. [*Chap. 6*]

Alexander, S., and Orbach, R. (1982). Density of states on fractals: fractons, *J. Phys. Lett., Paris* **43**, L625. [*Chap. 1, 5, 6*]

Alexander, S., and Pincus, P. (1978). Diffusion of labeled particles on one-dimensional systems, *Phys. Rev. B* **18**, 2011. [*Chap. 10*]

Alexander, S., Bernasconi, J., and Orbach, R. (1978). Low energy density of states for disordered chains, *J. Phys., Paris, C* **39**, 706. [*Chap. 5*]

Alexander, S., Bernasconi, J., Schneider, W. R., and Orbach, R. (1981). Excitation dynamics in random one-dimensional systems, *Rev. Mod. Phys.* **53**, 175. [*Chap. 5, 8*]

Alexander, S., Courtens, E., and Vacher, R. (1993). Vibrations of fractals: dynamic scaling, correlation functions and inelastic light scattering, *Physica A* **195**, 286. [*Chap. 6*]

Alexandrowicz, Z. (1980). Critically branched chains and percolation clusters, *Phys. Lett. A* **80**, 284. [*Chap. 6*]

Allen, C. E., and Seebauer, E. G. (1996). Surface diffusivities and reaction-rate constants: making a quantitative experimental connection, *J. Chem. Phys.* **104**, 2557. [*Chap. 13*]

Amar, J. G., and Family, F. (1990). Diffusion–annihilation in one dimension and kinetics of the Ising model at zero temperature, *Phys. Rev. E* **41**, 3258. [*Chap. 15*]

Amici, A., and Montuori, M. (1998). Scaling properties of discrete fractals, *e-preprint*: cond-mat/9811077. [*Chap. 1*]

Amitrano, C., Bunde, A., and Stanley, H. E. (1985). Diffusion of interacting particles on fractal aggregates, *J. Phys. A* **18**, L923. [*Chap. 10*]

Amitrano, C., Coniglio, A., and di Liberto, F. (1986). Growth probability distribution in kinetic aggregation processes, *Phys. Rev. Lett.* **57**, 1016. [*Chap. 6*]

Amritkar, R. E., and Roy, M. (1998). Percolation of finite-sized objects on a lattice, *Phys. Rev. E* **57**, 1269. [*Chap. 2*]

Anacker, L. W., and Kopelman, R. (1987). Steady state chemical kinetics on fractals: geminate and non-geminate generation of reactants, *J. Chem. Phys.* **91**, 5555. [*Chap. 13*]

Andrade Jr, J. S., Almeida, M. P., Mendes Filho, J., Havlin, S., Suki, B., and Stanley, H. E. (1997a). Fluid flow through porous media: the role of stagnant zones, *Phys. Rev. Lett.* **79**, 3901. [*Chap. 9*]

Andrade Jr, J. S., Street, D. A., Shibusa, Y., Havlin, S., and Stanley, H. E. (1997b). Diffusion and reaction in percolating pore networks, *Phys. Rev. E* **55**, 772. [*Chap. 13*]

Andrews, A. T. (1986). *Electrophoresis: Clinical Applications*, Oxford University Press, New York. [*Chap. 10*]

Anlauf, J. K. (1984). Asymptotically exact solution of the one-dimensional trapping problem, *Phys. Rev. Lett.* **52**, 1845. [*Chap. 12*]

Araujo, M., Havlin, S., and Stanley, H. E. (1991a). Anomalous fluctuations in tracer concentration in stratified media with random velocity-fields, *Phys. Rev. A* **44**, 6913. [*Chap. 4*]

Araujo, M., Havlin, S., Weiss, G. H., and Stanley, H. E. (1991b). Diffusion of walkers with persistent velocities, *Phys. Rev. A* **43**, 5207. [*Chap. 4*]

Araujo, M., Havlin, S., Larralde, H., and Stanley, H. E. (1992). Random force dominated reaction kinetics: reactants moving under random forces, *Phys. Rev. Lett.* **68**, 1791. [*Chap. 14*]

Araujo, M., Larralde, H., Havlin, S., and Stanley, H. E. (1993). Scaling anomalies in reaction front dynamics of confined systems, *Phys. Rev. Lett.* **71**, 3592. [*Chap. 14*]

Argyrakis, P., and Kopelman, R. (1990). Nearest-neighbor distance distributions of self-ordering in diffusion-controlled reactions. I. A + A simulations, *Phys. Rev. A* **41**, 2114 (1990). [*Chap. 13, 15*]

Argyrakis, P., and Kopelman, R. (1992). Diffusion-controlled binary reactions in low dimensions: refined simulations, *Phys. Rev. A* **45**, 5814. [*Chap. 13*]

Arratia, R. (1983). *Ann. Prob.* **11**, 362. [*Chap. 10*]

Avnir, D., ed. (1992). *The Fractal Approach to Heterogeneous Chemistry*, John Wiley, New York. [*Chap. 1*]

Avnir, D., and Kagan, M. (1984). Spatial structures generated by chemical reactions at interfaces, *Nature, Lond.* **307**, 717. [*Chap. 14*]

Babalievski, F. (1997). Comment on "Universal formulas for percolation thresholds. II. Extension to anisotropic and aperiodic lattices", *Phys. Rev. E* **56**, 332. [*Chap. 2*]

Babalievski, F. (1998). Cluster counting: the Hoshen–Kopelman algorithm vs. spanning tree approaches, *Int. J. Mod. Phys. C* **9**, 43. [*Chap. 2*]

Bak, P., Chen, K, and Tang, C. (1990). A forest-fire model and some thoughts on turbulence, *Phys. Lett. A* **147**, 297. [*Chap. 2*]

Bak, P. (1996). *How Nature Works*, Copernicus, New York. [*Chap. 1*]

Balagurov, B. Ya., and Vaks, V. G. (1973). Random walks of a particle on lattices with traps, *Sov. Phys. JETP* **38**, 968. [*Chap. 12*]

Balberg, I. (1987). Recent developments in continuum percolation, *Phil. Mag. B* **56**, 991. [*Chap. 2, 6*]

Balberg, I. (1998a). Limits on the continuum-percolation transport exponents, *Phys. Rev. B* **57**, 13351. [*Chap. 2, 6*]

Balberg, I. (1998b). Resistivity and electrical noise in continuum percolation, *Trends in Stat. Phys.* **2**, 39. [*Chap. 6*]

Balboni, D., Rey, P.-A., and Droz, M. (1995). Universality of a class of annihilation–coagulation models, *Phys. Rev. E* **52**, 6220. [*Chap. 13, 15*]

Ball, R. C., and Lee, J. R. (1996). Irreversible growth algorithm for branched polymers (lattice animals), and their relation to colloidal cluster–cluster aggregates. *J. Phys., Paris* **6**, 357. [*Chap. 7*]

Ball, R. C., and Witten, T. A. (1984). Causality bound on the density of aggregates, *Phys. Rev. A* **29**, 2966. [*Chap. 7*]

Ball, R. C., Havlin, S., and Weiss, G. H. (1987). Non-Gaussian random walks, *J. Phys. A* **20**, 4055. [*Chap. 7*]

Ball, Z., Phillips, H. M., Callahan, D. L., and Sauerbrey, R. (1994). Percolative metal-insulator transition in excimer-laser irradiated polyimide, *Phys. Rev. Lett.* **73**, 2099. [*Chap. 2*]

Ballesteros, H. G., Fernandez, L. A., Martin-Mayor, V., Muñoz Sudupe, A., Parisi, G., and Ruiz-Lorenzo, J. J. (1997). Measures of critical exponents in the four-dimensional site percolation, *Phys. Lett. B* **400**, 346. [*Chap. 2*]

Banavar, J. R., and Willemson, J. (1984). Probability density for diffusion on fractals, *Phys. Rev. B* **30**, 6778. [*Chap. 5, 6, 7*]

Barabasi, J. R., and Stanley, H. E. (1995). *Fractal Concepts in Surface Growth*, Cambridge University Press, Cambridge. [*Chap. 1, 4*]

Barat, K., and Chakrabarti, B. K. (1995). Statistics of self avoiding walks on random lattices, *Phys. Rep.* **258**, 377. [*Chap. 6, 10*]

Barber, M. N., and Ninham, B. W. (1970). *Random and Restricted Walks. Theory and Applications*, Gordon and Breach, New York. [*Chap. 3*]

Barkema, G. T., Howard, M. J., and Cardy, J. L. (1996). Reaction–diffusion front for A + B → 0 in one dimension, *Phys. Rev. E* **53**, R2017. [*Chap. 14*]

Barlow, M. T., and Bass, R. F. (1990). On the resistance of the Sierpinski carpet, *Proc. Roy. Soc. Lond. A, Mat.* **431**, 345. [*Chap. 5*]

Barlow, M. T., and Bass, R. F. (1992). Transitions densities for Brownian motion on the Sierpinski carpet, *Probability Theor. Related Fields* **91**, 307. [*Chap. 5*]

Barlow, M. T., and Perkins, E. A. (1988). Brownian motion on the Sierpinski gasket, *Probability Theor. Related Fields* **79**, 543. [*Chap. 5*]

Barlow, M. T., Bass, R. F., and Sherwood, J. D. (1990). Resistance and spectral dimension of Sierpinski carpets, *J. Phys. A* **23**, L253. [*Chap. 5*]

Barma, M., and Dhar, D. (1983). Directed diffusion in a percolation network, *J. Phys. C* **16**, 1451. [*Chap. 7, 9*]

Barma, M., and Ramaswamy, R. (1986). On backbends on percolation backbones, *J. Phys. A* **19**, L605. [*Chap. 9*]

Barma, M., and Ramaswamy, R. (1994). in *Nonlinearity and Breakdown in Soft Condensed Matter*, B. K. Chakrabarti, K. K. Bardhan, and A. Hansen, eds., Springer, Berlin. [*Chap. 9, 10*]

Barnsley, M. (1988). *Fractals Everywhere*, Academic Press, San Diego. [*Chap. 1*]

Barthelemy, M., Buldyrev, S. V., Havlin, S., and Stanley, H. E. (2000). Multifractal properties of the random resistor network *Phys. Rev. E* **61**, R3283. [*Chap. 6*]

Bassinthwaighte, J. B., Liebovitch, L, S., and West, B. J. (1994). *Fractal Physiology*, Oxford University Press, New York. [*Chap. 1*]

Batrouni, G. G., Hansen, A., and Roux, S. (1988). Negative moments of the current spectrum in the random-resistor network, *Phys. Rev. A* **38**, 3820. [*Chap. 6*]

Baumgärtner, A. (1995). *Polymers in Disordered Medium*, Habilitationsschrift, Gesamthochschule Duisburg. [*Chap. 10*]

Beeler, R. J., and Delaney, J. A. (1963). Order–disorder events produced by single vacancy migration, *Phys. Rev.* **130**, 926. [*Chap. 12*]

ben-Avraham, D. (1987). Discrete fluctuations and their influence on kinetics of reactions, *J. Stat. Phys.* **48**, 315. [*Chap. 13*]

ben-Avraham, D. (1993). Diffusion-limited three-body reactions in one dimension, *Phys. Rev. Lett.* **71**, 3733. [*Chap. 13, 19*]

ben-Avraham, D. (1995). The method of interparticle distribution functions for diffusion–reaction systems in one dimension, *Mod. Phys. Lett. B* **9**, 895. [*Chap. 15*]

ben-Avraham, D. (1997). The coalescence process, A + A → A, and the method of interparticle distribution functions, in *Nonequilibrium Statistical Mechanics in One Dimension*, V. Privman, ed., pp. 29–50, Cambridge University Press, Cambridge. [*Chap. 15*]

ben-Avraham, D. (1998a). Complete exact solution of diffusion-limited coalescence, A + A → A, *Phys. Rev. Lett.* **81**, 4756. [*Chap. 18*]

ben-Avraham, D. (1998b). Diffusion-limited coalescence, A + A ⇌ A, with a trap, *Phys. Rev. E* **58**, 4351. [*Chap. 18*]

ben-Avraham, D. (1998c). Fisher waves in the diffusion-limited coalescence process A + A ⇌ A, *Phys. Lett. A* **247**, 53. [*Chap. 18*]

ben-Avraham, D. (1998d). Inhomogeneous steady-states of diffusion-limited coalescence, A + A ⇌ A, *Phys. Lett. A* **249**, 415. [*Chap. 18*]

ben-Avraham, D. and Doering, C. R. (1988). Equilibrium of two-species annihilation with input, *Phys. Rev. A* **37**, 5007. [*Chap. 13*]

ben-Avraham, D., and Havlin, S. (1982). Diffusion on percolation clusters at criticality, *J. Phys. A* **15**, L691. [*Chap. 6*]

ben-Avraham, D., and Redner, S. (1986). Kinetics of *n*-species annihilation: mean-field and diffusion-controlled limits, *Phys. Rev. A* **34**, 501. [*Chap. 13*]

ben-Avraham, D., and Weiss, G. H. (1989). Statistical properties of nearest-neighbor distances in a diffusion–reaction model, *Phys. Rev. A* **39**, 6436. [*Chap. 12*]

ben-Avraham, D., and Zhong, D. (1993). Diffusion-limited many-body reactions in one dimension and the method of interparticle distribution functions, *Chem. Phys.* **180**, 329. [*Chap. 13, 19*]

ben-Avraham, D., Burschka, M., A., and Doering, C. R. (1990). Statics and dynamics of a diffusion-limited reaction: anomalous kinetics, nonequilibrium self-ordering, and a dynamic transition, *J. Stat. Phys.* **60**, 695. [*Chap. 13, 15, 16, 17*]

ben-Avraham, D., Taitelbaum, H., and Weis, G. H. (1991). Boundary conditions for a model of photon migration in a turbid medium, *Lasers in the Life Sciences* **4**, 29. [*Chap. 12*]

ben-Avraham, D., Leyvraz, F., and Redner, S. (1992). Superballistic motion in a "random walk" shear flow. *Phys. Rev. A* **45**, 2315. [*Chap. 4*]

ben-Avraham, D., Leyvraz, F., and Redner, S. (1994). Propagation and extinction in branching annihilating random walks, *Phys. Rev. E* **50**, 1843. [*Chap. 13*] ben-Avraham, D., Privman, V., and Zhong, D. (1995). Two-species annihilation with drift: a model with continuous concentration-decay exponents, *Phys. Rev. E* **52**, 6889. [*Chap. 10, 13*]

Benguigui, L. (1986). Lattice and continuum percolation transport exponents: experiments in two-dimensions, *Phys. Rev. B* **34**, 8177. [*Chap. 6*]

Ben-Mizrahi, A., and Bergman, D. J. (1981). Non-universal critical behaviour of random resistor networks with a singular distribution of conductances, *J. Phys. C* **14**, 909. [*Chap. 6*]

Ben-Naim, E., and Krapivsky, P. L. (1994). Cluster approximation for the contact process, *J. Phys. A* **27**, L481. [*Chap. 19*]

Ben-Naim, E., and Redner, S. (1992). Inhomogeneous two-species annihilation in the steady state, *J. Phys. A* **25**, L575. [*Chap. 14*]

Ben-Naim, E., Redner, S., and ben-Avraham, D. (1992). Bimodal diffusion in power-law shear flows. *Phys. Rev. A* **45**, 7207. [*Chap. 4*]

Ben-Naim, E., Redner, S., and Weiss, G. H. (1993). Partial absorption and "virtual" traps, *J. Stat. Phys.* **71**, 75. [*Chap. 12*]

Benson S. W. (1960). *The Foundations of Chemical Kinetics*, McGraw-Hill, New York. [*Chap. 11*]

Berezhkovskii, A. M., and Weiss, G. H. (1995). Some generalizations of the trapping problem, *Physica A* **215**, 40. [*Chap. 12*]

Berezhkovskii, A. M., Makhnovskii, Yu A., and Suris, R. A. (1989a). Fluctuation slow-down of the death of Brownian particles in the case of movable traps, *J. Phys. A* **22**, L615. [*Chap. 12*]

Berezhkovskii, A. M., Makhnovskii, A. Yu., and Suris, R. A. (1989b). Kinetics of diffusion-controlled reactions, *Chem. Phys.* **137**, 41. [*Chap. 13*]

Berezhkovskii, A. M., Bicout, D. J., and Weiss, G. H. (1999). Target and trapping problems: from the ballistic to the diffusive regime, *J. Chem. Phys.* **110**, 1112. [*Chap. 12*]

Berg, H. C. (1993). *Random Walks in Biology*, Princeton, NJ. [*Chap. 3*]

Berkolaiko, G., and Havlin, S. (1997). Territory covered by N Lévy flights on d-dimensional lattices, *Phys. Rev. E* **55**, 1395. [*Chap. 4, App. B*]

Berkolaiko, G., Havlin, S., Larralde, H., and Weiss, G. H. (1996). Expected number of distinct sites visited by N discrete Lévy flights on a one-dimensional lattice, *Phys. Rev. E* **53**, 5774. [*Chap. 4, App. B*]

Berkowitz, B., and Scher, H. (1998). Theory of anomalous chemical transport in random fracture networks, *Phys. Rev. E* **57**, 5858. [*Chap. 4*]

Berkowitz, B., Scher, H., and Sillman, S. E. (2000). Anomalous transport in laboratory-scale, heterogeneous porous media, *Water Resources Research* **36**, 149. [*Chap. 4*]

Bernasconi, J., Alexander, S., and Orbach, R. (1978). Classical diffusion in a one-dimensional disordered lattice. *Phys. Rev. Lett.* **41**, 185. [*Chap. 8*]

Bernasconi, J., Beyeler, J. U., Strassler, S., and Alexander, S. (1979). Anomalous frequency-dependent conductivity in disordered one-dimensional systems, *Phys. Rev. Lett.* **42**, 819. [*Chap. 8, 9*]

Bhatti, F. M, Brak, R., Essam, J. W. and Lookman, T. (1997). Series expansion analysis of the backbone properties of two-dimensional percolation clusters, *J. Phys. A* **30**, 6215. [*Chap. 2*]

Birdi, K. S. (1993). *Fractals in Chemistry, Geochemistry and Biophysics: an Introduction*, Plenum, New York. [*Chap. 1*]

Bixon, M., and Zwanzig, R. (1981). Diffusion in a medium with static traps, *J. Chem. Phys.* **75**, 2354. [*Chap. 12*]

Blumen, A., Klafter, J., and Zumofen, G. (1983). Trapping and reaction rates on fractals; A random-walk study, *Phys. Rev. B* **28**, 6112. [*Chap. 12*]

Blumen, A., Klafter, J., White, B. S., and Zumofen, G. (1984). Continuous-time random walks on fractals, *Phys. Rev. Lett.* **53**, 1301. [*Chap. 7*]

Blumen, A., Zumofen, G., and Klafter, J. (1986a). Random walks on ultrametric spaces: low temperature patterns, *J. Phys. A* **19**, L861. [*Chap. 8*]

Blumen, A., Klafter, J., and Zumofen, G. (1986b). in *Optical Spectroscopy of Glasses*, I. Zschokke, ed., p. 199, Reidel, Dordrecht. [*Chap. 12*]

Blumenfeld, R., and Aharony, A. (1985). Nonlinear resistor fractal networks, topological distances, singly connected bonds and fluctuations, *J. Phys. A* **18**, L443. [*Chap. 6*]

Bonner, R. F., Nossal, R., Havlin, S., and Weiss, H. G. (1987). Model for photon migration in turbid biological media, *J. Opt. Soc. Am. A* **4**, 423. [*Chap. 3, 12*]

Böttger, H., and Bryskin, V. V. (1980). Investigation of non-Ohmic hopping conduction by methods of percolation theory, *Phil. Mag. B* **42**, 297. [*Chap. 9*]

Böttger, H., and Bryskin, V. V. (1982). Hopping conductivity in ordered and disordered systems (III), *Phys. Stat. Sol. (b)* **113**, 9. [*Chap. 9*]

Bouchaud, J.-P., and Georges, A. (1990). Anomalous diffusion in disordered media: statistical mechanisms, models and physical applications, *Phys. Rep.* **195**, 128. [*Chap. 8*]

Brady, R. M., and Ball, R. C. (1984). Fractal growth of copper electrodeposits, *Nature, Lond.* **309**, 225. [*Chap. 7*]

Bramson, M., and Griffeath, D. (1980). Clustering and dispersion for interacting particles, *Ann. Prob.* **8**, 183. [*Chap. 13, 14, 15*]

Bramson, M., and Lebowitz, J. (1988). Asymptotic behavior of densities in diffusion-dominated annihilation reactions, *Phys. Rev. Lett.* **61**, 2397. [*Chap. 12, 13*]

Brandt, W. W. (1975). Use of percolation theory to estimate effective diffusion coefficients of particles migrating on various ordered lattices and in a random network structure, *J. Chem. Phys.* **63**, 5162. [*Chap. 6*]

Braunstein, L., Mártin, O. H., Grynberg, M. D., and Roman, H. E. (1992). Effects of probability of reaction on annihilation reactions in one dimension, *J. Phys. A* **25**, L255. [*Chap. 19*]

Brunet, E., and Derrida, B. (1997). Shift in the velocity of a front due to a cutoff, *Phys. Rev. E* **56**, 2597. [*Chap. 18*]

Bug, A. L. R., Grest, G. S., Cohen, M. H., and Webman, I. (1986). AC response near percolation threshold: transfer matrix calculation in 2D, *J. Phys. A* **19**, L323. [*Chap. 6, 9*]

Buldyrev, S. V., Goldberger, A.-L., Havlin, S., Peng, C.-K., Simons, M., and Stanley, H. E. (1993). Generalised Lévy-walk model for DNA nucleotide sequences, *Phys. Rev. E* **47**, 4514. [*Chap. 4*]

Buldyrev, S. V., Havlin, S., Kertesz, J., Sadr, R., Shehter, A., and Stanley, H. E. (1995a). Surface roughening with quenched disorder in high dimensions: exact results for the Cayley tree, *Phys. Rev. E* **52**, 373. [*Chap. 2*]

Buldyrev, S. V., Goldberger, A. L., Havlin, S., Mantegna, R. N., Matsa, M. E., Peng, C.-K., Simons, M., and Stanley, H. E. (1995b). Long-range correlation properties of coding and noncoding DNA-sequences genebank analysis, *Phys. Rev. E* **51**, 5084. [*Chap. 4, App. D*]

Bunde, A., and Havlin, S., eds. (1994). *Fractals in Science*, Springer, Berlin. [*Chap. 1*]

Bunde, A., and Havlin, S., eds. (1996). *Fractals and Disordered Systems*, 2nd ed., Springer, Berlin. [*Chap. 1, 2, 3, 5, 6*]

Bunde, A., and Havlin, S., eds. (1999). *Proceedings of the International Conference on Percolation and Disordered Systems: Theory and Applications*, *Physica A* **264**. [*Chap. 2*]

Bunde, A., Havlin, S., Nossal, R., and Stanley, H. E. (1985). Anomalous trapping: effect of interaction between diffusing particles, *Phys. Rev. B* **32**, 3367. [*Chap. 10*]

Bunde, A., Havlin, S., Stanley, H. E., Trus, B., and Weiss, G. H. (1986a). Diffusion in random structures with a topological bias, *Phys. Rev. B* **34**, 8129. [*Chap. 8, 9*]

Bunde, A., Harder, H., and Havlin, S. (1986b). Nonuniversality of diffusion exponents in percolation systems, *Phys. Rev. B* **34**, 3540. [*Chap. 6*]

Bunde, A., Mosely, L. L., Stanley, H. E., ben-Avraham, D., and Havlin, S. (1986c). Anomalously slow trapping of non-identical interacting particles by random sinks, *Phys. Rev. A* **34**, 2575. [*Chap. 10*]

Bunde, A., Havlin, S., and Roman, H. E. (1990). Multifractal features of random walks on random fractals, *Phys. Rev. A* **42**, R6274. [*Chap. 5, 6, 9*]

Bunde, A., Ingram, M. D., and Funke, K. (1994). The dynamic structure model for ionic transport in glass, *J. Non-Cryst. Solids* **172-174**, 1222. [*Chap. 2*]

Bunde, A., Havlin, S., Klafter, J., Graff, G., and Shehter, A. (1997). Anomalous size dependence of relaxational processes, *Phys. Rev. Let.* **78**, 3338. [*Chap. 12*]

Burioni, R., and Cassi, D. (1996). Universal properties of spectral dimension, *Phys. Rev. Lett.* **76**, 1091. [*Chap. 5*]

Burioni, R., Cassi, D., and Regina, S. (1999). Cutting-decimation renormalization for diffusive and vibrational dynamics on fractals, *Physica A* **265**, 323. [*Chap. 5*]

Burschka, M. A., Doering, C. R., and ben-Avraham, D. (1989). Transition in the relaxation dynamics of a reversible diffusion-limited reaction, *Phys. Rev. Lett.* **63**, 700. [*Chap. 17*]

Cardy, J. L. (1998). The number of incipient spanning clusters in two-dimensional percolation, *J. Phys. A* **31**, L105. [*Chap. 2, 6*]

Cardy, J. L., and Colaiori, F. (1999). Directed percolation and generalized friendly random walkers, *Phys. Rev. Lett.* **82**, 2232. [*Chap. 2*]

Cardy, J. L., and Grassberger, P. (1985). Epidemic models and percolation, *J. Phys. A* **18**, L267. [*Chap. 6*]

Castellani, C., and Peliti, L. (1986). Multifractal wavefunction at the localization threshold, *J. Phys. A* **19**, L429. [*Chap. 6*]

Cates, M. E. (1985). Range of validity of the Einstein reaction (comment and reply), *Phys. Rev. Lett.* **55**, 131. [*Chap. 5*]

Chakrabarti, B. K., and Kertesz, J. (1981). The statistics of self-avoiding walks on a disordered lattice, *Z. Phys. B* **44**, 221. [*Chap. 10*]

Chandrasekhar, S. (1943). Stochastic problems in physics and astronomy, *Rev. Mod. Phys.* **15**, 1. Reprinted in M. Wax, *Selected Papers in Noise and Stochastic Processes* (Dover, New York, 1954). [*Chap. 3*]

Cheon, M., Heo, M., Chang, I., and Stauffer, D. (1999). Fragmentation of percolation clusters in general dimensions, *Phys. Rev. E* **59**, R4733. [*Chap. 2*]

Chopard, B., Droz, M., Magnin, J., and Rácz, Z. (1997). Localization–delocalization transition of a reaction–diffusion front near a semipermeable wall, *Phys. Rev. E* **56**, 5343. [*Chap. 14*]

Chowdhury, D. (1985). Random walk on self-avoiding walk in external bias: diffusion, drift and trapping, *J. Phys. A* **18**, L761. [*Chap. 9*]

Cieplak, M., Maritan, A., and Banavar, J. R. (1994). Optimal paths and domain walls in the strong disorder limit, *Phys. Rev. Lett.* **72**, 2320. [*Chap. 2, 4*]

Cieplak, M., Maritan, A., Swift, M. R., Battacharya, A., Stella, A. L., and Banavar, J. R. (1995). Optimal paths and universality, *J. Phys. A* **28**, 5693. [*Chap. 4*]

Cieplak, M., Maritan, A., and Banavar, J. R. (1996). Invasion percolation and Eden growth: geometry and universality, *Phys. Rev. Lett.* **76**, 3754. [*Chap. 2, 7, 8*]

Clar, S., Schenk, K., and Schwabl, F. (1997). Phase transitions in a forest-fire model, *Phys. Rev. E* **55**, 2174. [*Chap. 2*]

Clément, E., Kopelman, R., and Sander, L. M. (1990). Bimolecular reaction $A + B \rightarrow 0$ at steady-state on fractals: anomalous rate law and reactant self-organization, *Chem. Phys.* **146**, 343. [*Chap. 13*]

Clément, E., Kopelman, R., and Sander, L. M. (1991). The diffusion-limited reaction $A + B \rightarrow 0$ on a fractal substrate, *J. Stat. Phys.* **65**, 919. [*Chap. 13*]

Clément, E., Kopelman, R., and Sander, L. M. (1994). The diffusion-limited reaction $A + A \rightarrow 0$ in the steady state: influence of correlations in the source, *Chem. Phys.* **180**, 337. [*Chap. 13, 16*]

Clerk, J. P., Giraud, G., Laugier, J. M., and Luck, J. M. (1990). The AC electrical conductance of binary disordered systms, percolation clusters, fractals and related models, *Adv. Phys.* **39**, 191. [*Chap. 9*]

Clifford, M. J., Roberts, E. P. L., and Cox, S. M. (1998). The influence of segregation on the yield for a series-parallel reaction, *Chem. Eng. Sci.* **53**, 1791. [*Chap. 14*]

Cole, B. J. (1995). Fractal time in animal behavior: the movement activity of drosophila, *J. Anim. Behav.* **50**, 1317. [*Chap. 4*]

Condat, C. A., Sibona, G., and Budde, C. E. (1995). Depletion region for diffusion-controlled reactions in a field, *Phys. Rev. E* **51**, 2839. [*Chap. 12*]

Coniglio, A. (1981). Thermal phase transition of the dilute q-state Potts and n-vector models at the percolation threshold, *Phys. Rev. Lett.* **46**, 250. [*Chap. 2*]

Coniglio, A. (1982). Cluster structure near the percolation threshold, *J. Phys. A* **15**, 3829. [*Chap. 2*]

Coppens, M. O., and Froment, G. F. (1995). Diffusion and reaction in a fractal catalyst pore: 1. Geometrical aspects, *Chem. Eng. Sci.* **50**, 1013. [*Chap. 13*]

Cornell, S., Droz, M., and Chopard, B. (1991). Role of fluctuations for inhomogeneous reaction–diffusion phenomena, *Phys. Rev. A* **44**, 4826. [*Chap. 14*]

Cox, S. M., Clifford, M. J., and Roberts, E. P. L. (1998). A two-stage reaction with initially separated reactants, *Physica A* **256**, 65. [*Chap. 14*]

Daoud, M. (1983). Spectral dimension and conductivity exponent of the percolating cluster, *J. Phys. Lett., Paris* **44**, 1925. [*Chap. 6*]

Daoud, M., Cotton, J. P., Farnoux, B., Jannik, G., Sarma, G., Benoit, H., Duplessix, R., Picot, C., and de Gennes, P. G. (1975). Solutions of flexible polymers. Neutron experiments and interpretation, *Macromolecules* **8**, 804. [*Chap. 10*]

Dayan, I., Gouyet, J. F., and Havlin, S. (1991). Percolation in multi-layer structures, *J. Phys. A* **24**, L287. [*Chap. 2*]

Dean, D. S. and Jansons, K. M. (1993). Brownian excursions on combs, *J. Stat. Phys.* **70**, 1313. [*Chap. 7*]

de Arcangelis, L., Redner, S., and Coniglio, A. (1985). Anomalous voltage distribution of random resistor networks and a new model for the backbone at the percolation threshold. *Phys. Rev. B* **31**, 4725. [*Chap. 6*]

de Arcangelis, L., Redner, S., and Coniglio, A. (1986a). Multiscaling approach in random resitor and random superconducting networks, *Phys. Rev. B* **34**, 4656. [*Chap. 6*]

de Arcangelis, L., Koplik, J., Redner, S., and Wilkinson, D. (1986b). Hydrodynamic dispersion in network models of porous media, *Phys. Rev. Lett.* **57**, 996. [*Chap. 9*]

de Arcangelis, L., Coniglio, A., and Redner, S. (1987). Multifractal structure of the incipient infinite percolating cluster, *Phys. Rev. B* **36**, 5631. [*Chap. 6*]

Dee, G. T. (1986). Patterns produced by percolation at moving reaction fronts, *Phys. Rev. Lett.* **57**, 275. [*Chap. 14*]

de Gennes, P. G. (1976a). La percolation: un concept unificateur, *La Recherche* **7**, 919. [*Chap. 6*]

de Gennes, P. G. (1976b). On the relation between percolation theory and the elasticity of gels, *J. Phys. Lett., Paris* **37**, L1. [*Chap. 6*]

de Gennes, P. G. (1979). *Scaling Concepts in Polymer Physics*, Cornell University Press, Ithaca. [*Chap. 6, 10*]

De Martino, A., and Giansanti, A. (1998a). Percolation and lack of self-averaging in a frustrated evolutionary model, *J. Phys. A* **31**, 8757. [*Chap. 2*]

De Martino, A., and Giansanti, A. (1998b). Critical percolation and lack of self-averaging in disordered models, *J. Phys. A* **31**, 8757. [*Chap. 2*] den Hollander, W. Th. F. (1984). Random walks on lattices with randomly distributed traps. I. The average number of steps until trapping, *J. Stat. Phys.* **37**, 331. [*Chap. 12*]

den Nijs, M. P. M. (1979). A relation between the temperature exponents of the eight-vertex and q-state Potts models, *J. Phys.* **12**, 1857. [*Chap. 2*]

Dentener, P. J. H., and Ernst, M. H. (1984). Diffusion in systems with static disorder, *Phys. Rev. B* **29**, 1755. [*Chap. 8*]

Derrida, B. (1982). A self-avoiding walk on random strips, *J. Phys. A* **15**, L119. [*Chap. 10*]

Derrida, B., and Pomeau, Y. (1982). Classical diffusion on a random chain, *Phys. Rev. Lett.* **48**, 627. [*Chap. 8*]

Derrida, B., and Vannimenus, J. (1982). A transfer-matrix approach to random resistor networks, *J. Phys. A* **15**, L557. [*Chap. 5*]

Derrida, B., and Zeitak, R. (1996). Distribution of domain sizes in the zero temperature Glauber dynamics of the one-dimensional Potts model, *Phys. Rev. E* **54**, 2513. [*Chap. 13, 15*]

Derrida, B., Zabolitzky, J. G., Vannimenus, J., and Stauffer, D. (1984). A transfer matrix program to calculate the conductivity of random resistor networks, *J. Stat. Phys.* **36**, 31. [*Chap. 5*]

des Cloizeaux, J., and Jannink, G. (1990). *Polymers in Solution: their Modelling and Structure*, Clarendon Press, Oxford. [*Chap. 10*]

De Turck, D., Gladney, L., and Pietrovito, A. (1994). More mechanisms: a model of enzyme kinetics, in *The Interactive Textbook of PFP*, Section 4.1.6, electronic web-site: http://dept.physics.upenn.edu/~www/gladney/mathphys/subsection4_1_6.html. [*Chap. 11*]

Dhar, D. (1978a). Self-avoiding random walks: some exactly soluble cases, *J. Math. Phys.* **19**, 5. [*Chap. 5*]

Dhar, D. (1978b). Entropy and phase transitions in partially ordered sets, *J. Math. Phys.* **19**, 1711. [*Chap. 5*]

Dhar, D. (1983). Exact solution of a directed-site animals-enumeration problem in three dimensions, *Phys. Rev. Lett.* **51**, 853. [*Chap. 7*]

Dhar, D. (1984). Diffusion and drift on percolation networks in an external field, *J. Phys. A* **17**, L257. [*Chap. 9*]

Dhar, D., and Ramaswamy, R. (1985). Classical diffusion on Eden trees, *Phys. Rev. Lett.* **54**, 1346. [*Chap. 5, 7*]

Dhar, D., and Stauffer, D. (1998). Drift and trapping in biased diffusion on disordered lattices, *Int. J. Mod. Phys. C* **9**, 349. [*Chap. 9*]

Djordjevic, Z. V., Havlin, S., Stanley, H. E., and Weiss, G. H. (1984). New method for growing branched polymers and large percolation clusters below p_c, *Phys. Rev. B* **30**, 478. [*Chap. 7*]

Dickinson, J. T., Langford, S. C., and Jensen, L. C. (1993). Recombination on fractal networks: photon and electron emission following fracture of materials, *J. Mater. Res.* **8**, 2921. [*Chap. 13*]

Digonnet, M. J. F., Davis, M. K., and Pantell, R. H. (1994). Rate equations for clusters in rare earth-doped fibers, *Opt. Fiber Tech.* **1**, 48. [*Chap. 11*]

Doering, C. R. (1992). Microscopic spatial correlations induced by external noise in a reaction–diffusion system, *Physica A* **188**, 386. [*Chap. 18*]

Doering, C. R., and ben-Avraham, D. (1988). Interparticle distribution functions and rate equations for diffusion-limited reactions, *Phys. Rev. A* **38**, 3035. [*Chap. 15, 16*]

Doering, C. R., and ben-Avraham, D. (1989). Diffusion-limited coagulation in the presence of particle input: exact results in one dimension, *Phys. Rev. Lett.* **62**, 2563. [*Chap. 16*]

Doering, C. R., and Burschka, M. A. (1990). Long crossover time in a finite system, *Phys. Rev. Lett.* **64**, 245. [*Chap. 17, 18*]

Doering, C. R., Burschka, M. A., and Horsthemke, W. (1991). Fluctuations and correlations in a diffusion–reaction system: exact hydrodynamics, *J. Stat. Phys.* **65**, 953. [*Chap. 18*]

Doi, M., and Edwards, S. F. (1986). *The Theory of Polymer Dynamics*, Clarendon Press, Oxford. [*Chap. 10*]

Dokholyan, N. V., Buldyrev, S. V., Havlin, S., King, P. R., Lee, Y., and Stanley, H. E. (1999). Distribution of shortest paths in percolation, *Physica A* **266**, 55. [*Chap. 2, 6*]

Domany, E., Alexander, S., Bensimon, D., and Kadanoff, L. P. (1983). Solutions to the Schrödinger equation on some fractal lattices, *Phys. Rev. B* **28**, 3110. [*Chap. 5*]

Domb, C. (1966). Crystal statistics with long-range forces: I. The equivalent neighbour model, *Proc. Phys. Soc.* **89**, 859. [*Chap. 2*]

Domb, C. (1969). Self-avoiding walks on lattices, *Adv. Chem. Phys.* **15**, 229. [*Chap. 5, 10*]

Donev, A., Rockwell, J., and ben-Avraham, D. (1999). Generalized von Smoluchowski model of reaction rates, with reacting particles and a mobile trap, *J. Stat. Phys.* **95**, 97. [*Chap. 18*]

Donsker, N. D., and Varadhan, S. R. S. (1979). On the number of distinct sites visited by a random walk, *Comm. Pure Appl. Math.* **32**, 721. [*Chap. 12*]

Drager, J., and Bunde, A. (1996). Multifractal features of random walks and localized vibrational excitations on random fractals: dependence on the averaging procedure, *Phys. Rev. E* **54**, 4596. [*Chap. 5, 6*]

Drager, J., and Bunde, A. (1999). Random walks on percolation with a topological bias: decay of the probability density, *Physica A* **266**, 62. [*Chap. 9*]

Drager, J., and Klafter, J. (1999). Sorting single events: mean arrival times of *N* random walkers, *Phys. Rev. E* **60**, 6503. [*App. B*]

Drossel, B., and Schwabl, F. (1992). Self-organized critical forest-fire model, *Phys. Rev. Lett.* **69**, 1629. [*Chap. 2*]

Dudek, M. R. (1995). Finite-size-scaling behavior of biased SAW at the percolation threshold, *Physica A* **219**, 114. [*Chap. 10*]

Dullien, F. A. L. (1979). *Porous Media, Fluid Transport and Pore Structure*, Academic Press, New York. [*Chap. 10*]

Dyson, F. J. (1953). The dynamics of disordered linear chains, *Phys. Rev.* **92**, 1331. [*Chap. 8*]

Ebert, U., and van Saarloos, W. (1998). Universal algebraic relaxation of fronts propagating into an unstable state and implications for moving boundary approximations. *Phys. Rev. Lett.* **80**, 1650. [*Chap. 18*]

Edwards, B. F., and Kerstein, A. R. (1985). Is there a lower critical dimension for chemical distance? *J. Phys. A* **18**, L1081. [*Chap. 6*]

Eizenberg, N. and Klafter, J. (1996). Critical exponents of self-avoiding walks in three dimensions, *Phys. Rev. E* **53**, 5078. [*Chap. 10*]

Elan, W. T., Kerstein, A. R., and Rehr, J. J. (1984). Critical properties of the critical dimension, *Phys. Rev. Lett.* **52**, 1516. [*Chap. 6*]

Elias-Kohav, T., Sheintuch, M., and Avnir, D. (1991). Steady-state diffusion and reactions in catalytic fractal porous media, *Chem. Eng. Sci.* **46**, 2787. [*Chap. 13*]

Elran, R., ben-Avraham, D., and Havlin, S. (1999). Diffusion exponents in percolation in two and three dimensions, *preprint*. [*Chap. 6*]

Essam, J. W. (1980). Percolation theory, *Rep. Prog. Phys.* **52**, 833. [*Chap. 2*]

Essam, J. W., and Bhatti, F. M. (1985). Series expansion evidence supporting the Alexander–Orbach conjecture in two dimensions, *J. Phys. A* **18**, 3577. [*Chap. 6*]

Essam, J. W., Gaunt, D. S., and Guttman, A. J. (1978). Percolation theory at the critical dimension, *J. Phys. A* **11**, 1983. [*Chap. 2*]

Essam, J. W., Lookman, T., and DeBell, K. (1996). New series expansion data for surface and bulk resistivity and conductivity in two-dimensional percolation, *J. Phys. A*, **29**, L143. [*Chap. 6*]

Even, C., Russ, S., Repain, V., Pieranski, P., and Sapoval, B. (1999). Localizations in fractal drums: an experimental study, *Phys. Rev. Lett.* **83**, 726. [*Chap. 5*]

Family, F., and Coniglio, A. (1984). Geometrical arguments against the Alexander–Orbach conjecture for lattice animals and diffusion limited aggregates, *J. Phys. A* **17**, L285. [*Chap. 6*]

Family, F., and Landau, D. P., eds. (1984). *Proceedings of the International Topical Conference on Kinetics of Aggregation and Gelation, Athens, GA*, North-Holland, Amsterdam. [*Chap. 8*]

Family, F., and Meakin, P. (1989). Kinetics of droplet growth processes: simulations, theory, and experiments, *Phys. Rev. A* **40**, 3836. [*Chap. 13*]

Fedders, P. A. (1978). Two-point correlation functions for a distinguishable particle hopping on a uniform one-dimensional chain, *Phys. Rev. B* **17**, 40. [*Chap. 10*]

Feder, J. (1988). *Fractals*, Plenum, New York. [*Chap. 1*]

Feng, S., Halperin, B. I., and Sen, P. N. (1987). Transport properties of continuum systems near the percolation threshold, *Phys. Rev. B* **35**, 197. [*Chap. 6*]

Field, G. B., and Saslow, W. C. (1965). *Astrophys. J.* **142**, 568. [*Chap. 13*]

Fischer, L. (1969). *An Introduction to Gel Chromatography*, North Holland, Amsterdam. [*Chap. 9*]

Fisher, M. E. (1966). Shape of a self-avoiding walk or polymer chain, *J. Chem. Phys.* **44**, 616. [*Chap. 5, 10*]

Fisher, M. E., and Burford, R. J. (1967). Theory of critical-point scattering and correlations. I. The Ising model. *Phys. Rev.* **156**, 583. [*Chap. 5*]

Fixman, M. (1984). Quenching by static traps: initial-value and steady-state problems, *Phys. Rev. Lett.* **52**, 791. [*Chap. 12*]

Flory, P. J. (1949). The configuration of real polymer chains, *J. Chem. Phys.* **17**, 303. [*Chap. 10*]

Flory, P. J. (1971). *Principles of Polymer Chemistry*, Cornell University Press, Ithaca, NY. [*Chap. 10*]

Fournier, J., Boiteux, G., and Seytre, G. (1997). Fractal analysis of the percolation network in epoxy–polypyrrol composites, *Phys. Rev. B* **56**, 5207. [*Chap. 2*]

Friedlander, S. K. (1977). *Smoke Dust and Haze*, John Wiley, New York. [*Chap. 11, 13*]

Frisch, U. (1995). *Turbulence*, Cambridge University Press, Cambridge. [*Chap. 1*]

Frojdh, P., and den Nijs, M. (1997). Crossover from isotropic to directed percolation, *Phys. Rev. Lett.* **78**, 1850. [*Chap. 2*]

Galam, S., and Mauger, A. (1997). Universal formulas for percolation thresholds. II. Extension to anisotropic and aperiodic lattices, *Phys. Rev. E* **56**, 322. [*Chap. 2*]

Galam, S., and Mauger, A. (1998). Topology invariance in percolation thresholds, *Eur. Phys. J. B* **1**, 255. [*Chap. 2*]

Gálfi, L., and Rácz, Z. (1988). Properties of the reaction front in an $A + B \rightarrow C$ type reaction–diffusion process, *Phys. Rev. A* **38**, 3151. [*Chap. 14*]

Gandjbakhche, A. H., Gannot, I., and Bonner, R. F. (1996). Photon migration theory applied to 3D optical imaging of tissue, *Proceedings of the 1996 9th Annual Meeting of IEEE Lasers and Electro-Optics Society*, Part 1, Vol. 1, p. 395. [*Chap. 12*]

Gawlinski, E. T., and Stanley, H. E. (1981). Continuum percolation in two dimensions: Monte Carlo tests of scaling and universality for non-interacting discs, *J. Phys. A* **14**, L291. [*Chap. 6*]

Gefen, Y., and Goldhirsch, I. (1985). Biased diffusion on random networks: mean first passage time and DC conductivity, *J. Phys. A* **18**, L1037. [*Chap. 9*]

Gefen, Y., and Goldhirsch, I. (1987). Relation between the classical resistance of inhomogeneous networks and diffusion, *Phys. Rev. B* **35**, 8639. [*Chap. 5*]

Gefen, Y., Mandelbrot. B. B., and Aharony, A. (1980). Critical phenomena on fractal lattices, *Phys. Rev. Lett.* **45**, 855. [*Chap. 1*]

Gefen, Y., Aharony, A., Mandelbrot, B. B., and Kirkpatrick, S. (1981). Solvable fractal family and its possible relation to the backbone at percolation, *Phys. Rev. Lett.* **47**, 1771. [*Chap. 1, 5, 6*]

Gefen, Y., Aharony, A., and Mandelbrot, B. B. (1983a). Phase transitions on fractals: I. Quasi-linear lattices, *J. Phys. A* **16**, 1267. [*Chap. 5*]

Gefen, Y., Aharony, A., and Alexander, S. (1983b). Anomalous diffusion on percolation clusters, *Phys. Rev. Lett.* **50**, 77. [*Chap. 6, 9*]

Gefen, Y., Aharony, A., Shapir, Y., and Mandelbrot, B. B. (1984). Phase transitions on fractals: II. Sierpinski gaskets, *J. Phys. A* **17**, 435. [*Chap. 5*]

Geisel, T., Nierwetberg, J., and Zacherl, A. (1985). Accelerated diffusion in Josephson junctions and related chaotic systems, *Phys. Rev. Lett.* **54**, 616. [*Chap. 4*]

Giacometti, A., and Murthy, K. P. N. (1996). Diffusion and trapping on a one-dimensional lattice, *Phys. Rev. E* **53**, 5647. [*Chap. 12*]

Gillis, J. E., and Weiss, G. H. (1970). Expected number of distinct sites visited by a random walk with an infinite variance, *J. Math. Phys.* **11**, 1308. [*Chap. 4, App. B*]

Giona, M. (1992). First-order reaction–diffusion kinetics in complex fractal media, *Chem. Eng. Sci.* **47**, 1503. [*Chap. 13*]

Giona, M. (1994). An energy barrier model of biased transport in disordered systems, *Chaos Solitons Fractals* **4**, 461. [*Chap. 9*]

Given, J. A., and Mandelbrot, B. B. (1983). Diffusion on fractal lattices and fractal Einstein relation, *J. Phys. A* **16**, L565. [*Chap. 5*]

Glasser, M. L., and Agmon, N. (1987). Scavenging of one-dimensional diffusion with random traps, *J. Chem. Phys.* **86**, 5104. [*Chap. 12*]

Goldhirsch, I., and Gefen, Y. (1987). Biased random walk on networks, *Phys. Rev. A* **35**, 1317. [*Chap. 7, 9*]

Gonzalez, A. P., Pereyra, V. D., Milchev, A., and Zgrablich, G. (1995). Kinetics of simple reactions in a dichotomic barrier model, *Phys. Rev. Lett.* **75**, 3954. [*Chap. 13*]

Gould, H., and Kohin, R. P. (1984). Diffusion on lattice animals and percolation clusters: a renormalisation group approach. *J. Phys. A* **17**, L159. [*Chap. 5*]

Gouyet, J.-F. (1992). *Physique et structures fractales*, Masson, Paris. [*Chap. 1*]

Grassberger, P. (1983). On the behavior of the general epidemic process and dynamical percolation, *Math. Biosci.* **63**, 157. [*Chap. 2*]

Grassberger, P. (1985). On the spreading of two-dimensional percolation, *J. Phys. A* **18**, L215. [*Chap. 6*]

Grassberger, P. (1986). Spreading of percolation in three and four dimensions, *J. Phys. A* **19**, 1681. [*Chap. 2*]

Grassberger, P. (1992a). Numerical studies of critical percolation in three dimensions, *J. Phys. A* **25**, 5867. [*Chap. 2*]

Grassberger, P. (1992b). Spreading and backbone dimensions of 2D percolation, *J. Phys. A* **25**, 5475. [*Chap. 2*]

Grassberger, P. (1993). Recursive sampling of random walks: self avoiding walks in disordered media, *J. Phys. A* **26**, 1023. [*Chap. 10*]

Grassberger, P. (1999a). Conductivity exponent and backbone dimension in 2-d percolation, *Physica A* **262**, 251. [*Chap. 2, 6*]

Grassberger, P. (1999b). Pair connectedness and shortest path scaling in critical percolation, *J. Phys. A* **32**, 6233. [*Chap. 6*]

Grassberger, P., and Procaccia, I. (1982). The long time properties of diffusion in a medium with static traps, *J. Chem. Phys.* **77**, 6281. [*Chap. 12*]

Grimmett, G. R. (1989). *Percolation*, Springer, New York. [*Chap. 2*]

Grossmann, S., Wegner, F., and Hoffmann, K. H. (1985). Anomalous diffusion on a self-similar hierarchical structure, *J. Phys. Lett., Paris* **46**, L575. [*Chap. 8*]

Gutfraind, R., and Sheintuch, M. (1992). Scaling approach to study diffusion and reaction processes on fractal catalysts, *Chem. Eng. Sci.* **47**, 4425. [*Chap. 13*]

Guyer, R. A. (1984). Diffusion on the Sierpinski gasket: a random walker on a fractally structured object, *Phys. Rev. A* **29**, 2751. [*Chap. 5*]

Hahn, K., Kärger, J., and Kukla, V. (1996). Single-file diffusion observation, *Phys. Rev. Lett.* **76**, 2762. [*Chap. 10*]

Haken, H. (1978). *Synergetics*, Springer, Berlin. [*Chap. 13*]

Haken, H. (1983). *Advanced Synergetics: Instability Hierarchies of Self-Organizing Systems and Devices*, Springer, Berlin. [*Chap. 13*]

Halperin, B. I., Feng, S., and Sen, P. N. (1985). Differences between lattice and continuum percolation transport exponents, *Phys. Rev. Lett.* **54**, 2391. [*Chap. 6, 8*]

Halsey. T. L., Meakin, P., and Procaccia, I. (1986). Scaling structure of the surface layer of diffusion-limited aggregates, *Phys. Rev. Lett.* **56**, 854. [*Chap. 6*]

Hamilton, B., Jacobs, J., Hill, D. A., Pettifer, R. F., Teehan, D., and Canham, L. T. (1998). Size-controlled percolation pathways for electrical conduction in porous silicon, *Nature, Lond.* **393** 443. [*Chap. 6*]

Harder, H., Bunde, A., and Havlin, S. (1986). Non-linear response in percolation systems, *J. Phys. A* **19**, L927. [*Chap. 9*]

Harder, H., Bunde, A., and Havlin, S. (1987). Diffusion on fractals with singular waiting time distribution, *Phys. Rev. B* **36**, 3874. [*Chap. 8*]

Harris, A. B. (1983). Self-avoiding walks on random lattices, *Z. Phys. B* **49**, 347. [*Chap. 10*]

Harris, A. B. (1987). Resistance correlations in random structures, *Phil. Mag. B* **56**, 835. [*Chap. 6*]

Harris, A. B., and Lubensky, T. C. (1984). Diluted continuous spin models near the percolation threshold, *J. Phys. A* **17**, L609. [*Chap. 6*]

Harris, A. B., Kim, S., and Lubensky, T. C. (1984). Expansion for the conductivity of a random resistor network, *Phys. Rev. Lett.* **53**, 743. [*Chap. 6*]

Harris, A. B., Meir, Y., and Aharony, A. (1987). Diffusion on percolation clusters, in *Time-Dependent Effects in Disordered Materials. Proceedings of a NATO Advanced Study Institute*, Plenum, New York, pp. 213–16. [*Chap. 6*]

Harris, E. J. (1960). *Transport and Accumulation in Biological Systems*, Butterworths, London. [*Chap. 10*]

Harris, T. E. (1965). Diffusion with "collisions" between particles, *J. Appl. Prob.* **2**, 323. [*Chap. 10*]

Hattori, K., Hattori, T., and Kusuoka, S. (1993). Self-avoiding paths on the 3-dimensional Sierpinski gasket. *Publ. Res. I Math. Sci.* **29**, 455. [*Chap. 10*]

Haus, J. W., and Kehr, K. W. (1987). Diffusion in regular and disordered lattices, *Phys. Rep.* **150**, 141. [*Chap. 8*]

Havlin, S. (1984a). Comment on the Aharony–Stauffer conjecture, *Phys. Rev. Lett.* **53**, 1705. [*Chap. 6, 7*]

Havlin, S. (1984b). Intrinsic properties of percolation clusters and branched polymers, in *Kinetics of Aggregation and Gelation*, F. Family and D. P. Landau, eds., Elsevier, Amsterdam. [*Chap. 6, 7*]

Havlin, S. (1985). Range of validity of the Einstein relation, *Phys. Rev. Lett.* **55**, 130. [*Chap. 5*]

Havlin, S., and ben-Avraham, D. (1982a). Fractal dimensionality of polymer chains, *J. Phys. A* **15**, L311. [*Chap. 5*]

Havlin, S., and ben-Avraham, D. (1982b). Theoretical and numerical study of fractal dimensionality in self-avoiding walks, *Phys. Rev. A* **26**, 1728. [*Chap. 5*]

Havlin, S., and ben-Avraham, D. (1983). Diffusion and fracton dimensionality on fractals and on percolation clusters, *J. Phys. A* **16**, L483. [*Chap. 5, 6*]

Havlin, S., and ben-Avraham, D. (1987). Diffusion in disordered media, *Adv. Phys.* **36**, 695. [*Chap. 6, 7*]

Havlin, S., and Bunde, A. (1989). Probability densities of random walks in random systems, *Physica D* **38**, 184. [*Chap. 5*]

Havlin, S., and Nossal, R. (1984). Topological properties of percolation clusters, *J. Phys. A* **17**, L427. [*Chap. 1, 5, 6*]

Havlin, S., and Weissman, H. (1986). Mapping between hopping on hierarchical structures and diffusion on a family of fractals, *J. Phys. A* **19**, L1021. [*Chap. 8*]

Havlin, S., ben-Avraham, D., and Movshovitz, D. (1983a). Percolation on fractal lattices, *Phys. Rev. Lett.* **51**, 2347. [*Chap. 2*]

Havlin, S., ben-Avraham, D., and Sompolinsky, H. (1983b). Scaling behavior of diffusion on percolation clusters, *Phys. Rev. A* **27**, 1730. [*Chap. 6*]

Havlin, S., Djordjevic, Z. V., Majid, I., Stanley, H. E., and Weiss, G. H. (1984a). Relation between dynamic transport properties and static topological structure for the lattice-animal model of branched polymers, *Phys. Rev. Lett.* **53**, 178. [*Chap. 5, 7*]

Havlin, S., Weiss, G. H., ben-Avraham, D., and Movshovitz, D. (1984b). Structure of clusters generated by random walks, *J. Phys. A* **17**, L849. [*Chap. 5*]

Havlin, S., Nossal, R., Trus, B., and Weiss, G. H. (1984c). Universal substrates of percolation clusters: the skeleton, *J. Phys. A* **17**, 1957. [*Chap. 7*]

Havlin, S., Dishon, M., Kiefer, J. E., and Weiss, G. H. (1984d). Trapping of random walks in two and three dimensions, *Phys. Rev. Lett.* **53**, 407. [*Chap. 12*]

Havlin, S., Weiss, G. H., Kiefer, J. E., and Dishon, M. (1984e). Exact enumeration of random walks with traps, *J. Phys. A* **17**, L347. [*Chap. 12*]

Havlin, S., Kiefer, J. E., Weiss, G. H., ben-Avraham, D., and Glazer, Y. (1985a). Properties of the skeleton of aggregates grown on a Cayley tree, *J. Stat. Phys.* **41**, 489. [*Chap. 7*]

Havlin, S., Trus, B. L., Weiss, G. H., and ben-Avraham, D. (1985b). The chemical distance distribution in percolation clusters, *J. Phys. A* **18**, L247. [*Chap. 5, 6*]

Havlin, S., Movshovitz, D., Trus, B. L., and Weiss, G. H. (1985c). Probability densities for the displacement of random walks on percolation clusters, *J. Phys. A* **18**, L719. [*Chap. 5, 6*]

Havlin, S., Nossal, R., Trus, B., and Weiss, G. H. (1985d). Diffusion on treelike clusters, *Phys. Rev. B* **31**, 7497. [*Chap. 5, 7*]

Havlin, S., Nossal, R., and Trus, B. (1985e). Cluster growth model for treelike structures, *Phys. Rev. A* **32**, 3829. [*Chap. 7*]

Havlin, S., Bunde, A., Glaser, Y., and Stanley, H. E. (1986a). Diffusion with a topological bias on random structures with a power-law distribution of dangling ends, *Phys. Rev. A*, **34**. [*Chap. 9*]

Havlin, S., Bunde, A., Stanley, H. E., and Movshovitz, D. (1986b). Diffusion on percolation clusters with a bias in topological space: non-universal behavior, *J. Phys. A* **19**, L693. [*Chap. 8, 9*]

Havlin, S., Trus, B. L., and Weiss, G. H. (1986c). A phase transition in the dynamics of an exact model for hopping transport, *J. Phys. A* **19**, L817. [*Chap. 8*]

Havlin, S., Bunde, A., Weissman, H., and Aharony, A. (1987a). Nonuniversal transport exponents in quasi-one-dimensional systems with a power-law distribution of conductances, *Phys. Rev. B* **35**, 397. [*Chap. 8*]

Havlin, S., Kiefer, J. E., and Weiss, G. H. (1987b). Anomalous diffusion on random combs, *Phys. Rev. A* **35**, 1403. [*Chap. 7, 8*]

Havlin, S., Blumberg-Selinger, R., Schwartz, M., Stanley, H. E., and Bunde, A. (1988). Random multiplicative processes and transport in structures with correlated disorder, *Phys. Rev. Lett.* **61**, 1438. [*Chap. 2*]

Havlin, S., Schwartz, M., Blumberg-Selinger, R., Stanley, H. E., and Bunde, A. (1989). Universality classes for diffusion in the presence of correlated spatial disorder, *Phys. Rev. A* **40**, 1717. [*Chap. 2*]

Havlin, S., Kopelman, R., Schoonover, R., and Weiss, G. H. (1991). Diffusive motion in a fractal medium in the presence of a trap, *Phys. Rev. A* **43**, 5228. [*Chap. 6*]

Havlin, S., Larralde, H., Trunfio, P., Kiefer, J. E., Stanley, H. E., and Weiss, G. H. (1992). Number of distinct sites visited by N particles diffusing on a fractal, *Phys. Rev. A* **46**, R1717. [*Chap. 5, 14, App. B*]

Havlin, S., Araujo, M., Lereah, Y., Larralde, H., Shehter, A., Stanley, H. E., Trunfio, P., and Vilensky, B. (1995). Complex dynamics in initially separated reaction–diffusion systems, *Physica A* **221**, 1. [*Chap. 14*]

Heckmann, K. (1972). in *Passive Permeability of Cell Membranes, Biomembranes*, F. Kreuzer and J. F. G. Segers, eds., Plenum, New York, Vol. 3, p. 127. [*Chap. 10*]

Heidel, B., Knobler, C. M., Hilfer, R., and Bruinsma, R. (1986). Pattern formation at liquid interfaces, *Phys. Rev. Lett.* **60**, 2492. [*Chap. 14*]

Henkel, M., Orlandini, E., and Schütz, G. M. (1995). Equivalence between stochastic systems, *J. Phys. A* **28**, 6335. [*Chap. 13, 15*]

Henkel, M., Orlandini, E., and Santos, J. (1997). Reaction–diffusion processes from equivalent integrable quantum chains, *Ann. Phys., NY* **259**, 163. [*Chap. 13, 15*]

Herrmann, H. J. (1986). Geometrical cluster growth models and kinetic gelation, *Phys. Rep.* **136**, 153. [*Chap. 7*]

Herrmann, H. J., and Roux, J., eds. (1990). *Statistical Models for the Fracture of Disordered Media*, North Holland, Amsterdam. [*Chap. 2*]

Herrmann, H. J., and Stanley, H. E. (1988). The fractal dimension of the minimum path in two- and three-dimensional percolation, *J. Phys. A* **21**, L829. [*Chap. 2*]

Herrmann, H. J., Derrida, B., and Vannimenus, J. (1984a). Superconductivity exponents in two- and three-dimensional percolation, *Phys. Rev. B* **30**, 4080. [*Chap. 6*]

Herrmann, H. J., Hong, D. C., and Stanley, H. E. (1984b). Backbone and elastic backbone of percolation clusters obtained by the new method of "burning", *J. Phys. A* **17**, L261. [*Chap. 2, 6*]

Hinrichsen, H. (1996). Matrix product ground states for asymmetric exclusion processes with parallel dynamics, *J. Phys. A* **29**, 3659. [*Chap. 13, 15*]

Hinrichsen, H., Krebs, K., and Pfanmüller, M. P. (1995). Finite-size scaling of one-dimensional reaction–diffusion systems. Part I: analytic results, *J. Stat. Phys.* **78**, 1429. [*Chap. 13, 15*]

Hinrichsen, H., Sandow, S., and Peschel, I. (1996a). On matrix product ground states for reaction–diffusion models, *J. Phys. A* **29** 2643. [*Chap. 13, 15*]

Hinrichsen, H., Krebs, K., and Peschel, I. (1996b). Solution of a one-dimensional reaction–diffusion model with spatial asymmetry, *Z. Phys. B* **100**, 105. [*Chap. 13, 15*]

Hong, D. C., and Stanley, H. E. (1983a). Cumulant renormalisation group and its application to the incipient infinite cluster in percolation, *J. Phys. A* **16**, 1475. [*Chap. 2, 6*]

Hong, D. C., and Stanley, H. E. (1983b). Cumulant renormalisation group and its application to the incipient infinite cluster in percolation, *J. Phys. A* **16**, L525. [*Chap. 6*]

Hong, D. C., Havlin, S., Hermann, H. J., and Stanley, H. E. (1984). Breakdown of Alexander–Orbach conjecture for percolation: exact enumeration of random walks on percolation backbones, *Phys. Rev. B* **30**, 4083. [*Chap. 5, 6*]

Hovi, J. P., and Aharony, A. (1997a). Renormalization group calculation of distribution functions: structural properties for percolation clusters, *Phys. Rev. E* **56**, 172. [*Chap. 2*]

Hovi, J. P., and Aharony, A. (1997b). Different self-avoiding walks on percolation clusters: a small-cell real-space renormalization-group study, *J. Stat. Phys.* **86**, 1163. [*Chap. 10*]

Howard, M., and Cardy, J. (1995). Fluctuation effects and multiscaling of the reaction–diffusion front for $A + B \to 0$, *J. Phys. A* **28**, 3599. [*Chap. 14*]

Hoyuelos, M., and Mártin, H. O. (1993a). Annihilation reactions in 2-dimensional percolation clusters: effects of short-range interactions, *Phys. Rev. E* **48**, 71. [*Chap. 13*]

Hoyuelos, M., and Mártin, H. O. (1993b). Rate equation of the $A + A \to A$ reaction with probability of reaction and diffusion, *Phys. Rev. E* **48**, 3309. [*Chap. 19*]

Hoyuelos, M., and Mártin, H. O. (1994). Annihilation reaction $A + A \to 0$ with diffusion and interaction between particles in disordered structures, *Phys. Rev. E* **50**, 600. [*Chap. 13*]

Hoyuelos, M., and Mártin, H. O. (1995). Annihilation reactions: crossover from mean-field to anomalous behaviors, *Chaos, Solitons Fractals* **6**, 213. [*Chap. 13*]

Hoyuelos, M., and Mártin, H. O. (1996). Annihilation and coagulation reactions in low-dimensional substrata: effects of probability of reaction and short-range interactions, *Langmuir* **12**, 61. [*Chap. 13*]

Huber, G., Jensen, M. H., and Sneppen, K. (1995). A dimension formula for self-similar and self-affine fractals, *Fractals* **3**, 525. [*Chap. 7*]

Huberman, B. A., and Kerszberg, M. (1985). Ultradiffusion: the relaxation of hierarchical systems, *J. Phys. A* **18**, L331. [*Chap. 8*]

Huckestein, B., and Schweitzer, L. (1994). Relation between the correlation dimensions of multifractal wave functions and spectral measures in integer quantum Hall systems, *Phys. Rev. Lett.* **72**, 713. [*Chap. 5*]

Hughes, B. D. (1995). *Random Walks and Random Environments*, Clarendon Press, Oxford. [*Chap. 3*]

Ikeda, H., Itoh, S., and Adams, M. A. (1997). Anomalous diffusion in percolating magnets with fractal geometry, *Physica B* **241**, 585. [*Chap. 6*]

Inaguma, Y., and Itoh, M. (1996). Influences of carrier concentration and site percolation on lithium ion conductivity in perovskite-type oxides, *Solid State Ionics* **86–8**, 257. [*Chap. 6*]

Ispolatov, I., Krapivsky, P. L., and Redner, S. (1995). Kinetics of $A + B \to 0$ with driven diffusive motion, *Phys. Rev. E* **52**, 2540. [*Chap. 10, 13*]

Ivanov, P. Ch., Rosenblum, M. G., Peng, C.-K., Mietus, J., Havlin, S., Stanley, H. E., and Goldberger, A. L. (1996). Scaling behaviour of heartbeat intervals obtained by wavelet-based time-series analysis, *Nature, Lond.* **383**, 323. [*App. D*]

Jacobs, D. J., Mukherjee, S., and Nakanishi, H. (1994). Diffusion on a DLA cluster in two and three dimensions, *J. Phys. A* **27**, 4341. [*Chap. 7*]

Jain, N. C., and Pruitt, W. E. (1971). *J. d'Analyse Math.* **24**, 369. [*App. B*]

Jan, N. and Stauffer, D. (1998). Random site percolation in three dimensions, *Int. J. Mod. Phys. C* **9**, 341. [*Chap. 2*]

Jan, N., Hong, D. C., and Stanley, H. E. (1985). The fractal dimension and other percolation exponents in four and five dimensions, *J. Phys. A* **18**, L935. [*Chap. 2*]

Janowsky, S. A. (1995a). Asymptotic behavior of $A + B \to$ inert for particles with a drift, *Phys. Rev. E* **51**, 1858. [*Chap. 10, 13*]

Janowsky, S. A. (1995b). Spatial-organization in the reaction $A + B \to$ inert for particles with drift, *Phys. Rev. E* **52**, 2535. [*Chap. 10, 13*]

Janssen, H. K. (1985). Renormalized field theory of dynamical percolation, *Z. Phys. B* **58**, 311. [*Chap. 2, 6*]

Janssen, H. K., and Stenull, O. (1999). Diluted networks of nonlinear resistors and fractal dimensions of percolation clusters, *e-preprint*: cond-mat/9910427. [*Chap. 6*]

Janssen, H. K., Stenull, O., and Oerding, K. (1999). Resistance of Feynman diagrams and the percolation backbone dimension, *Phys. Rev. E* **59**, R6239. [*Chap. 2, 6*]

Jensen, I. (1993). Conservation laws and universality in branching annihilating random walks, *J. Phys. A* **26**, 3921. [*Chap. 13*]

Jensen, I. (1994). Critical exponents for branching annihilating random walks with an even number of offspring, *Phys. Rev. E* **50**, 3623. [*Chap. 13*]

Jensen, P., Barabási, A.-L., Larralde, H., Havlin, S., and Stanley, H. E. (1994). Controlling nanostructures, *Nature, Lond.* **368**, 22. [*Chap. 7*]

Jiang, Z., and Ebner, C. (1990). Simulation study of reaction fronts, *Phys. Rev. A* **42**, 7483. [*Chap. 14*]

Jones, J. D., and Luby-Phelps, K. (1996). Tracer difusion through F-actin: effect of filament length and cross-linking, *Biophys. J.* **71**, 2742. [*Chap. 6*]

Kang, K., and Redner, S. (1984). Fluctuation-dominated kinetics in diffusion-controlled reactions, *Phys. Rev. Lett.* **52**, 955. [*Chap. 13*]

Kantelhardt, J. W., and Bunde, A. (1997). Electrons and fractons on percolation structures at criticality: sublocalization and superlocalization, *Phys. Rev. E* **56**, 6693. [*Chap. 5, 6*]

Kantelhardt, J. W., Bunde, A., and Schweitzer, L. (1998). Extended fractons and localized phonons on percolation clusters, *Phys. Rev. Lett.* **81**, 4907. [*Chap. 6*]

Kanter, I. (1990). Synchronous or asynchronous parallel dynamics. Which is more efficient?, *Physica D* **42**, 273. [*Chap. 13, 15*]

Kärger, J., Haberlandt, R., and Heitjans, P., eds. (1998). *Diffusion in Condensed Matter*, Vieweg, Berlin. [*Chap. 3*]

Kehr, K. W., and Binder. K. (1987). Simulation of diffusion in lattice gases and related kinetic phenomena, in *Applications of the Monte Carlo Method in Statistical Physics*, K. Binder, ed., Springer, Berlin. [*Chap. 10*]

Keizer, J. (1985). Theory of rapid bimolecular reactions in solution and membranes, *Acc. Chem. Res.* **18**, 235. [*Chap. 13*]

Keizer, J. (1987). Diffusion effects on rapid bimolecular chemical reactions, *Chem. Rev.* **87**, 167. [*Chap. 13*]

Kertesz, J. (1981). Percolation of holes between overlapping spheres: Monte Carlo calculation of the critical volume fraction, *J. Phys., Paris* **42**, L393. [*Chap. 2*]

Kessler, D. A., Ner, Z., and Sander, L. M. (1998). Front propagation: precursors, cutoffs, and structural stability, *Phys. Rev. E* **58**, 107. [*Chap. 18*]

Kesten, H. (1982). *Percolation Theory for Mathematicians*, Birkhauser, Boston. [*Chap. 2*]

Ketzmerick, R., Kruse, K., Kraut, S., and Geisel, T. (1997). What determines the spreading of a wave packet? *Phys. Rev. Lett.* **79**, 1959. [*Chap. 5*]

Keyser, R. F., and Hubbard, J. B. (1983). Diffusion in a medium with a random distribution of static traps, *Phys. Rev. Lett.* **51**, 79. [*Chap. 12*]

Keyser, R. F., and Hubbard, J. B. (1984). Reaction diffusion in a medium containing a random distribution of nonoverlapping traps, *J. Chem. Phys.* **80**, 1127. [*Chap. 12*]

Kim, D., and Kahng, B. (1985). Comment on 'Self avoiding walks on finitely ramified Fractals', *Phys. Rev. A* **31**, 1193. [*Chap. 10*]

King, P. R., Andrade, J. S., Buldyrev, S. V., Dokholyan, N., Lee, Y., Havlin, S., and Stanley, H. E. (1999). Predicting oil recovery using percolation, *Physica A* **266**, 107. [*Chap. 2*]

Kinzel, W. (1983). Directed percolation, in *Percolation Structures and Processes*, G. Deutscher, R. Zallen, and J. Adler, eds., *Annals of the Israeli Physics Society*, Vol. 5, p. 425, Adam Hilger, Bristol. [*Chap. 2*]

Kirkpatrick, S. (1979). in *Inhomogeneous Superconductors*, D. U. Gubser, T. L. Francarilla, J. R. Leibowitz, and S. A. Wolf, eds., AIP Conference Proceedings **58**, American Institute of Physics, New York. [*Chap. 5*]

Kirkpatrick, T. R. (1982). Time dependent transport in a fluid with static traps, *J. Chem. Phys.* **76**, 4225. [*Chap. 12*]

Kirsch, A. (1998). Phase transition in two-dimensional biased diffusion, *Int. J. Mod. Phys. C* **9**, 1021. [*Chap. 9*]

Kivelson, S. (1980). Hopping conduction and the continuous-time random-walk model, *Phys. Rev. B* **21**, 5755. [*Chap. 8*]

Klafter, J., and Silbey, R. (1980). Derivation of the continuous-time random-walk equation, *Phys. Rev. Lett.* **44**, 55. [*Chap. 8*]

Klafter, J., Zumofen, G., and Blumen, A. (1984). Long time properties of trapping on fractals, *J. Phys. Lett., Paris* **45**, L49. [*Chap. 12*]

Klafter, J., Shlesinger, M. F., and Zumofen, G. (1996). Beyond Brownian motion, *Phys. Today* **49**, 33. [*Chap. 4*]

Kleban, P., and Ziff, R. M. (1998). Exact results at the 2-D percolation point, *Phys. Rev. B* **57**, R8075. [*Chap. 2*]

Klein, D. J., and Seitz, W. A. (1984). Self-interacting self-avoiding walk on the Sierpinski gasket. *J. Phys., Paris* **45**, L241. [*Chap. 10*]

Klug, A. (1958). *Acta Crystallogr.* **11**, 515. [*Chap. 3*]

Kogut, P. M., and Straley, J. P. (1979). Distribution-induced non-universality of the percolation conductivity exponents, *J. Phys. C* **12**, 2151. [*Chap. 6, 8*]

Koo, Y. E., and Kopelman, R. (1991). Space- and time-resolved diffusion-limited binary reaction kinetics in capillaries: experimental observation of segregation, anomalous exponents, and depletion zone, *J. Stat. Phys.* **65**, 893. [*Chap. 14*]

Koo, Y. E., Li, L., and Kopelman, R. (1990). Reaction front dynamics in diffusion-controlled particle–antiparticle annihilation: experiments and simulations, *Mol. Cryst. Liq. Cryst.* **183**, 187. [*Chap. 14*]

Kopelman, R. (1976). Exciton percolation in molecular alloys and aggregates, in *Topics in Applied Physics* Vol. 15, F. K. Fang, ed., Springer, Heidelberg. [*Chap. 6*]

Kopelman, R. (1986). Rate processes on fractals: theory, simulations, and experiments, *J. Stat. Phys.* **42**, 185. [*Chap. 15*]

Kopelman, R. (1987) Low-dimensional exciton reactions, *Phil. Mag. B* **56**, 717. [*Chap. 12, 13*]

Kopelman, R. (1988). Fractal reaction-kinetics, *Science* **241**, 1620. [*Chap. 15*]

Kopelman, R., Parus, S. J., and Prasad, J. (1988). Exciton reactions in ultrathin molecular wires, filaments, and pores: a case study of kinetics and self-ordering in low dimensions, *Chem. Phys.* **128**, 209. [*Chap. 13, 15*]

Koplik, J., Redner, S., and Wilkinson, D. (1988). Transport and dispersion in random networks with percolation disorder, *Phys. Rev. A* **37**, 2619. [*Chap. 6, 9*]

Koscielny-Bunde, E., Bunde, A., Havlin, S., Roman, H. E., Goldreich, Y., and Schellnhuber, H. J. (1998). Indication of a universal persistence law governing atmospheric variability, *Phys. Rev. Lett.* **81**, 729. [*Chap. 4*]

Koza, Z. (1998). Asymptotic properties of the A + B → 0 reaction–diffusion front, *Phil. Mag. B* **77**, 1437. [*Chap. 14*]

Koza, Z., and Taitelbaum, H. (1998). Spatiotemporal properties of diffusive systems with a mobile imperfect trap, *Phys. Rev. E* **57**, 237. [*Chap. 12*]

Koza, Z., Yanir, T., and Taitelbaum, H. (1998). Nearest-neighbor distance at a single mobile trap, *Phys. Rev. E* **58**, 6821. [*Chap. 12*]

Kozak, J. J. (1997). Random walks on the Menger sponge, *Chem. Phys. Lett.* **275**, 199. [*Chap. 12*]

Krapivsky, P. L. (1993). Aggregation–annihilation processes with injection, *Physica A* **198**, 157. [*Chap. 13, 15*]

Krapivsky, P. L. (1994). Diffusion-limited aggregation process with 3-particle elementary reactions, *Phys. Rev. E* **49**, 3233. [*Chap. 13*]

Kremer, K. (1981). Self-avoiding walks (SAW's) on diluted lattices, a Monte-Carlo analysis, *Z. Phys. B* **45**, 148. [*Chap. 10*]

Kumar, S., Singh, Y., and Joshi, Y. P. (1990a). Critical exponents of self-avoiding walks on a family of truncated n-simplex lattices, *J. Phys. A* **23**, 2987. [*Chap. 10*]

Kumar, S., Singh, Y., and Joshi, Y. P. (1990b). Addendum: critical exponents of self-avoiding walks on a family of truncated n-simplex lattices, *J. Phys. A* **23**, 5115. [*Chap. 10*]

Kutasov, D., Aharoni, A., Domany, E., and Kinzel, W. (1986). Dynamic transition in a hierarchical Ising system, *Phys. Rev. Lett.* **56**, 2229. [*Chap. 8*]

Kutner, R., and Maass, P. (1998). Lévy flights with quenched noise amplitudes, *J. Phys. A* **31**, 2603. [*Chap. 4*]

Kuzovkov, V., and Kotomin, E. (1988). Kinetics of bimolecular reactions in condensed media: critical phenomena and microscopic self-organization, *Rep. Prog. Phys.* **51**, 1479. [*Chap. 13*]

Kuzovkov, V., and Kotomin, E. (1992). Self-organization in the A + B → 0 reaction of charged particles, *Physica A* **191**, 172. [*Chap. 13*]

Kuzovkov, V., and Kotomin, E. (1993). Dynamic particle aggregation in the bimolecular A + B → 0 reaction, *J. Chem. Phys.* **98**, 9107. [*Chap. 13*]

Labini, F. S., Montuori, M., and Pietronero, L. (1998a). Multifractal properties of galaxy distribution, *J. Phys. IV, Paris* **8**, 115. [*Chap. 1*]

Labini, F. S., Montuori, M., and Pietronero, L. (1998b). Scale-invariance of galaxy clustering, *Phys. Rep.* **293**, 62. [*Chap. 1*]

Laibowitz, R. G., and Gefen, Y. (1984). Dynamic scaling near the percolation threshold in thin Au films, *Phys. Rev. Lett.* **53**, 380. [*Chap. 9*]

Laidler, K. J. (1965). *Chemical Kinetics*, McGraw-Hill, New York. [*Chap. 11*]

Lam, P. M., and Zhang, Z. Q. (1984). Self-avoiding walks on percolation clusters at criticality and lattice animals, *Z. Phys. B* **56**, 155. [*Chap. 10*]

Langlands, R. (1994). Conformal invariance in two-dimensional percolation, *Bull. Am. Math. Soc.* **30**, 1. [*Chap. 2*]

Larralde, H. and Weiss, G. H. (1995). A generating function for the 2nd moment of the distinct number of sites visited by an *n*-step lattice random walk, *J. Phys. A* **28**, 5217. [*Chap. 3*]

Larralde, H., Trunfio, P., Havlin, S., Stanley, H. E., and Weiss, G. H. (1992a). Territory covered by *N* diffusing particles, *Nature, Lond.* **355**, 423. [*Chap. 3, App. B*]

Larralde, H., Trunfio, P., Havlin, S., Stanley, H. E., and Weiss, G. H. (1992b). Number of distinct sites visited by *N* random walkers, *Phys. Rev. A* **45**, 7128. [*Chap. 3, App. B*]

Larralde, H., Araujo, M., Havlin, S., and Stanley, H. E. (1992c). Reaction front for A + B $\rightarrow$ C diffusion–reaction systems with initially separated reactants, *Phys. Rev. A* **46**, 855. [*Chap. 14*]

Larralde, H., Araujo, M., Havlin, S., and Stanley, H. E. (1992d). Diffusion–reaction kinetics for A + B(static) $\rightarrow$ C(inert) for one-dimensional system with initially separated reactants, *Phys. Rev. A* **46**, R6121. [*Chap. 14*]

Larralde, H., Lereah, Y., Trunfio, P., Dror, J., Havlin, S., Rosenbaum, R., and Stanley, H. E. (1993). Reaction kinetics of diffusing particles injected into a *d*-dimensional reactive substrate, *Phys. Rev. Lett.* **70**, 1461. [*Chap. 14*]

Lavenda, B. H. (1985). Brownian motion, *Sci. Am.*, Feb., 70. [*Chap. 3*]

Lea, E. J. A. (1963). *J. Theor. Biol.* **5**, 102. [*Chap. 10*]

Lee, B. P. (1994). Renormalization group calculation for the reaction kA $\rightarrow$ 0, *J. Phys. A* **27**, 2633. [*Chap. 13*]

Lee, B. P., and Cardy, J. (1994). Scaling of reaction zones in the A + B $\rightarrow$ 0 diffusion-limited reaction, *Phys. Rev. E* **50**, R3287. [*Chap. 13*]

Lee, B. P., and Cardy, J. (1997). Renormalization group study of the A + B $\rightarrow$ 0 diffusion-limited reaction (vol. 80, pg. 971, 1995), *J. Stat. Phys.* **87**, 951. [*Chap. 13*]

Lee, J. W. (1997). Monte Carlo studies on three-species two-particle diffusion-limited reactions, *Physica A* **256**, 351. [*Chap. 10*]

Lee, J. W. (1998). Self-attracting walk on lattices, *J. Phys. A* **31**, 3929. [*Chap. 3*]

Lee, S. B. (1996). Critical behavior and crossover scaling of self-avoiding walks on diluted lattices, *J. Korean Phys. Soc.* **29**, 1. [*Chap. 10*]

Lee, S. B., and Woo, K. Y. (1995). True self-avoiding walks on fractal lattices above the upper marginal dimension, *J. Phys. A* **28**, 7065. [*Chap. 10*]

Lee, Y., Andrade, J. S., Buldyrev, S. V., Dokholyan, N. V., Havlin, S., King, P. R., Paul, G., and Stanley, H. E. (1999). Traveling time and traveling length for flow in porous media, *Phys. Rev. E* **60**, 3425. [*Chap. 2, 6*]

Levitt, D. G. (1973). Dynamics of a single-file pore: non Fickian behavior, *Phys. Rev. A* **8**, 3050. [*Chap. 10*]

Leyvraz, F., and Stanley, H. E. (1983). To what class of fractals does the Alexander–Orbach conjecture apply?, *Phys. Rev. Lett.* **51**, 2048. [*Chap. 6*]

Li, B., Madras, N., and Sokal, A. D. (1995). Critical exponents, hyperscaling, and universal amplitude ratios for 2-dimensional and 3-dimensional self-avoiding walks, *J. Stat. Phys.* **80**, 661. [*Chap. 10*]

Lianos, P., and Duportail, G. (1992). Reactions in lipid vesicles: pyrene excimer formation in restricted geometries: effect of temperature and concentration, *Eur. Biol. J.* **21**, 29. [*Chap. 13*]

Liesegang, R. E., (1896). *Naturwiss. Wochensch.* **11**, 353. [*Chap. 14*]

Liggett, T. M. (1985). *Interacting Particle Systems*, Springer, New York. [*Chap. 11, 13*]

Lin, J.-C. (1991). Closure schemes for joint density functions in diffusion-limited reactions, *Phys. Rev. A* **44**, 6706. [*Chap. 13*]

Lin, Z. Q, Yang, Z. R., and Qin, Y. (1997). Renormalization group approach to bond percolation on Sierpinski carpets, *Acta Phys. Sin. – Ov. Ed.* **6**, 257. [*Chap. 2*]

Lindenberg, K., Sheu, W.-S., and Kopelman, R. (1991). Scaling properties of diffusion-limited reactions: fractals, *Phys. Rev. A* **43**, 7070. [*Chap. 13*]

Lobb, C. J., and Forrester, M. G. (1987). Measurement of nonuniversal critical behavior in a two-dimensional continuum percolating system, *Phys. Rev. B* **35**, 1899. [*Chap. 6*]

Lobb, C. J., and Frank, D. J. (1984). Percolative conduction and the Alexander–Orbach conjecture in two dimensions, *Phys. Rev. B* **30**, 4090. [*Chap. 6*]

Lorenz, C. M., and Ziff, R. M. (1998). Precise determination of the bond percolation thresholds and finite-size scaling corrections for the sc, fcc and bcc lattices, *Phys. Rev. E* **57**, 230. [*Chap. 2*]

Lubensky, T. C., and Isaacson, J. (1981). Critical behavior of branched polymers and the Lee–Xang edge singularity, *Phys. Rev. Lett.* **46**, 871. [*Chap. 7*]

Lubensky, T. C., and Tremblay, A. M. J. (1986). ε-expansion for transport exponents of continuum percolating systems, *Phys. Rev. B* **34**, 3408. [*Chap. 6*]

Lubkin, G. B. (1998). Small-world networks can provide a new tool to study diverse systems, *Phys. Today* **51**, 21. [*Chap. 2*]

Lucena, L. S., Araujo, J. M., Tavares, D. M., da Silva, L. R., and Tsallis, C. (1994). Ramified polymerization in dirty media: a new critical phenomenon, *Phys. Rev. Lett.* **72**, 230. [*Chap. 7*]

Luck, J. M. (1985). A real-space renormalisation group approach to electrical and noise properties of percolation clusters, *J. Phys. A* **18**, 2061. [*Chap. 9*]

Lushnikov, A. A. (1987). Binary reaction $1 + 1 \to 0$ in one dimension, *Phys. Lett. A* **120**, 135. [*Chap. 13, 15*]

Machta, J. (1985). Random walks on site disordered lattices, *J. Phys. A* **18**, L531. [*Chap. 8*]

Machta, J., Guyer, R. A., and Moore, S. M. (1986). Conductivity in percolation networks with broad distributions of resistances, *Phys. Rev. B* **33**, 4818. [*Chap. 6*]

Mai, J., Sokolov, I. M., and Blumen, A. (1996). Front propagation and local ordering in one-dimensional irreversible autocatalytic reactions, *Phys. Rev. Lett.* **77**, 4462. [*Chap. 18*]

Majid, I., ben-Avraham, D., Havlin, S., and Stanley, H. E. (1984). Exact-enumeration approach to random walks on percolation clusters in two dimensions, *Phys. Rev. B* **30**, 1626. [*Chap. 5, 6, App. C*]

Makhnovskii, Ya., Maslova, M. E., and Berezhkovskii, A. M. (1998a). Many-body effects in the trapping problem with a field, *J. Chem. Phys.* **108**, 6431. [*Chap. 12*]

Makhnovskii, Ya., Yang, D. Y., Berezhkovskii, A. M., Sheu, S. Y., and Lin, S. H. (1998b). Effect of trap clustering on Brownian particle trapping rate, *Phys. Rev. E* **58**, 4340. [*Chap. 12*]

Makse, H., Havlin, S., and Stanley, H. E. (1995). Modeling urban growth patterns, *Nature, Lond.* **377**, 608. [*Chap. 2*]

Makse, H., Havlin, S., Stanley, H. E., and Schwartz, M. (1996). Method for generating long-range correlations for large systems, *Phys. Rev. E* **53**, 5445. [*Chap. 2, 4, App. D*]

Malcai, O., Lidar, D. A., Biham, O., and Avnir, D. (1997). Scaling range and cutoffs in empirical fractals, *Phys. Rev. E* **56**, 2817. [*Chap. 1*]

Mandelbrot, B. B. (1974). Intermittent turbulence in self-similar cascades: divergence of high moments and dimension of the carrier, *J. Fluid Mech.,* **62**, 331. [*Chap. 6*]

Mandelbrot, B. B. (1977). *Fractals: Form, Chance and Dimension*, Freeman, San Francisco. [*Chap. 1*]

Mandelbrot, B. B. (1982). *The Fractal Geometry of Nature*, Freeman, San Francisco. [*Chap. 1, 4, 5, 6*]

Manna, S. S., and Dhar, D. (1996). Fractal dimension of backbone of Eden trees, *Phys. Rev. E* **54**, R3063. [*Chap. 7*]

Mantegna, R. N., and Stanley, H. E. (1995). Scaling behavior in the dynamics of an economic index, *Nature, Lond.* **376**, 46. [*Chap. 4*]

Margolin, G., Berkowitz, B., and Scher, H. (1998). Structure, flow, and generalized conductivity scaling in fracture networks, *Water Resources Research* **34**, 2103. [*Chap. 4*]

Margolina, A. (1985). The fractal dimension of cluster perimeters generated by a kinetic walk, *J. Phys. A* **18**, L651. [*Chap. 2*]

Marro, J., and Dickman, R. (1999). *Nonequilibrium Phase Transitions in Lattice Models*, Cambridge University Press, Cambridge. [*Chap. 11*]

Mártin, H. O., and Braunstein, L. (1993). Study of $A + A \to 0$ with probability of reaction and diffusion in one-dimension and in fractal substrata, *Z. Phys. B* **91**, 521. [*Chap. 13*]

Mártin, H. O., Iguain, J. L., and Hoyuelos, M. (1995). Steady state of imperfect annihilation and coagulation reactions, *J. Phys. A* **28**, 5227. [*Chap. 19*]

Martin, J. (1972). *Phase Transitions and Critical Phenomena*, Vol. 3, C. Domb and M. S. Green, eds., Academic, New York. [*Chap. 7*]

Mattis, D. C., and Glasser, M. L. (1998). The uses of quantum field-theory in diffusion-limited reactions, *Rev. Mod. Phys.* **70**, 979. [*Chap. 13, 15*]

McKenzie, D. S. (1976). Polymers and scaling, *Phys. Rep.* **27**, 2. [*Chap. 5*]

McKenzie, D. S., and Moore, M. A. (1971). Shape of a self-avoiding walk or polymer chain, *J. Phys. A* **4**, L82. [*Chap. 5*]

Meakin, P. (1983a). Diffusion-controlled cluster formation in two, three and four dimensions, *Phys. Rev. A* **27**, 604. [*Chap. 7*]

Meakin, P. (1983b). Diffusion controlled clusters in 2–6 dimensional space, *Phys. Rev. A* **27**, 1495. [*Chap. 7*]

Meakin, P. (1998). *Fractals, Scaling and Growth far from Equilibrium*, Cambridge University Press, Cambridge. [*Chap. 1, 7*]

Meakin, P., and Stanley, H. E. (1983). Spectral dimension for the diffusion-limited aggregation model of colloid growth, *Phys. Rev. Lett.* **51**, 1457. [*Chap. 5, 7*]

Meakin, P., and Stanley, H. E. (1984). Novel dimension-independent behavior for diffusive annihilation on percolation fractals. *J. Phys. A* **17**, L173. [*Chap. 13*]

Meakin, P., Majid, I., Havlin, S., and Stanley, H. E. (1984). Topological properties of diffusion limited aggregation and cluster–cluster aggregation, *J. Phys. A* **17**, L975. [*Chap. 5, 7*]

Meakin, P., Stanley, H. E., Coniglio, A., and Witten, T. A. (1985). Surfaces, interfaces and screening of fractal structures, *Phys. Rev. A* **32**, 2364. [*Chap. 6*]

Meakin, P., Coniglio, A., Stanley, H. E., and Witten, T. A. (1986). Scaling properties for surfaces of fractal and nonfractal objects: an infinite hierarchy of critical exponents, *Phys. Rev. A* **34**, 3325. [*Chap. 6*]

Meneveau, C. M., and Sreenivasan, K. R. (1987). The multifractal spectrum of the dissipation field in turbulent flows, *Nucl. Phys. B Proc. Suppl.* **2**, 49. [*Chap. 1*]

Meneveau, C. M., and Sreenivasan, K. R. (1991). The multifractal nature of turbulent energy dissipation, *J. Fluid Mech.* **224**, 429. [*Chap. 1*]

Menyhárd, N. (1994). One-dimensional non-equilibrium kinetic Ising models with branching annihilating random walks, *J. Phys. A* **27**, 6139. [*Chap. 13*]

Metz, V. (1995). Renormalization of finitely-ramified fractals, *Proc. Roy. Soc. Edinburgh A, Mat.* **125**, 1085. [*Chap. 5*]

Meyer, M., Jaenisch, V., Maass, P., and Bunde, A. (1996a). Mixed alkali effects in crystals of β- and β=EF=EF-alumina structure, *Phys. Rev. Lett.* **76**, 2338. [*Chap. 2*]

Meyer, M., Bunde, A., and Havlin, S. (1996b). Clustering of independently diffusing individuals by birth and death processes, *Phys. Rev. E* **54**, 5567. [*Chap. 3*]

Meyer, M., Nicoloso, N., and Jaenisch, V. (1997). Percolation model for the anomalous conductivity of fluorite-related oxides, *Phys. Rev. B* **56**, 5961. [*Chap. 10*]

Middlemiss, K. M., Whittington, S. G., and Gaunt, D. C. (1980). Monte-Carlo study of the percolating cluster for the square lattice site problem, *J. Phys. A* **13**, 1835. [*Chap. 6*]

Milosevic, S., and Zivic, I. (1991). Self-avoiding walks on fractals studied by the Monte-Carlo renormalization-group, *J. Phys. A* **15**, L833. [*Chap. 10*]

Milosevic, S., and Zivic, I. (1993). Universal crossing of the self-avoiding walk critical exponent ν at the Euclidean value 3/4 for several different fractal families, *J. Phys. A* **24**, 7263. [*Chap. 10*]

Mitescu, C. D., and Roussenq, J. (1976). An ant in a labyrinth: diffusion in a percolation system, *Comptes Rendus Hebdomadaires des Séances de l'Académie des Sciences A* **283**, 999. [*Chap. 6*]

Mitescu, C. D., Ottavi, H., and Roussenq, J. (1979). Diffusion on percolation lattices: the labyrinthine ant, in *Electrical Transport and Optical Properties of Inhomogeneous Media*, AlP Conference Proceedings Vol. 40, p. 377, J. C. Garland and D. B. Tanner, eds., AlP, New York. [*Chap. 6*]

Montroll, E. W. and Weiss, G. H. (1965). Random walks on lattices. II, *J. Math. Phys.* **6**, 167. [*Chap. 3*]

Montroll, E. W., and Shlesinger, M. F. (1984). On the wonderful world of random walks, in *Nonequilibrium Phenomena II: From Stochastics to Hydrodynamics*, J. L. Lebowitz and E. W. Montroll eds., North Holland, Amsterdam. [*Chap. 3*]

Moukarzel, C. F. (1998). A fast algorithm for backbones, *Int. J. Mod. Phys. C* **8**, 887. [*Chap. 2*]

Moukarzel, C. F., Duxbury, P. M., and Leath, P. L. (1997). Infinite-cluster geometry in central-force networks, *Phys. Rev. Lett.* **78**, 1480. [*Chap. 2*]

Movshovitz, D., and Havlin, S. (1988). Structural and dynamical properties of random-walk clusters, *J. Phys. A* **21**, 2761. [*Chap. 3*]

Muthukumar, M., and Cukier, R. I. (1981). Concentration dependence of diffusion-controlled processes among stationary reactive sinks, *J. Stat. Phys.* **26**, 453. [*Chap. 12*]

Muzy, J. F., Bacry, E., and Arneodo, A. (1991). Wavelets and multifractal formalism for singular signals: application to turbulence data. *Phys. Rev. Lett.* **67**, 3515. [*App. D*]

Nakanishi, H. (1994). Random and self-avoiding walks in disordered media, in *Annual Reviews of Computer Physics I*, D. Stauffer, ed., Singapore, World Scientific. [*Chap. 10*]

Nakanishi, H., and Herrmann, H. J. (1993). Diffusion and spectral dimension on Eden tree, *J. Phys. A* **26**, 4513. [*Chap. 7*]

Nakanishi, H., and Moon, J. (1992). Self-avoiding walk on critical percolation cluster, *Physica A* **191**, 309. [*Chap. 10*]

Nakayama, T., Yakubo, K., and Orbach, R. L. (1994). Dynamical properties of fractal networks: scaling, numerical simulations, and physical realizations, *Rev. Mod. Phys.* **66**, 381. [*Chap. 6*]

Narasimhan, S. L. (1996). Kinetic self-avoiding walks on randomly diluted lattices at the percolation threshold, *Phys. Rev. E* **53**, 1986. [*Chap. 10*]

Nelson, D. R., and Shnerb, N. (1998). Non-Hermitian localization and population biology, *Phys. Rev. E* **58**, 1383. [*Chap. 13*]

Nicolis, G., and Prigogine, I. (1980). *Self-Organization in Non-Equilibrium Systems*, John Wiley, New York. [*Chap. 13*]

Niemeyer, L., and Pinnekamp, F. (1982). in *Proceedings of the International Symposium on Gaseous Dielectrics, Knoxville*, Pergamon, New York, p. 379. [*Chap. 7*]

Niemeyer, L., Pietronero, L., and Wiesmann, H. J. (1984). Fractal dimension of dielectric breakdown, *Phys. Rev. Lett.* **52**, 1033. [*Chap. 7*]

Nienhuis, B. (1982). Exact critical point and critical exponents of $O(n)$ models in two dimensions, *Phys. Rev. Lett.* **49**, 1062. [*Chap. 2, 10*]

Nittmann, J., and Stanley, H. E. (1986). Tip splitting without interfacial tension and dendritic growth patterns arising from molecular anisotropy, *Nature, Lond.* **321**, 663. [*Chap. 7*]

Nittmann, J., Daccord, G., and Stanley, H. E. (1985). Fractal growth of viscous fingers; quantitative characterization of a fluid instability phenomenon, *Nature, Lond.* **314**, 141. [*Chap. 7*]

Normand, J. M., and Herrmann, H. J. (1995). Precise determination of the conductivity exponent of 3D percolation using "Percola", *Int. J. Mod. Phys. C* **6**, 813. [*Chap. 6*]

Oerding, K. (1996). The $A + B \to 0$ annihilation reaction in a quenched random velocity field, *J. Phys. A* **29**, 7051. [*Chap. 13*]

Ogston, A. G., Preston, B. N., and Wells, J. D. (1973). On the transport of compact particles through solutions of chain polymers, *Proc. Roy. Soc. Lond.* **333**, 297. [*Chap. 6*]

Ohtsuki, T. (1982). Diffusion of classical particles in a random medium. II. Nonlinear response, *J. Phys. Soc. Japan* **51**, 1493. [*Chap. 9*]

Ohtsuki, T. (1985). Diffusion-controlled trapping by extended traps, *Phys. Rev. A* **32**, 699. [*Chap. 12*]

Ohtsuki, T., and Keyes, T. (1984). Mobility and linear response theory on percolation lattices, *Phys. Rev. Lett.* **52**, 1177. [*Chap. 7, 9*]

Okazaki, A., Maruyama, K., Okumura, K., Hasegawa, Y., and Miyazima, S. (1996). Critical exponents of a two-dimensional continuum percolation system, *Phys. Rev. E* **54**, 3389. [*Chap. 2*]

Oppenheim, I., Schuler, K. E., and Weiss, G. H. (1977). *Stochastic Processes in Chemical Physics*, MIT Press, Cambridge, MA. [*Chap. 11, 13*]

Orbach, R. (1986). Dynamics of fractal networks, *Science* **231**, 814. [*Chap. 7*]

Ordemann, A., Berkolaiko, G., Havlin, S., and Bunde, A. (2000). Swelling–collapse transition of self-attracting walks, *Phys. Rev. E* **61**, R1005. [*Chap. 10*]

Oshanin, G., and Blumen, A. (1998). Kinetic description of diffusion-limited reactions in random catalytic media, *J. Chem. Phys.* **108**, 1140. [*Chap. 13*]

Oshanin, G., Sokolov, I. M., Argyrakis, P., and Blumen, A. (1996). Fluctuation-dominated $A + B \to 0$ kinetics under short-ranged interparticle interactions, *J. Chem. Phys.* **105**, 6304. [*Chap. 13*]

O'Shaughnessy, B., and Procaccia, I. (1985a). Analytical solutions for diffusion on fractal objects, *Phys. Rev. Lett.* **54**, 455. [*Chap. 5, 6, 7*]

O'Shaughnessy, B., and Procaccia, I. (1985b). Diffusion on fractals, *Phys. Rev.* **32**, 3073. [*Chap. 6*]

Pandey, R. B. (1984). Classical diffusion drift and trapping in random percolating systems, *Phys. Rev. B* **30**, 489. [*Chap. 9*]

Pandey, R. B. (1986). Disorder-induced transport: I. Simple cubic lattice, *J. Phys. A* **19**, 3925. [*Chap. 8*]

Pandey, R. B. (1992). Breakdown of diffusive motion in an interacting lattice gas, *Physica A* **187**, 77. [*Chap. 10*]

Pandey, R. B., and Stauffer, D. (1983). Confirmation of dynamical scaling at the percolation threshold, *Phys. Rev. Lett.* **51**, 527. [*Chap. 5, 6*]

Parichha, T. K., and Talapatra, G. B. (1996). Energy transfer and trapping in a mixed molecular solid: a Monte Carlo simulation study, *Chem. Phys. Lett.* **257**, 622. [*Chap. 12*]

Parisi, G., and Sourlas, N. (1981). Crtitical behavior of branched polymers and the Lee–Yang edge singularity, *Phys. Rev. Lett.* **46**, 871. [*Chap. 7*]

Patterson, L. (1984). Diffusion-limited aggregation and two-fluid displacements in porous media, *Phys. Rev. Lett.* **52**, 1621. [*Chap. 7*]

Peitgen, H.-O. and Richter, P., H. (1986). *The Beauty of Fractals*, Springer, Heidelberg. [*Chap. 1*]

Peitgen, H.-O., Jürgens, H., and Saupe, D. (1992). *Chaos and Fractals*, Springer, New York. [*Chap. 1*]

Peliti, L. (1985). Path integral approach to birth–death processes on a lattice, *J. Phys., Paris* **46**, 1469. [*Chap. 13, 15*]

Peng, C.-K., Havlin, S., Schwartz, M., and Stanley, H. E. (1991). Directed polymer and ballistic deposition growth with correlated noise, *Phys. Rev. A* **44**, R2239. [*Chap. 4*]

Peng, C.-K., Buldyrev, S. V., Goldberger, A. L., Havlin, S., Sciortino, F., Simons, M., and Stanley, H. E. (1992a). Long-range correlations in nucleotide sequences, *Nature, Lond.* **356**, 168. [*Chap. 4*]

Peng, C.-K., Buldyrev, S. V., Goldberger, A. L., Havlin, S., Sciortino, F., Simons, M., and Stanley, H. E. (1992b). Fractal landscapes of DNA walk, *Physica A* **191**, 25. [*Chap. 4*]

Peng, C.-K., Mietus, J., Hausdorff, J. M., Havlin, S., Stanley, H. E., and Goldberger, A. L. (1993). Long-range anticorrelations and non-Gaussian behavior of the heartbeat, *Phys. Rev. Lett.* **70**, 1343. [*Chap. 4*]

Peng, C.-K., Buldyrev, S. V., Havlin, S., Simons, M., Stanley, H. E., and Goldberger, A. L. (1994). On the mosaic organization of DNA nucleotides, *Phys. Rev. E* **49**, 1685. [*App. D*]

Perondi, L. F., Elliott, R. J., and Kaski, K. (1997). Tracer diffusion in bond-disordered square lattices. 1, *J. Phys. Cond. Matter* **9**, 7933. [*Chap. 10*]

Petersen, J., Roman, H. E., Bunde, A., and Dieterich, W. (1989). Nonuniversality of transport exponents in continuum percolation systems: effects of finite jump distance, *Phys. Rev. B* **39**, 893. [*Chap. 6*]

Peyrelasse, J., and Boned, C. (1990). Conductivity, dielectric relaxation, and viscosity of ternary microemulsions: the role of the experimental path and the point of view of percolation theory, *Phys. Rev. A* **41**, 938. [*Chap. 6*]

Phillips, J. C. (1996). Stretched exponential relaxation in molecular and electronic glasses, *Rep. Prog. Phys.* **59**, 1133. [*Chap. 12*]

Phillips, J. C., Rasaiah, J. C., and Hubbard, J. B. (1998). Comment on "Anomalous size dependence of relaxational processes", *Phys. Rev. Lett.* **80**, 5453. [*Chap. 12*]

Pike, R., and Stanley, H. E. (1981). Order propagation near the percolation threshold, *J. Phys. A* **14**, L169. [*Chap. 6*]

Porto, M., Shehter, A., Bunde, A., and Havlin, S. (1996). Branched polymers in the presence of impurities, *Phys. Rev. E* **54** 1742. [*Chap. 7*]

Porto, M., Havlin, S., Schwarzer, S., and Bunde, A. (1997a). Optimal path in strong disorder and shortest path in invasion percolation with trapping, *Phys. Rev. Lett.* **79**, 4060. [*Chap. 2, 8, 10*]

Porto, M., Bunde, A., Havlin, S., and Roman, H. E. (1997b). Structural and dynamical properties of the percolation backbone in two dimensions, *Phys. Rev. E* **56**, 1667. [*Chap. 2, 6, 8*]

Porto, M., Havlin, S., Roman, H. E., and Bunde, A. (1998). Probability distribution of the shortest path on the percolation cluster, its backbone, and skeleton, *Phys. Rev. E* **58**, R5205. [*Chap. 6*]

Porto, M., Bunde, A., and Havlin, S. (1999). Distribution of dangling ends on the incipient percolation cluster, *Physica A* **266**, 96. [*Chap. 2, 7, 8*]

Pottier, N. (1995). Diffusion on random comb-like structures – field-induced trapping effects, *Physica A* **217**, 440. [*Chap. 7*]

Prakash, S., Havlin, S., Schwartz, M., and Stanley, H. E. (1992). Structural and dynamical properties of long-range correlated percolation, *Phys. Rev. A* **46**, R1724. [*Chap. 2*]

Privman, V. (1992). Model of cluster growth and phase separation: exact results in one dimension, *J. Stat. Phys.* **69**, 629. [*Chap. 13, 15*]

Privman, V. (1994a). Exact results for diffusion-limited reactions with synchronous dynamics, *Phys. Rev. E* **50**, 50. [*Chap. 13, 15*]

Privman, V. (1994b). Exact results for 1D conserved order parameter model, *Mod. Phys. Lett. B* **8**, 143. [*Chap. 13, 15*]

Privman, V., ed. (1997). *Nonequilibrium Statistical Mechanics in One Dimension*, Cambridge University Press, Cambridge. [*Chap. 13*]

Privman, V., and Grynberg, M. D. (1992). Fast-diffusion mean-field theory for k-body reactions in one dimension, *J. Phys. A* **25**, 6567. [*Chap. 13*]

Privman, V., Doering, C. R., and Frisch, H. L. (1993). Crossover from rate-equation to diffusion-controlled kinetics in two-particle coagulation, *Phys. Rev. E* **48**, 846. [*Chap. 19*]

Privman, V., Cadilhe, A. M. R., and Glasser, M. L. (1995). Exact solutions of anisotropic diffusion-limited reactions with coagulation and annihilation, *J. Stat. Phys.* **81**, 881. [*Chap. 13, 15*]

Privman, V., Cadilhe, A. M. R., and Glasser, M. L. (1996). Anisotropic diffusion-limited reactions with coagulation and annihilation, *Phys. Rev. E* **53**, 739. [*Chap. 13, 15*]

Rabinovich, S., Roman, H. E., Havlin, S., and Bunde, A. (1996). Critical dimensions for random walks on random-walk chains, *Phys. Rev. E* **54**, 3606. [*Chap. 5*]

Rácz, Z. (1985). Diffusion-controlled annihilation in the presence of particle sources: exact results in one dimension, *Phys. Rev. Lett.* **55**, 1707. [*Chap. 13, 15, 16*]

Rácz, Z., and Plischke, M. (1987). Correlations in a nonequilibrium steady state: exact results for a generalized kinetic Ising model, *Acta Phys. Hungarica* **62**, 203. [*Chap. 13*]

Ragot, B. R., and Kirk, J. G. (1997). Anomalous transport of cosmic ray electrons, *Astron. Astrophys.* **327**, 432. [*Chap. 12*]

Ramaswamy, R., and Barma, M. (1987). Transport in random networks in a field: interacting particles, *J. Phys. A* **20**, 2973. [*Chap. 9, 10*]

Rammal, R., and Toulouse, G. (1982). Spectrum of the Schrödinger equation on a self-similar structure, *Phys. Rev. Lett.* **49**, 1194. [*Chap. 5*]

Rammal, R., and Toulouse, G. (1983). Random walks on fractal structures and percolation clusters, *J. Phys. Lett., Paris* **44**, L13. [*Chap. 1, 5, 6*]

Rammal, R., Angles d'Auriac, J. C., and Benoit, A. (1984a). Universality of the spectral dimension of percolation clusters, *Phys. Rev. B* **30**, 4087. [*Chap. 6*]

Rammal, R., Toulouse, G., and Vannimenus, J. (1984b). Self-avoiding walks on fractal spaces: exact results and Flory approximation, *J. Phys., Paris* **45**, 389. [*Chap. 10*]

Rammal, R., Tannous, C., and Tremblay, A. M. S. (1985). $1/f$ noise in random resistor networks: fractals and percolating systems, *Phys. Rev. A* **31**, 2662. [*Chap. 6*]

Rammal, R., Toulouse, G., and Virasoro, M. A. (1986). Ultrametricity for physicists, *Rev. Mod. Phys.* **58**, 765. [*Chap. 8*]

Rant, R. (1997). Diffusion-limited reaction rates on self-affine fractals, *J. Phys. Chem. B* **101**, 3781. [*Chap. 13*]

Redner, S. (1982). A Fortran program for cluster enumeration, *J. Stat. Phys.* **29**, 309. [*Chap. 7*]

Redner, S. (1990). Superdiffusion in random velocity-fields, *Physica A* **168**, 551. [*Chap. 4*]

Redner, S. (1997). Scaling theories of bimolecular reactions, in *Nonequilibrium Statistical Mechanics in One Dimension*, V. Privman, ed., Cambridge University Press, Cambridge. [*Chap. 13*]

Redner, S., and ben-Avraham, D. (1990). Nearest-neighbour distance of diffusing particles from a single trap, *J. Phys. A* **23**, L1169. [*Chap. 12*]

Redner, S., and Kang, K. (1983). Asymptotic solution of interacting walks in one dimension, *Phys. Rev. Lett.* **81**, 1729. [*Chap. 12*]

Redner, S., and Kang, K. (1984). Kinetics of the 'scavenger' reaction, *J. Phys. A* **17**, L451. [*Chap. 12*]

Redner, S., Koplik, J., and Wilkinson, D. (1987). Hydrodynamic dispersion in a self-similar geometry, *J. Phys. A* **20**, 1543. [*Chap. 9*]

Reigada, R., Sagues, F., Sokolov, I. M., Sancho, J. M., and Blumen, A. (1997). Fluctuation-dominated kinetics under stirring, *Phys. Rev. Lett.* **78**, 741. [*Chap. 11*]

Reis, F. D. A. A. (1996a). Diffusion on regular random fractals, *J. Phys. A* **29**, 7803. [*Chap. 5*]

Reis, F. D. A. A. (1996b). Scaling for random walks on Eden trees, *Phys. Rev. E* **54**, R3079. [*Chap. 7*]

Reis, F. D. A. A., and Riera, R. (1995). Directed self-avoiding walks on Sierpinski carpets: series results, *J. Phys. A* **28**, 1257. [*Chap. 10*]

Revathi, S., Balakrishnan, V., Lakshmibala, S., and Murthy, K. P. N. (1996). Validity of the mean-field approximation for diffusion on a random comb, *Phys. Rev. E* **54**, 2298. [*Chap. 7*]

Rey, P.-A., and Droz, M. (1997). A renormalization group study of reaction–diffusion models with particles input, *J. Phys. A* **30**, 1101. [*Chap. 13, 15, 16*]

Reynolds, P. J., Stanley, H. E., and Klein, W. (1980). Large-cell MC renormalization group for percolation, *Phys. Rev. B* **21**, 1223. [*Chap. 2*]

Rice, S. A. (1985). *Diffusion-Limited Reactions*, Elsevier, Amsterdam. [*Chap. 11, 13*]

Richards, P. M. (1977). Theory of one-dimensional hopping conductivity and diffusion, *Phys. Rev. B* **16**, 1393. [*Chap. 10*]

Rigby, S. P., and Gladden, L. F. (1998). Influence of structural heterogeneity on selectivity in fractal catalyst structures, *J. Catal.* **80**, 44. [*Chap. 13*]

Rintoul, M. D. and Torquato, S. (1997). Precise determination of the critical threshold and exponents in a three-dimensional continuum percolation model, *J. Phys. A* **30**, L585. [*Chap. 2*]

Rintoul, M. D., Moon, J., and Nakanishi, H. (1994). Statistics of self-avoiding walks on randomly diluted lattices, *Phys. Rev. E* **49**, 2790. [*Chap. 10*]

Riordan, J., Doering, C. R., and ben-Avraham, D. (1995). Fluctuations and stability of Fisher waves, *Phys. Rev. Lett.* **75**, 565. [*Chap. 18*]

Ritzenberg, A. L., and Cohen, R. I. (1984). First passage percolation: scaling and critical exponents, *Phys. Rev. B* **30**, 4036. [*Chap. 6*]

Robillard, S., and Tremblay, A. M. S. (1986). Anomalous diffusion on fractal lattices with site disorder, *J. Phys. A* **19**, 2171. [*Chap. 8*]

Roman, H. E., Bunde, A., and Dieterich, W. (1986). Conductivity of dispersed ionic conductors: a percolation model with two critical points, *Phys. Rev. B* **34**, 3439. [*Chap. 2*]

Roman, H. E., Dräger, J., Bunde, A., Havlin, S., and Stauffer, D. (1995). Distributions of polymers in disordered structures, *Phys. Rev. E* **52**, 6303. [*Chap. 10*]

Roman, H. E., Ordemann, A., Porto, M., Bunde, A., and Havlin, S., (1998). Structure of self-avoiding walks on percolation clusters at criticality, *Phil. Mag. B* **77**, 1357. [*Chap. 6, 10*]

Rosenstock, B. (1969). Luminescent emission from an organic solid with traps, *Phys. Rev.* **187**, 1166. [*Chap. 12*]

Roux, S. (1985). Flory calculation of the fractal dimensionality of the shortest path in a percolation cluster, *J. Phys. A* **18**, L395. [*Chap. 6*]

Roux, S., Mitescu, C., Charlaix, E., and Baudet, C. (1986). Transfer matrix algorithm for convection-biased diffusion, *J. Phys. A* **19**, L687. [*Chap. 9*]

Roy, A. K., and Chakrabarti, B. K. (1987). Scaling theory for the statistics of self-avoiding walks on random lattices, *J. Phys. A* **20**, 215. [*Chap. 10*]

Rubin, Z., Sunshine, S. A., Heaney, M. B., Bloom, I., and Balberg, I. (1999). Critical behavior of the electrical transport properties in a tunneling-percolation system, *Phys. Rev. B* **59**, 12196. [*Chap. 2*]

Safonov, V. P., Shalaev, V. M., Markel, V. A., Danilova, Y. E., Lepeshkin, N. N., Kim, W., Rautian, S. G., and Armstrong, R. L. (1998). Spectral dependence of selective photomodification in fractal aggregates of colloidal particles, *Phys. Rev. Lett.* **80**, 1102. [*Chap. 6*]

Sahimi, M. (1994). *Applications of Percolation Theory*, Taylor and Francis, London. [*Chap. 2*]

Saleur, H., and Duplantier, B. (1987). Exact determination of the percolation hull exponent in two-dimensions, *Phys. Rev. Lett.* **58**, 2325. [*Chap. 2*]

Saleur, H., Sammis, C. G., and Sornette, D. (1996). Discrete scale invariance, complex fractal dimensions, and log-periodic fluctuations in seismicity, *J. Geol. Res.* **101**, 17661. [*Chap. 1, 9*]

Samorodnitsky, G., and Taqqu, M. S. (1994). *Stable Non-Gaussian Random Processes*, Chapman and Hall, New York. [*Chap. 4*]

Sánchez, A. D. (1999). Trapping reactions for mobile particles and a trap in the laboratory frame, *Phys. Rev. E* **59**, 5021. [*Chap. 12*]

Sánchez, A. D., Rodriguez, M. A., and Wio, H. S. (1998). Results in trapping reactions for mobile particles and a single trap, *Phys. Rev. E* **57**, 6390. [*Chap. 12*]

Sankey, D. F., and Fedders, P. H. (1977). The generalized atomic hopping problem – particle correlation functions, *Phys. Rev. B* **15**, 3586. [*Chap. 10*]

Santos, J. E., Schütz, G. M., and Stinchcombe, R. B. (1996). Diffusion–annihilation dynamics in one spatial dimension, *J. Chem. Phys.* **105**, 2399. [*Chap. 15*]

Sapoval, B., Rosso, M., and Gouyet, J. F. (1985). The fractal nature of a diffusion front and the relation to percolation, *J. Phys. Lett., Paris* **46**, L149. [*Chap. 2*]

Sapoval, B., Gobron, Th., and Margolina, A. (1991). Vibrations of fractal drums, *Phys. Rev. Lett.* **67**, 2974. [*Chap. 5*]

Sapozhnikov, V. B. (1998). Reply to the comment by J. W. Lee. Self-attracting walk: are the exponents universal? *J. Phys. A* **31**, 3935. [*Chap. 3*]

Sartoni, G., and Stella, A. L. (1997). Finite-size scaling analysis of biased diffusion on fractals, *Physica A* **241**, 453. [*Chap. 9*]

Saxton, M. J. (1994). Anomalous difusion due to obstacles: a Monte Carlo study, *Biophys. J.* **66**, 394. [*Chap. 6*]

Scher, H., and Lax, M. (1973). Stochastic transport in a disordered solid. I. Theory, *Phys. Rev. B* **7**, 4491. [*Chap. 3, 8, 9*]

Scher, H., and Montroll, E. (1975). Anomalous transit-time dispersion in amorphous solids, *Phys. Rev. B* **12**, 2455. [*Chap. 3, 8*]

Schiff, E. A. (1995). Diffusion-controlled bimolecular recombination of electrons and holes in a-Si : H, *J. Non-Cryst. Solids* **190**, 1. [*Chap. 13*]

Schlögl, F. (1972). Chemical reaction model for nonequilibrium phase transition, *Z. Phys.* **253**, 147. [*Chap. 11*]

Schnörer, H., Kuzovkov, V., and Blumen, A. (1989). Segregation in annihilation reactions without diffusion: analysis of corelations, *Phys. Rev. Lett.* **63**, 805. [*Chap. 13*]

Schoonover, R., ben-Avraham, D., Havlin, S., Kopelman, R., and Weiss, G. H. (1991). Nearest-neighbor distances in diffusion-controlled reactions modelled by a single mobile trap, *Physica A* **171**, 232. [*Chap. 12*]

Schreiber, M., and Grussbach, H. (1991). Multifractal wave functions at the Anderson transition, *Phys. Rev. Lett.* **67**, 607. [*Chap. 1*]

Schütz, G. M. (1995). Diffusion–annihilation in the presence of a driving field, *J. Phys. A* **28**, 3405. [*Chap. 13, 15, 18*]

Schütz, G. M. (1997). Diffusion-limited annihilation in inhomogeneous environments, *Z. Phys. B* **104**, 583. [*Chap. 18*]

Schütz, G. M., and Mussawisade, K. (1998). Annihilating random walks in one-dimensional disordered media, *Phys. Rev. E* **57**, 2563. [*Chap. 18*]

Schwartz, N., Nazaryev, A. L., and Havlin, S. (1998). Optimal path in two and three dimensions, *Phys. Rev. E* **58**, 7642. [*Chap. 2*]

Schwartz, N., Porto, M., Havlin, S., and Bunde, A. (1999). Optimal path in weak and strong disorder, *Physica A* **266**, 317. [*Chap. 4*]

Schwarzer, S., Havlin, S., and Bunde, A. (1999). Structural properties of invasion percolation with and without trapping: shortest path and distributions, *Phys. Rev. E* **59**, 3262. [*Chap. 2*]

Seifert, E., and Suessenbach, M. (1984). Test of universality for percolative diffusion, *J. Phys. A* **17**, L703. [*Chap. 9*]

Seksek, O., Biwersi, J., and Verkman, A. S. (1997). Translational diffusion of macromolecule-sized solutes in cytoplasm and nucleus, *J. Cell Biol.* **138**, 131. [*Chap. 6*]

Sen, P. (1997). Statistics of red sites on elastic and full backbone, *Physica A* **238**, 39. [*Chap. 2*]

Sen, P. N., Roberts, J. N., and Halperin, B. I. (1985). Nonuniversal critical exponents for transport in percolating systems with a distribution of bond strengths, *Phys. Rev. B* **32**, 3306. [*Chap. 6*]

Shchur, I. N., and Kosyakov, S. S. (1998). Probability of incipient spanning clusters in critical two-dimensional percolation, *Nucl. Phys. B – Proc. Suppl.* **63**, 664. [*Chap. 2*]

Shenkel, A., Zhang, J., and Zhang, Y. C. (1993). Long range correlations in human writings, *Fractals* **1**, 47. [*Chap. 4*]

Shimshoni, E., Auerbach, A., and Kapitulnik, A. (1998). Transport through quantum melts, *Phys. Rev. Lett.* **80**, 3352. [*Chap. 6*]

Shlesinger, M. F. (1974). Asymptotic solutions of continuous-time random walks, *J. Stat. Phys.* **10**, 421. [*Chap. 7*]

Shlesinger, M. F., and Klafter, J. (1986). Lévy walks versus Lévy flights, in *On Growth and Form*, H. E. Stanley and N. Ostrowsky, eds., Martinus Nijhoff, Dordrecht. [*Chap. 4*]

Shlesinger, M. F., and Montroll, E. W. (1984). On the Williams–Watts function of dielectric relaxation, *Proc. Natl. Acad. Sci. U.S.A.* **81**, 1280. [*Chap. 12*]

Shlesinger, M. F., West, B., and Klafter, J. (1987). Lévy dynamics of enhanced diffusion: application to turbulence, *Phys. Rev. Lett.* **58**, 1100. [*Chap. 4*]

Simon, H. (1995). Concentration for one and two species one-dimensional reaction–diffusion systems, *J. Phys. A* **28**, 6585. [*Chap. 13, 15*]

Sinai, Ya. (1982). *Proceedings of the Berlin Conference on Mathematical Problems in Theoretical Physics*, R. Schrader, R. Seiler and D. A. Uhlenbroch, eds., Springer, Berlin, p. 12. [*Chap. 8*]

Sinder, M., and Pelleg, J. (1999). Properties of the crossover from nonclassical chemical kinetics in a reversible $A + B \rightleftharpoons C$ reaction–diffusion process, *Phys. Rev. E* **60**, R6259. [*Chap. 14*]

Skal, A. S., and Shklovskii, B. I. (1975). Topology of an infinite cluster in the theory of percolation and its relationship to the theory of hopping conduction, *Soviet Phys. Semicond.* **8**, 1029. [*Chap. 6*]

Sokolov, I. M., and Blumen, A. (1999). Scaling and patterns in surface fragmentation processes, *Physica A* **266**, 299. [*Chap. 2*]

Sokolov, P. T., and Kauffmann, H. F. (1998). Excitation trapping in dynamically disordered polymers, *Macromolecules* **31**, 2521. [*Chap. 12*]

Solomon, T., Weeks, E., and Swinney, H. (1993). Observation of anomalous diffusion and Lévy flights in a two-dimensional rotating flow, *Phys. Rev. Lett.* **71**, 3975. [*Chap. 4*]

Solomon, T. H., Weeks, E. R., and Swinney, H. L. (1994). Chaotic advection in a two-dimensional flow: Lévy flights and anomalous diffusion, *Physica D* **76**, 70. [*Chap. 4*]

Sompolinsky, H. (1981). Time-dependent order parameters in spin-glasses, *Phys. Rev. Lett.* **47**, 935. [*Chap. 8*]

Sornette, D. (1998). Discrete scale invariance and complex dimensions, *Phys. Rep.* **297**, 239. [*Chap. 9*]

Sornette, D., and Johansen, A. (1997). Large financial crashes, *Physica A* **245**, 411. [*Chap. 9*]

Sornette, D., Johansen, A., Arneodo, A., Muzy, J. F., and Saleur, H. (1996a). Complex fractal dimensions describe the hierarchical structure of diffusion-limited-aggregation clusters, *Phys. Rev. Lett.* **76**, 251. [*Chap. 1*]

Sornette, D., Johansen, A., and Bouchaud, J.-P. (1996b). Stock market crashes, precursors and replicas, *J. Phys. I, Paris* **6**, 167. [*Chap. 9*]

Spitzer, F. (1970). *Adv. Math.* **5**, 246. [*Chap. 10*]

Spitzer, F. (1976). *Principles of Random Walks*, Springer, New York. [*Chap. 3*]

Spouge, J. L. (1988). Exact solutions for a diffusion–reaction process in one dimension, *Phys. Rev. Lett.* **60**, 871. [*Chap. 13, 15*]

Stanley, H. E. (1984). Application of fractal concepts to polymer statistics and to anomalous transport in randomly porous media, *J. Stat. Phys.* **36**, 843. [*Chap. 6*]

Stanley, H. E., and Coniglio, A. (1984). Flow in porous media: the "backbone" fractal at the percolation treshold, *Phys. Rev. B* **29**, 522. [*Chap. 6*]

Stanley, H. E., and Meakin, P. (1988). Multifractal phenomena in physics and chemistry, *Nature, Lond.*, **335**, 405. [*Chap. 6*]

Stanley, H. E., and Ostrowsky, N., eds. (1990). *Correlations and Connectivity: Geometric Aspects of Physics, Chemistry and Biology*, Kluwer, Dordrecht. [*Chap. 1*]

Stanley, H. E., Majid, I., Margolina, A., and Bunde, A. (1984). Direct tests of the Aharony–Stauffer argument, *Phys. Rev. Lett.* **53**, 1706. [*Chap. 6*]

Stauffer, D. (1979). Scaling theory of percolation clusters, *Phys. Rep.* **54**, 1. [*Chap. 6*]

Stauffer, D. (1985a). *Introduction to Percolation Theory*, Taylor and Francis, London. [*Chap. 2*]

Stauffer, D. (1985b). Monte Carlo study of biased diffusion at the percolation threshold, *J. Phys. A* **18**, 1827. [*Chap. 9*]

Stauffer, D. (1999). New simulations on old biased diffusion, *Physica A* **266**, 35. [*Chap. 9*]

Stauffer, D., and Aharony, A. (1994). *Introduction to Percolation Theory*, Taylor & Francis, London. [*Chap. 2*]

Stauffer, D., and Sornette, D. (1998). Log-periodic oscillations for biased diffusion in 3D random lattices, *Physica A* **252**, 271. [*Chap. 9*]

Stephen, M. J., and Kariotis, R. (1982). Diffusion in a one-dimensional disordered system, *Phys. Rev. B* **26**, 2917. [*Chap. 8*]

Stoll, E. (1998). A fast cluster counting algorithm for percolation on and off lattices, *Computer Phys. Comm.* **109**,1. [*Chap. 2*]

Straley, J. P. (1982). Threshold behaviour of random resistor networks: a synthesis of theoretical approaches, *J. Phys. C* **15**, 2333. [*Chap. 6, 8*]

Strenski, P. N., Bradley, R. M., and Debierre, J. M. (1991). Scaling behavior of percolation surfaces in three dimensions, *Phys. Rev. Lett.* **66**, 1330. [*Chap. 2*]

Sudbury, A. (1990). The branching annihilating process: an interacting particle system, *Ann. Prob.* **18**, 581. [*Chap. 13*]

Suding, P. N., and Ziff, R. M. (1999). Site percolation thresholds for Archimedean lattices, *Phys. Rev. E* **60**, 275. [*Chap. 2*]

Sugihara, G., and May R. M. (1990). Application of fractals in ecology, *Trends Ecol. Evol.* **5**, 79. [*Chap. 1*]

Taitelbaum, H. (1991). Nearest-neighbor distances at an imperfect trap in two dimensions, *Phys. Rev. A* **43**, 6592. [*Chap. 12*]

Taitelbaum, H., and Koza, Z. (1998). Anomalous kinetics of reaction–diffusion fronts, *Phil. Mag. B* **77**, 1389. [*Chap. 14*]

Taitelbaum, H., and Weiss, G. H. (1994). Anomalous segregation at a single trap in disordered chains, *Phys. Rev. E* **50**, 2357. [*Chap. 12*]

Taitelbaum, H., Kopelman, R., Weiss, G. H., and Havlin, S. (1990). Statistical properties of nearest-neighbor distances at an imperfect trap, *Phys. Rev. A* **41**, 3116. [*Chap. 12*]

Taitelbaum, H., Havlin, S., Kiefer, J., Trus, B. L., and Weiss, G. H. (1991). Some properties of the $A + B \to C$ reaction–diffusion system with initially separated components, *J. Stat. Phys.* **65**, 873. [*Chap. 14*]

Taitelbaum, H., Koo, Y. E., Havlin, S, Kopelman, R., and Weiss, G. H. (1992). Exotic behavior of the reaction front in the $A + B \to C$ reaction–diffusion system, *Phys. Rev. A* **46**, 2151. [*Chap. 14*]

Taitelbaum, H., Vilensky, B., Lin A., Yen, A., Koo, Y. E. L., and Kopelman, R. (1996). Competing reactions with initially separated components, *Phys. Rev. Lett.* **77**, 1640. [*Chap. 14*]

Takayasu, H. (1990). *Fractals in the Physical Sciences*, John Wiley, New York. [*Chap. 1, App. D*]

Takayasu, H., and Tretyakov, A. Yu. (1992). Extinction, survival, and dynamical phase transition of branching annihilating random walk, *Phys. Rev. Lett.* **68**, 3060. [*Chap. 13*]

Teitel, S., and Domany, E. (1985). Dynamical phase transitions in hierarchical structures, *Phys. Rev. Lett.* **55**, 2176. [*Chap. 5, 8*]

Terao, T., Yamaya, A. and Nakayama, T. (1998). Vibrational dynamics of cluster–cluster aggregations, *Phys. Rev. E* **57**, 4426. [*Chap. 6*]

Torney, D. C., and McConnell. H. E. (1983). Diffusion-limited reactions in one dimension, *Proc. R. Soc., Lond. A* **387**, 147. [*Chap. 13, 15*]

Toussaint, D., and Wilczek, F. (1983). Particle–antiparticle annihilation in diffusive motion, *J. Chem. Phys.* **78**, 2642. [*Chap. 13*]

Turcotte, D. L. (1992). *Fractals and Chaos in Geology and Geophysics*, Cambridge University Press, Cambridge. [*Chap. 1*]

Tzschichholtz, F., Bunde, A., and Havlin, S. (1989). Loopless percolation clusters, *Phys. Rev. A* **39**, R5470. [*Chap. 7*]

van Beijeren, H., and Kehr, K. W. (1986). Correlation factor, velocity autocorrelation function and frequency-dependent tracer diffusion coefficient, *J. Phys. C* **19**, 1319. [*Chap. 10*]

van Beijeren, H., and Kutner, R. (1988). Tracer diffusion in concentrated lattice gas models. Rectangular lattices with anisotropic jump rates, *J. Stat. Phys.* **49**, 1043. [*Chap. 10*]

van Beijeren, H., Kehr, K. W., and Kutner, R. (1983). Diffusion in concentrated lattice gases. III. Tracer diffusion on a one-dimensional lattice, *Phys. Rev. B* **28**, 5711. [*Chap. 10*]

van Damme, H., *et al.* (1986). in *Chemical Reactions in Organic and Inorganic Constrained Systems*, J. J. Fripiat and P. Sinay, eds., Reidel, Dordrecht. [*Chap. 12*]

van der Marck, S. C. (1998). Calculation of percolation thresholds in high dimensions for fcc, bcc, and diamond lattices, *Int. J. Mod. Phys. C* **9**, 529. [*Chap. 2*]

van der Meer, M., Schuchardt, R., and Keiper, R. (1982). Strong-field hopping in disordered semiconductors. A problem of directed percolation, *Phys. Stat. Sol. B* **110**, 571. [*Chap. 9*]

Vanderzande, C., and Komoda, A. (1992). Critical behavior of self-avoiding walks on percolation clusters, *Phys. Rev. A* **45**, R5335. [*Chap. 10*]

van Kampen, N. G. (1981). *Stochastic Processes in Physics and Chemistry*, North-Holland, Amsterdam. [*Chap. 11, 13*]

van Lien, N., and Shklovskii, B. I. (1981). Hopping conduction in strong electric fields and directed percolation, *Solid State Comm.* **38**, 99. [*Chap. 9*]

Vannimenus, J., Nadal, J. P., and Martin, H. (1984). On the spreading dimension of percolation and directed percolation clusters, *J. Phys. A* **17**, L351. [*Chap. 6*]

Vicsek, T. (1981). Random walks on bond percolation clusters: AC hopping conductivity below the threshold, *Z. Phys. B* **45**, 153. [*Chap. 6*]

Vicsek, T. (1991). *Fractal Growth Phenomena*, 2nd ed., World Scientific, Singapore. [*Chap. 1, 7, App. D*]

Vicsek, T., and Barabási, A. L. (1991). Multi-affine model for the velocity distribution in fully turbulent flows. *J. Phys. A* **24**, L845. [*App. D*]

Vicsek, T., and Kertesz, J. (1981). Monte-Carlo renormalisation-group approach to percolation on a continuum: test of universality, *J. Phys. A* **14**, L31. [*Chap. 2*]

Viswanathan, G. W., Afanasyev, V., Buldyrev, S. V., Murphy, E. J., Prince, P. A., and Stanley, H. E. (1996). Lévy flight search patterns of wandering albatrosses, *Nature, Lond.* **381**, 413. [*Chap. 4*]

von Smoluchowski, R. (1917). *Z. Phys. Chem.* **29**, 129. [*Chap. 12, 14*]

Voss, R. (1991). Random fractals forgeries, in *Fundamental Algorithms in Computer Graphics*, R. Earnshaw, ed., Springer, Berlin, pp. 805–35. [*Chap. 4*]

Wang, J., and Lubensky, T. C. (1986). Percolation conductivity exponent t to second order in $\varepsilon = 6 - d$, *Phys. Rev. B* **33**, 4998. [*Chap. 6*]

Watts, D. J., and Strogatz, S. H. (1998). Collective dynamics of 'small-world' networks, *Nature, Lond.* **393**, 440. [*Chap. 2*]

Webman, I. (1981). Effective-medium approximation for diffusion on a random lattice, *Phys. Rev. Lett.* **47**, 1497. [*Chap. 6*]

Webman, I. (1984a). Diffusion and trapping of excitations on fractals, *Phys. Rev. Lett.* **52**, 220. [*Chap. 12*]

Webman, I. (1984b). Propagation and trapping of excitations on percolation clusters, *J. Stat. Phys.* **26**, 603. [*Chap. 12*]

Weeks, E. R., Urbach, J. S., and Swinney, H. L. (1996). Anomalous diffusion in asymmetric random walks with a quasi-geostrophic flow example, *Physica D* **97**, 291. [*Chap. 4*]

Wehefritz, B., Krebs, K., and Pfanmüller, M. P. (1995). Finite-size scaling of one-dimensional reaction–diffusion systems. Part II: numerical methods, *J. Stat. Phys.* **78**, 1471. [*Chap. 13*]

Weigert, S., Eicke, H. F., and Meier, W. (1997). Electric conductivity near the percolation transition of a nonionic water-in-oil microemulsion, *Physica A* **242**, 95. [*Chap. 6*]

Weiss, G. H. (1993). Nearest-neighbour distance to a trap in a one-dimensional Smoluchowski model, *Physica A* **192**, 617. [*Chap. 12*]

Weiss, G. H. (1994). *Aspects and Applications of the Random Walk*, North Holland, Amsterdam. [*Chap. 3, App. B*]

Weiss, G. H., and Havlin, S. (1984). Trapping of random walks on the line, *J. Stat. Phys.* **37**, 17. [*Chap. 12*]

Weiss, G. H., and Havlin, S. (1985). Non-Markovian reaction sites and trapping, *J. Chem. Phys.* **83**, 5670. [*Chap. 12*]

Weiss, G. H., and Havlin, S. (1986). Some properties of a random walk on a comb structure, *Physica A* **134**, 474. [*Chap. 7*]

Weiss, G. H., and Havlin, S. (1987). Conductivity in hierarchical networks with a broad distribution of resistors, *Phys. Rev. B* **36**, 807. [*Chap. 6*]

Weiss, G. H., and Rubin, R. J. (1983). Random walks: theory and selected applications, *Adv. Chem. Phys.* **52**, 363. [*Chap. 3*]

Weiss, G. H., Havlin, S., and Bunde, A. (1985). On the survival probability of a random walk in a finite lattice with a single trap, *J. Stat. Phys.* **40**, 191. [*Chap. 12*]

Weissman, H., Weiss, G. H., and Havlin, S. (1989). Transport properties of the CTRW with a long-tailed waiting-time density, *J. Stat. Phys.* **57**, 301. [*Chap. 4*]

West, B. J. (1990). *Fractal Physiology and Chaos in Medicine*, World Scientific, Singapore. [*Chap. 1*]

West, B. J., and Deering, W. (1994). Fractal physiology for physicists: Lévy statistics, *Phys. Rep.* **246**, 1. [*Chap. 1*]

White, S. R., and Barma, M. (1984). Field-induced drift and trapping in percolation networks, *J. Phys. A* **17**, 2995. [*Chap. 7, 9*]

Wilke, S., Gefen, Y., Ilkovic, V., Aharony, A., and Stauffer, D. (1984). Diffusion on random clusters and the parasite problem, *J. Phys. A* **17**, 647. [*Chap. 5, 7*]

Wilkinson, D., and Willemsen, J. F. (1983). Invasion percolation, a new form of percolation theory, *J. Phys. A* **16**, 3365. [*Chap. 2*]

Winter, C. L., Newman, C. M., and Neuman, S. P. (1984). A perturbation expansion for diffusion in a random velocity field, *SIAM J. Appl. Math.* **44**, 411. [*Chap. 8*]

Witten, T. A., and Cates, M. E. (1986). Tenuous structures from disorderly growth processes, *Science* **232**, 1607. [*Chap. 7*]

Witten, T. A., and Sander, L. M. (1981). Diffusion-limited aggregation, a kinetic critical phenomenon, *Phys. Rev. Lett.* **47**, 1400. [*Chap. 7*]

Witten, T. A., and Sander. L. M. (1983). Diffusion-limited aggregation, *Phys. Rev. B* **27**, 5686. [*Chap. 7*]

Wolfling, S., and Kantor, Y. (1999). Loops in one-dimensional random walks, *Eur. Phys. J. B* **12**, 569. [*Chap. 3*]

Woo, K. Y., and Lee, S. B. (1991). Monte Carlo study of self-avoiding walks on a percolation cluster, *Phys. Rev. A* **44**, 999. [*Chap. 10*]

Wu, J. J., and McLachlan, D. S. (1997). Percolation exponents and thresholds obtained from the nearly ideal continuum percolation system graphite–boron nitride, *Phys. Rev. B* **56**, 1236. [*Chap. 2, 6*]

Yen, A., Koo, Y-E. L., and Kopelman, R. (1996). Experimental study of a crossover from nonclassical to classical chemical kinetics: an elementary and reversible $A + B \rightleftharpoons C$ reaction–diffusion process in a capillary, *Phys. Rev. E* **54**, 2447. [*Chap. 14*]

Yuste, S. B. (1997). Escape times of n random walkers from a fractal labyrinth, *Phys. Rev. Lett.* **79**, 3565. [*Chap. 5*]

Yuste, S. B. (1998). Order statistics of diffusion on fractals, *Phys. Rev. E* **57**, 6327. [*Chap. 5*]

Zabolitzky, J. G. (1984). Monte-Carlo evidence against the Alexander–Orbach conjecture for percolation conductivity, *Phys. Rev. B* **30**, 4077. [*Chap. 5, 6*]

Zel'dovich, Ya. B., and Ovchinnikov, A. A. (1977). Asymptotic form of the approach to equilibrium and concentration fluctuations, *JETP Lett.* **26**, 440. [*Chap. 13*]

Zhong, D., and ben-Avraham, D. (1995). Diffusion-limited coalescence with finite reaction rates in one dimension, *J. Phys. A* **28**, 33. [*Chap. 19*]

Ziff, R. M. (1992). Spanning probability in 2D percolation, *Phys. Rev. Lett.* **69**, 2670. [*Chap. 2*]

Ziff, R. M. (1999). Exact critical exponent for the shortest-path scaling function in percolation, *J. Phys. A* **32**, L457. [*Chap. 6*]

Ziff, R. M., and Sapoval, B. (1987). The efficient determination of the percolation threshold by a frontier-generating walk in a gradient, *J. Phys. A* **19**, L1169. [*Chap. 2*]

Zimm, B. H., and Stockmayer, W. H. (1949). The dimensions of chain molecules containing branches and rings, *J. Chem. Phys.* **17**, 1301. [*Chap. 7*]

Zivic, I., Milosevic, S., and Stanley, H. E. (1998). Comparative study of self-avoiding trails and self-avoiding walks on a family of compact fractals, *Phys. Rev. E* **58**, 5376. [*Chap. 10*]

Zumofen, G., Blumen, A., and Klafter, J. (1984). Scaling behaviour for excitation trapping on fractals, *J. Phys. A* **17**, L479. [*Chap. 12*]

Zumofen, G., Klafter, J., and Blumen, A. (1990). Enhanced diffusion in random velocity fields, *Phys. Rev. A* **42**, 4601. [*Chap. 4*]

Zumofen, G., Klafter, J., and Blumen, A. (1991a). Transient $A + B \rightarrow 0$ reaction on fractals: stochastic and deterministic aspects, *J. Stat. Phys.* **65**, 1015. [*Chap. 13*]

Zumofen, G., Klafter, J., and Blumen, A. (1991b). Interdomain gaps in transient $A+B \rightarrow 0$ reactions on fractals, *Phys. Rev. A* **44**, 8394. [*Chap. 13*]

Zumofen, G., Klafter, J., and Blumen, A. (1991c). Scaling properties of diffusion-limited reactions: simulation results, *Phys. Rev. A* **43**, 7068. [*Chap. 13*]

Zumofen, G., Klafter, J., and Shlesinger, M. F. (1997). Mixing in reaction dynamics under enhanced diffusion conditions, *Phys. Rep.* **290**, 157. [*Chap. 11*]

Index

active front 198
affinity transformation 259
Airy equation 214
Anderson localization 82
annihilation
    n-species 188
    one-species 179, 215, 257
    two-species 182, 225
    two-species, with drift 189
    two-species, stochastic 186
anomalous diffusion 35, 46, 60, 68, 107
    dimension – *see* walk exponent
    experiments 96
ant in the labyrinth xiii, 74, 263
Alexander–Orbach conjecture 79, 95
Alexandrowicz–Leath algorithm 24, 29, 83, 94, 265
autocorrelation function 137

backbends 139
backbone 21, 30, 90, 92, 99, 106, 150
ballistic particles (motion) 177, 187
barriers 115, 177
bias
    Cartesian 127, 133
    topological 127, 132
birth (of particles) 207, 212
blind ant 264
blobs 22
Boltzmann distribution/factor 41, 47, 119
box-counting algorithm 8
box dimension 11, 258
branched polymers – *see* lattice animals
Branching-annihilating walks (BAW) 188
branching Koch curve 70, 149
Brownian motion – *see* diffusion, random walk
brushes 111, 132

Cantor set 11, 237
Cauchy–Schwartz inequality 225
Cayley tree 25, 29, 88, 100, 133, 136, 216
central limit theorem 36
characteristic function 35
chemical path/length/distance 7, 23, 68, 83, 262

chemical exponent d_{min} 7, 26, 83
chemical dimension d_ℓ 23, 83, 99
chemical reaction 159
chemical shell 24
classical rate equations 160, 180, 223, 233
    autonomous 225, 227
coalescence process 212
    generalized 188
    irreversible 226
    reversible 229
    single-species 163, 177
    with a trap 248
    with bias 215
combs 106
    hierarchical 108
    infinitely long teeth 106
    varying teeth lengths 108, 129
concentration (density)
    global 157, 228
    local 163, 228
concentration front 233
concentration/reaction profile 194
conductance 117
conductivity 39, 63, 77, 89, 92
    ac 136
conductivity exponent $\tilde{\mu}$ 63
conservation laws 164
contact process 165, 257
continuity equation 39
continuous-time random walk (CTRW) 41, 46, 88, 106, 187
correlation length ξ 15, 27, 74, 83
    in chemical distance, ξ_ℓ 83
correlation length exponent ν 15
creeping motion 124
critical field 162
critical threshold (of percolation) 13, 30
crossing probability 31, 84
crossover behavior 20, 75, 78, 82, 112, 176, 235, 253, 261
cutoff length
    lower 5
    upper 5, 19, 28

dangling ends 22, 90, 106, 125, 131
data collapse 38, 40, 79, 85, 87, 88, 89, 94
Democrats/Republicans 185
density–density correlation function 9
density of states 66, 82
detrended fluctuation function 267
dielectric constant 137
difference equation 37, 213
diffusion 37, 207, 212
    with a drift 39
diffusion coefficient/constant 35, 110, 118, 142
diffusion equation 37, 65, 174
diffusion exponent – *see* walk exponent
diffusion front 81, 103
diffusion-limited aggregates (DLAs) 100, 104, 115
diffusion-limited/controlled process 158, 260
diffusion/reaction time 157, 249
dilation symmetry – *see* self-similarity
dimensional analysis 222
directed percolation 56
discrete fluctuations 165, 185
distance to nearest particle 172
DNA 49, 54
Donsker–Varadhan result 170
dynamical phase transition 161, 228, 231
dynamical exponents 76, 79, 82, 92, 95
    bounds 86, 94
    in chemical space 84
    in loopless fractals 103

Eden growth model 100
Eden trees 100
effective order of reaction 225
effective rate equation 181
Einstein's relation 39, 41, 44, 63, 77, 103, 118, 136
    for the drag coefficient 44
elastic backbone 23
end-to-end exponent ν_{SAW} 145
enrichment technique 103, 104
    for SAWs 146
entropy 231
epidemics 24
epsilon expansion (ϵ-expansion) 81, 92
equilibrium 230, 236, 244
evolution equation 210, 212, 239
exact enumeration 59, 88, 104, 110, 131, 263
excluded-volume interaction 133, 141

filtering 269
finite-size effect 91, 96, 199, *see also* cutoff length
first-passage time 48, 61, 128, 140, 260
Fisher wave 246
Flory theory 147
Flory-type argument 84, 86, 104
forest-fire model 24, 31
fractals
    deterministic 3
    finitely/infinitely ramified 7, 61, 71, 216
    random 7, 74
    self-affine 9, 53
fractal carpets 32

fractal dimension d_f 3
    complex 12
    local, $d_{f,local}$ 10
    global, $d_{f,global}$ 10
    directional, d_f^x, etc. 10, 108
fractional Brownian motion (FBM) 50, 56
fractons 65, 82
fracton dimension d_s 7, 66, 71, 79, 84, 103, 171
fragmentation 32
front profile 202, 242
fugacity 149

gap exponent 70, 90, 95
geminate (initial conditions) 185
generator 4, 71, 95
graph dimension – *see* chemical dimension
Green function 217
growth sites – *see* Rammal–Toulouse equation

Hausdorff dimension 258
Hausdorff measure 258
heartbeat fluctuations 55, 269
heat equation 247
hull 22
Hurst exponent 268

incipient infinite cluster 13, 30, 74, 82
initiator/genus 4, 71, 116
injected particles 200
input 208, 212, 222, 254
    cooperative 227
interparticle distribution function (IPDF) 208, 255
invariance under dilation – *see* self-similarity
invasion percolation 32
invertase 165
ionic transport 32

kinetic equations – *see* evolution equations
kinetic phase transition – *see* dynamical phase
    transition
Kirkwood approximation 250
Koch curve 3, 6, 23, 68
Koch snowflake – *see* Koch curve
KPZ model 112
Kramers–Kronig relation 137

lacunarity 7
landscapes – *see* self-affine path
Laplacian operator 65, 67
lattice animals 100, 103
    on Cayley trees 104
law of mass-action 160
Leath algorithm – *see* Alexandrowicz–Leath
    algorithm
Lévy distribution 48
Lévy flight 48, 166, 187, 260, 262
Lévy walk 50
localization–delocalization transition 82
log-periodic oscillations 133, 140
long-range correlated walks 50

loopless fractal/lattice 25, 98
  infinitely ramified 113

Markovian (non-Markovian) process 33, 60, 67, 100, 175
master equation 43, 114, 263
mean field 109, 142, 181, 183, 242, 247
Menger sponge – *see* Sierpinski sponge
metal–insulator transition 32
method of empty intervals 211
minimal path – *see* chemical path/length/distance
moments (of a distribution) 42, 43, 67, 89
modified Koch curve 24
multiaffinity 268
multifractals 70, 90, 96, 124, 268
multifractal resistors chain 95
myopic ant 264

nonequilibrium state 220
number of distinct sites visited 44, 48, 72, 80, 108, 120, 168
  Lévy flights 56

Ohm's law 39, 63
oil recovery 32, 127
optimal path 31, 56, 125
order parameter 14
order parameter critical exponent β 14, 163

particle input 181
percolation 13, 74, 100, 106, 121
  anisotropic 29, 30
  bootstrap 15
  bond 13
  continuum 15, 30, 92, 264
  fractal dimension 18, 74, 83
  loopless 101
  site 15
  transition 13, 74
perimeter – *see* hull
periodic distribution 219
persistence length 55, 72
polymers – *see* self-avoiding walks
polynomial trees 100
Potts model 30
power spectrum 266
probability current 39
probability density 67
  chemical space 84
  CTRW 42
  diffusion in combs 107, 108
  diffusion in percolation 87
  diffusion in SAWs 70
  discrete random walks 35
  multifractal 90
  SAW 68
probability of return (to the origin) 64, 66

quasistatic approximation 172

radiating boundary conditions 175
Rammal–Toulouse equation 80, 103

random resistor networks 97
random-void model – *see* Swiss-cheese model
random walk 26
  asymmetric/biased 39, 43, 246, 260
  attracting 153
  long-range correlated 50
  persistent 43, 72
  recurrent (nonrecurrent) 61
  self-attracting/repelling 44
  simple 33
rate equations – *see* classical rate equations
reaction constant 159
reaction–diffusion equation 163, 192
reaction front 193
reaction kinetics 157
reaction-limited/controlled process 158
reactor tank 157
recursion relation 244
red bonds 22, 86, 89
relaxation 226, 232
renormalization group 7, 31, 61, 135, 148
resistance 63, 86, 101
resistance exponent $\tilde{\zeta}$ 63, 64, 92, 118
response function 270
restricted/unrestricted ensemble 75, 197
Rosenstock approximation 168
roughness 53
roughness exponent 54, 112

sand-box algorithm 8
scale invariance – *see* self-similarity
scaling analysis 39, 77
scaling distribution 219
scaling function/form 27, 68, 78, 83, 90, 200, 219, 226
scaling interparticle distance 219
scaling relation 28, 148
scavenger problem 176
Schlögl's model
  first 161
  second 165
Schrödinger equation 65, 67
segregation 183, 192, 225
self-affine path 11, 53, 267
self-averaging 31
self-avoiding walks (SAWs) 44, 68, 84, 88, 95, 144
  growing SAWs (GSAWs) 151
  end-to-end exponent 70
self-ordering 220
self-organized criticality 228
self-similar 3, 6, 258
  percolation clusters 19
  random walks 55
self-similar dimension 258
series expansions 31
shape exponent δ 47, 68, 88, 95
shielding 238
shortest path exponent – *see* chemical exponent
Sierpinski carpet 7
  random 7, 19
Sierpinski gasket 4, 6, 20, 23, 59, 69, 94, 121, 149

Sierpinski sponge 4, 7
similarity transformation 258
Sinai problem 119, 122
single-file diffusion 153
skeleton – *see* backbone
small-world network model 32
Smoluchowski's model 167
sliding box 267
space–time diagram 218, 223, 230
span (of random walks) 48, 75, 110
specific heat exponent α 29
spectral dimension – *see* fracton dimension
spin glass 115
stability analysis 162
stable/unstable phase 240
step probability density 35
step structure function 36
Sterling approximation 35
stirring 159, 166
stoichiometric coefficients 159
structure factor $S(q)$ 9
super-diffusion 124
superionic conductor (hollandite) 115, 137, 139, 153
surface growth 10
survival probability 48, 168, 227, 260
susceptibility 18
susceptibility exponent γ 18, 143
Swiss-cheese model 92, 115

telegrapher's equation 43
three-point correlation 245
topological dimension – *see* chemical dimension

tracer diffusion 141, 257
  in DLAs 143
  in percolation 144
transport exponents – *see* dynamical exponents
turbulent diffusion 50
two-point correlation 244

ultrametricity 116
universality 15
universality class 15, 56
upper critical dimension 25
  diffusion-limited coalescence 181
  DLA 105
  Fisher waves (reversible coalescence) 242, 248
  percolation 25
  reaction fronts 193
  RW with barriers and wells 120
  two-species annihilation 183

voters model 185
vibrations in fractals – *see* fractons

waiting time (distribution) 41, 47, 50, 107, 130
walk dimension d_w 46, 93, 197
wavelet transform 268
weather fluctuations 55
wells 115
Wiener–Khinchine theorem 136
world-line – *see* space–time diagram

Zwanzig formalism 114, 118, 121

Printed in the United States
By Bookmasters